REVIEWS IN FOOD AND NUTRITION TOXICITY

Volume 3

REVIEWS IN FOOD AND NUTRITION TOXICITY

Edited by
Victor R. Preedy and Ronald R. Watson

Volume 1
Volume 2
Volume 3

REVIEWS IN FOOD AND NUTRITION TOXICITY

Volume 3

Edited by

Victor R. Preedy and Ronald R. Watson

CRC Press is an imprint of the
Taylor & Francis Group, an **informa** business
A TAYLOR & FRANCIS BOOK

First published 2005 by Taylor & Francis

Published 2019 by CRC Press
Taylor & Francis Group
6000 Broken Sound Parkway NW, Suite 300
Boca Raton, FL 33487-2742

First issued in paperback 2019

ISBN-13: 978-0-367-45419-7 (pbk)
ISBN-13: 978-0-8493-3516-7 (hbk)

Visit the Taylor & Francis Web site at
http://www.taylorandfrancis.com

and the CRC Press Web site at
http://www.crcpress.com

Library of Congress Cataloging-in-Publication Data

Catalog record is available from the Library of Congress

Preface

In this third volume of reviews, we continue the scientific dialogue of introducing state-of-the-art chapters on food and nutrition toxicity. This addresses a pressing need by public health workers, nutritionists, food technologists, toxicologists, epidemiologists, and dieticians in general. The book serves those working within the industry as well as those in academia. The chapters are written by internationally recognized authors who have distinguished scientific backgrounds.

The first three chapters describe the toxic and pathological aspects of specific substances, namely methyleugenol, methyl mercury, and uranium. Chapter 4 is more broad-based in its scientific coverage and deals with polychlorinated biphenyls (i.e., PCBs) in general. Chapter 5 deals with cholesterol oxidation products. Chapters 6 and 7 cover targets of both scientific and media interest, namely nutraceuticals and genetically modified (i.e., GM) foods. Chapter 8 covers food contaminants relevant to children and Chapter 9 discusses aspects of dietary gluten with reference to celiac disease. All chapters are comprehensively referenced and illustrated with relevant figures and tables.

Food production processes and nutritional or dietary habits are continually changing and it is important to learn from past lessons and embrace a multidisciplinary approach. For example, some cellular mechanisms elucidated by studying one toxin may also be relevant to other areas of food pathology and it is our intention to impart such comprehensive information in a single series. We thus anticipate that the coverage in Volume 3 will stimulate specific and broad-based interests with applicability to other foods or nutritional substances.

Victor R. Preedy and Ronald Ross Watson

The Editors

Victor R. Preedy, Ph.D., D.Sc., F.R.C.Path., is a professor in the Department of Nutrition and Dietetics, King's College, London. He directs studies regarding protein turnover, cardiology, nutrition, and, in particular, the biochemical aspects of alcoholism. Dr. Preedy graduated in 1974 from the University of Aston with a combined honors degree in biology and physiology with pharmacology. He received his Ph.D. in 1981 in the field of nutrition and metabolism, specializing in protein turnover. In 1992, he received membership in the Royal College of Pathologists based on his published works, and in 1993 a D.Sc. degree for his outstanding contribution to the study of protein metabolism. At the time, he was one of the university's youngest recipients of this distinguished award. Dr. Preedy was elected a fellow of the Royal College of Pathologists in 2000. He has published more than 475 articles, which include more than 150 peer-reviewed manuscripts based on original research, and 70 reviews. His current major research interests include the role of alcohol in enteral nutrition and the molecular mechanisms responsible for alcoholic muscle damage.

Ronald R. Watson, Ph.D., attended the University of Idaho but graduated from Brigham Young University in Provo, Utah, with a degree in chemistry in 1966. He earned his Ph.D. in biochemistry from Michigan State University in 1971. His postdoctoral schooling in nutrition and microbiology was completed at the Harvard School of Public Health, where he gained two years of postdoctoral research experience in immunology.

From 1973 to 1974, Dr. Watson was assistant professor of immunology and performed research at the University of Mississippi Medical Center in Jackson. He was assistant professor of microbiology and immunology at the Indiana University Medical School from 1974 to 1978 and associate professor at Purdue University in the Department of Food and Nutrition from 1978 to 1982. In 1982, Dr. Watson joined the faculty at the University of Arizona Health Sciences Center in the Department of Family and Community Medicine of the School of Medicine. He is currently professor of health promotion sciences in the Mel and Enid Zuckerman Arizona College of Public Health.

Dr. Watson is a member of several national and international nutrition, immunology, cancer, and alcoholism research societies. He is presently funded by the National Heart Blood and Lung Institute to study nutrition and heart disease in mice with AIDS. Dr. Watson has edited more than 35 books on nutrition and 53 scientific books and has contributed to more than 500 research and review articles.

Contributor List

Kamal M. Abdo
National Institutes of Health
Research Triangle Park, North Carolina

Edmond J. Barrata
U.S. Food and Drug Administration
Winchester, Massachusetts

Janusz Z. Byczkowski
Consultant
Fairborn, Ohio

E. Davies
University of Manchester
Manchester, England

Alessio Fasano
University of Maryland
Baltimore, Maryland

Lynn R. Goldman
Johns Hopkins University
Baltimore, Maryland

Felicity Goodyear-Smith
University of Auckland
Auckland, New Zealand

D. Greenacre
University of Manchester
Manchester, England

Bernhard Hennig
University of Kentucky
Lexington, Kentucky

Gabriella Leonarduzzi
University of Torino
Torino, Italy

G.B. Lockwood
University of Manchester
Manchester, England

Jerry D. Johnson
Battelle Memorial Institute
Columbus, Ohio

Gabriele Ludewig
University of Iowa
Iowa City, Iowa

Michelle Maria Pietzak
University of Southern California
Los Angeles, California

Giuseppe Poli
University of Torino
Torino, Italy

Pachaikani Ramadass
University of Kentucky
Lexington, Kentucky

Larry W. Robertson
University of Iowa
Iowa City, Iowa

Barbara Sottero
University of Torino
Torino, Italy

Michal Toborek
University of Kentucky
Lexington, Kentucky

Veronica Verde
University of Torino
Torino, Italy

Table of Contents

1 Methyleugenol in the Diet: Toxic and Pathological Aspects

Jerry D. Johnson and Kamal M. Abdo

CONTENTS

0-8493-3516-7/05/$0.00+$1.50

ABSTRACT

Methyleugenol (MEG), an allylalkoxybenzene compound, produces a favorable taste and elicits a pleasant smell. These properties have promoted its use as a flavoring substance in foods and drinks, as well as an additive in body bath products to enhance their fragrance. MEG, included in the list of generally recognized as safe (GRAS) substances, is found naturally in plants but is also synthetically produced for commercial purposes. It has relatively large production and usage estimates, a regular rate of intake by humans, and is structurally related to other substances shown to possess carcinogenic activity. Recently, 2-year rodent bioassay studies indicated that there was clear evidence of carcinogenic activity for MEG under the conditions used for testing. Organs affected by MEG exposure included the liver, stomach, blood, testes and uterus, kidney, peritoneum, and skin. The MEG-related toxic and pathological findings in these organs, as well as the etiology, are described and reviewed in this chapter. The toxicity of MEG is attributed to the formation of a reactive metabolite. Bioactivation occurs when the liver metabolizes MEG to 1-hydroxymethyleugenol, which is then sulfated to form 1-sulfoxymethyleugenol, a sulfate ester that is unstable. Upon hydrolysis, a carbonium ion intermediate is formed. This highly reactive electrophile combines with proteins and other macromolecules such as DNA, which may progress to cellular damage and genotoxicity. The results from testing MEG in a variety of genotoxic assays are reviewed. In summary, the toxicity spectrum for MEG is dose dependent. Although relatively high doses do not result in sudden and severe life-threatening responses, prolonged exposure to MEG at relatively low doses may produce deleterious toxic and pathological effects.

INTRODUCTION

An old adage, familiar to many, can be paraphrased as "Too much of a good thing can kill you." This expression is often used in reference to overindulgence in certain types of food or alcohol. Using the latter as an example, it is easy enough to find research that claims that a small volume of alcohol taken on a daily basis can be beneficial to one's health but excessive consumption of alcohol, acutely or chronically, is most often detrimental. The list of items that this old saying applies to is long, and this review may add one more — methyleugenol (MEG). In small amounts, MEG produces a favorable taste and pleasant smell, but there is a threshold that when crossed over can lead to toxicity and death. In short, a little is OK, a lot is not.

MEG, chemically classified as an allylalkoxybenzene, is a naturally occurring substance that is used primarily as a flavoring agent for foods and drinks. Interest in the toxicity of MEG exposure has mounted steadily in the past few years. To a large part, renewed interest was generated by the outcome of a 2-yr rodent gavage study (National Toxicology Program [NTP], 1998) that led analysts to conclude that there was clear evidence of carcinogenic activity under the conditions used for testing. In the years prior to this landmark study, MEG was primarily being studied in academic laboratories for its pharmacological and biochemical effects, with some attention given to toxicity. As information began to accumulate regarding the carcinogenicity of related compounds, e.g., safrole, it became only a matter of time before plans commenced for formal testing of MEG. In the early 1990s, MEG was nominated by the National Cancer Institute and the U.S. Food and Drug Administration (FDA) for toxicity and carcinogenicity testing by the NTP, based on the high potential for human exposure through its use as a flavoring agent and its structural resemblance to the known carcinogen safrole (International Agency for Research on Cancer [IARC], 1976) and eugenol.

This chapter provides a review of the toxicological and pathological effects of MEG as well as the essential supportive and supplemental information needed to put these effects into perspective for the development of nutritional products when MEG is being considered as a flavoring agent or component. Early sections of this review provide general background information (uses of MEG, estimates and factors that affect human ingestion, federal regulations, environmental concerns and considerations, and relevant physical–chemical properties) as well as a review of the pharmacological effects of MEG from investigators who used varying *in vivo*, *in vitro*, or *in situ* methods. After these sections, the effects of MEG on endpoints routinely collected during the in-life phase of animal studies are described (survival, body weight (BW), and general toxicity), followed by a description of the toxicological and pathological findings by target organ. In addition, within each target organ section, an etiology for a given finding is described. Finally, a review of the literature regarding MEG genotoxic findings is presented.

GENERAL BACKGROUND INFORMATION

USES

MEG produces a favorable taste and elicits a pleasant smell. It provides tastes described as spice, cinnamon, clove, mouth tingle, fresh, peppery, and woody. As a palate pleaser, MEG is used as a flavoring substance to enhance the taste of ice cream, pudding, relish, baked goods (cookies, pies, gingerbread), nonalcoholic beverages (soft drinks), egg nog, patés, ketchup, chutney, apple butter, cigarettes, chewing gum, hard and soft candy, and jellies. The concentration of MEG in these products ranges from approximately 5 ppm (ice cream) to 52 ppm (jellies) (Hall and Oser, 1965; Furia and Bellanca, 1975). Also, because MEG is a component of commonly used spices and essential oils (see the following sources), and these substances are used extensively to add flavor to foods, it becomes incorporated into a variety of foods when such spices are added to enhance flavor. MEG is also used to modify spice, heavy fruit, root beer, and anise-type flavors in meats and condiments (Smith et al., 2002). Although its good taste is the basis for including MEG in foods and beverages, its pleasant fragrance is the basis for including it in a variety of body bath products. MEG is used in perfumes, creams, lotions, sunscreens, soaps, and detergents. Bath products contain concentrations of MEG that include as little as 0.01% in creams and lotions to as much as 0.8% found in perfumes (Opdyke, 1979; Radian, 1991).

Another use for MEG also takes advantage of its aromatic and fragrant properties. Howlett (1912, 1915) first reported that certain fruit flies are attracted to MEG, including the male oriental fruit fly (*Dacus dorsalis*). The usefulness of MEG as a lure and method for controlling oriental fruit fly infestations was investigated extensively by Steiner and coworkers (Steiner, 1952; Steiner et al., 1965, 1970). In combination with malathion, MEG was used to control an oriental fruit fly outbreak in the U.S. (Hays and Laws, 1991).

Human Ingestion

Human exposure to MEG occurs when certain types of foods or foods that have been altered by flavor additives are ingested. A variety of fruits naturally contain MEG — e.g., oranges and bananas — and even a vegetable (black pepper) has been reported to contain it. Certain types of spices and essential oils also contain natural quantities of MEG. Foods that do not naturally contain MEG are often flavored with spices that do, and it is the addition of MEG-containing spices to foods that accounts for the predominant source of MEG exposure for the vast portion of the population. A list of foods, spices, and essential oils reported to contain MEG is provided in Table 1.1.

The highest intake of MEG in a food preparation occurs by consumption of an Italian sauce called pesto. Those individuals who ingest pesto regularly and in sizable portions represent a specialized eating group referred to as "pesto-eaters." Pesto is prepared from fresh sweet basil, and the most common basil cultivar used in pesto production is *Ocimum basilicum* cv. Genovese Gigante. Pesto-eaters may ingest up to 10 g of basil in a single serving, and basil from northwestern Italy contains more than 40% MEG oil (Miele et al., 2001a). The content of MEG in basil depends on the growth stage the plant has reached. Miele and coworkers (2001a, 2001b) reported that plants analyzed 4 and 6 weeks after sowing contained MEG and eugenol as the main constituents. MEG was the predominant constituent in the plants until it reached

TABLE 1.1
Foods (Fruits, Vegetables, and Spices) and Essential Oils Reported to Contain MEG

Source	MEG Conc.[a]	MEG (mg)/tsp[b]	Reference
	Fruits[c]		
Apple (juice)	—[d]		
Banana	—		
Bilberries	—		
Grapefruit (juice)	—		
Orange (oil and juice)	240 ppm (oil)	1.2	MacGregor et al., 1974
	42 ppb (juice)	0.21	
	Vegetables[e]		
Black peppers	—		
	Spices[f]		
Anise	0.0001%	0.005	Smith et al., 2002
Basil (fresh/dried)	0.0011/0.026%	0.055–1.3	Smith et al., 2002
Mace	0.002%	0.1	Smith et al., 2002
Nutmeg	0.008%	0.4	Smith et al., 2002
Pimento berry (allspice)	0.13%	6.5	Smith et al., 2002
Tarragon	0.008%	0.4	Smith et al., 2002
	Essential Oils[g]		
Anise	400 ppm	2	Furia and Bellanca, 1975
Banana oil	0.00001%	0.0005	Smith et al., 2002
Bay	—		
Betel	—		
Sweet basil	2.0%	100	de Vincenzi et al., 2000
Blackberry (essence)	—		
Cajeput	—		
Cardamom	0.1%	5	de Vincenzi et al., 2000
Cinnamon (leaves)	—		
Citronella	—		
Clove	—		
Hyacinth	—		
Hyssop	0.54%	27	de Vincenzi et al., 2000
Japanese calamus	—		
Laurel (fruits and leaves)	—		
Lemongrass	2.0%	100	de Vincenzi et al., 2000
Mace	—		
Myrtle	2.3%	120	de Vincenzi et al., 2000
Nutmeg	0.1–17.9%	5–895	de Vincenzi et al., 2000
Pimento	5.0–8.8%	250–440	de Vincenzi et al., 2000
Pixuri seeds	520 ppm	260	Carlini et al., 1983
Rose	1.5%	75	Mookherjee and Wilson, 1996

TABLE 1.1 (continued)
Foods (Fruits, Vegetables, and Spices) and Essential Oils Reported to Contain MEG

Source	MEG Conc.[a]	MEG (mg)/tsp[b]	Reference
Rosemary	1.9–2.0%	95–100	Mookherjee and Wilson, 1996
Sweet flag	1.0%	50	de Vincenzi et al., 2000
Tarragon	—		

[a] Values expressed as parts per million (ppm), parts per billion (ppb), or grams/100 ml (%).
[b] Quantity of MEG in a level teaspoon (5 g or 5 ml) of a given food, spice, or essential oil.
[c] WHO, 1981; MacGregor et al., 1974.
[d] MEG concentration not found in the literature.
[e] WHO, 1981.
[f] Smith et al., 2002.
[g] FEMA 1978; Furia and Bellanca, 1975; Carlini et al., 1983; *Farm Chemical Handbook*, 1992; Mookherjee and Wilson, 1996; WHO, 1981; de Vincenzi et al., 2000.

a height of 6.5 to 10 cm; afterwards, eugenol predominated. The content of MEG or eugenol was related more to plant height than age. In addition, at any given height, MEG was more prevalent in the lower part of the plant, whereas eugenol was more likely to be localized in the upper part. Other factors that affect plant MEG content include geographical location, plant type, plant part used, processing and extraction methods, harvest time, and drying and storage conditions. The interested reader can obtain further information about the influence of these factors on MEG content in the following references: Tsai and Sheen, 1987; Bobin et al., 1991; Lawrence and Shu, 1993. Thus, a myriad of cultivation factors can affect the MEG content in a plant.

Production estimates for MEG include the amount produced by direct synthesis and natural sources. MEG is synthesized by methylation of eugenol (Opdyke, 1979). In 1990, the annual production of synthetic MEG in the U.S. was estimated to be 11,500 kg (Stanford Research Institute [SRI International], 1990). For natural sources, the production estimate for MEG is a function of the MEG concentration in the source and the annual production of the food, spice, or essential oil. For example, banana oil contains approximately 0.00001% MEG, and the 1999 annual production volume for banana oil was estimated to be 58,000,000 lb (Smith et al., 2002). Thus, banana oil production contributed 5.8 lb of MEG to the annual production volume. Similar calculations can be made for all MEG-containing foods, spices, and essential oils and, when totaled, provide total annual production estimates for MEG. It is the total production of MEG, together with other parameters, that provides a method for estimating human intake.

The primary method for estimating human intake is by determining the ingestion rate of MEG-containing foods or foods flavored with substances that contain MEG. Other sources of exposure, e.g., occupational or environmental, and routes of exposure, e.g., dermal or inhalation, occur to a minimal extent. For example, a National Occupational Exposure Survey by NIOSH from 1981 to 1983 estimated that approximately

TABLE 1.2
Summary of Human Exposure Levels for MEG

Parameter	Value	Reference
Total average daily intake	1–10 μg/kg BW/d (total diet)	Smith et al., 2002
	0.5 μg/kg BW/d (from traditional foods/spices)	
	0.16 μg/kg BW/d (from essential oils)	
	≥10–100 μg/kg BW/d ("pesto-eater")	
	0.26 μg/kg BW/d	Stofberg and Grundschober, 1987; NAS, 1989
Daily per capita intake	73 μg/d	WHO, 1981
	6 μg/d	Miller et al., 1983

2800 workers were potentially exposed to MEG (NIOSH, 1990). Because occupational and environmental exposure to MEG occur to such a small extent, and MEG is not even a minor air or water contaminant, these sources of MEG are not included in calculations for the level of human exposure. Consequently, the average daily intake for MEG is calculated using the concentrations of MEG found in foods, spices, and essential oils; the annual production of a given food, spice, or essential oil (assumed to represent the amount consumed); the size of the population ingesting the annual production level; average adult BW; and the number of days in a year. Table 1.2 summarizes reported average daily intake values for MEG. As is obvious from the table values, the daily intake of MEG occurs in relatively minute quantities, i.e., μg/kg/d. Human intake estimates are valuable for many reasons, and with respect to the present focus on the toxicological and pathological effects of MEG, they contribute necessary information to the risk assessment evaluation as well as assist in setting doses for the subchronic and chronic bioassay studies. These estimates are also essential in a thorough assessment of the risk to exposure.

Human serum concentrations are another method used for evaluating human exposure. By determining the concentration range of an analyte (or its primary metabolite) in the circulating blood, which is referred to as the internal dose, researchers are able to estimate a reference range for a chemical in the general population. The reference range provides information about the prevalence of incidental exposure to an analyte and human background levels, and allows increasing or decreasing trends that occur as a result of changing exposure climates to be ascertained (Needham et al., 1996). Information in the literature about human MEG background levels is scant. However, as part of a Centers for Disease Control (CDC)/National Institute of Environmental Health Sciences (NIEHS) investigation to estimate the MEG concentration in the general U.S. population, Barr et al. (2000) developed a simple solid-phase extraction procedure followed by an isotope-dilution gas-chromatograph, high-resolution mass-spectrometry method that was sensitive enough to quantify MEG in human serum. In 206 adult subjects, 98% had detectable serum concentrations of MEG (limit of detection was <3.1 pg/g). The mean MEG

TABLE 1.3
Serum MEG Concentrations for Selected Demographic Factors

Demographic Factor	Group	No. of Subjects	MEG Concentration (pg/g) Geometric Mean	Minimum	Maximum
Gender	Male	64	17	3.9	110
	Female	142	16	2.2	390
Race/ethnicity	Non-Hispanic white	53	19	4.9	84
	Non-Hispanic black	96	14	2.2	390
	Mexican American	41	19	5.4	110
	Other	16	16	5.2	78
Age	18–26 yr	48	16	2.2	390
	27–34 yr	54	16	4.1	95
	35–42 yr	43	16	2.2	110
	43+ yr	61	17	2.2	160
Smoking exposure status	Nonsmoker, no smoker at home	107	5.6	2.2	390
	Smoker or living with 1 smoker	69	17	2.2	160
	Smoker or living with ≥2 smokers	28	19	4.9	110
Diet	≤1 cake/cola per d	114	16	2.2	160
	≥1 cake/cola per d	92	17	2.2	390
Fasting	≤9 h for males	23	19	4.3	110
	≥9 h for males	41	16	3.7	84
	≤9 h for females	56	14	2.2	54
	≥9 h for females	86	18	2.2	390
Drinks bottled water	Yes	29	19	5.8	95
	No	175	16	2.2	390

Source: Barr, D.B., Barr, J.R., Bailey, S.L., Lapeza, C.R., Jr., Beeson, M.D., Caudill, S.P., Maggio, V.L., Schecter, A., Masten, S.A., Lucier, G.W., Needham, L.L., and Sampson, E.J. (2000), Levels of methyleugenol in a subset of adults in the general U.S. population as determined by high resolution mass spectrometry, *Environmental Health Perspectives*, 108: 1–6.

serum concentration was 24 pg/g (median 16 pg/g; maximum 390 pg/g). Demographic variables were evaluated, and the results are summarized in Table 1.3. Males fasting longer than 9 h produced slightly lower serum MEG concentrations than those who fasted for less than 9 h; the opposite result was obtained for females. In a separate study (Centers for Disease Control and Prevention/ National Institute of Environmental Health Sciences [CDCP/NIEHS, 2000]), serum MEG concentrations of 100 pg/g were obtained in five fasted adults given a meal containing 60 µg of MEG. The results of this investigation indicated that incidental MEG exposure of the population is widespread, but only in some demographic variables were slight differences observed between MEG concentration and age, race or ethnicity, gender, fasting, smoking or nonsmoking, and diet.

U.S. Regulations

MEG is regulated by the FDA and the U.S. Environmental Protection Agency (EPA). FDA regulatory action allows MEG to be in food, provided the minimum quantity required to produce the intended effect is used and processing occurs in accordance with all principles of good manufacturing practice (Code of Federal Regulations [CFR], 1977, 1996). EPA regulatory action exempts MEG from the requirement of tolerances on all raw agricultural commodities when used in combination with oriental fruit fly eradication programs (CFR, 1971; 1982). MEG is included in the GRAS (generally recognized as safe) list of substances. It received this status in 1965, and more recent evaluations, in 1979 and 2001, substantiated that MEG should retain its GRAS status. Thus, with respect to human consumption, MEG is permitted for intended use as a synthetic flavoring substance and adjuvant in foods (Smith et al., 2002).

Environmental Considerations

MEG is a negligible, and could be considered a relatively nonexistent, environmental contaminant. A few studies have reported measurable concentrations of MEG in the environment. One such study found MEG in the wastewater of a paper mill effluent, which had MEG concentrations of 0.001 to 0.002 mg/l (Keith, 1976). In another study, MEG and malathion were used to control fruit fly infestations (Hays and Laws, 1991). Although only a proposal, a plan was conceived, according to which MEG would be saturated onto cigarette filters and the filters aerially distributed over areas infested with fruit flies to act as a lure prior to spraying the congregating insects with malathion (Shaver and Bull, 1980). This plan prompted environmental experiments to be conducted to determine the fate of MEG in the environment.

Experiments by Shaver and Bull (1980) examined MEG's persistence in soil and water, leaching from soil, and fate in field-grown tomatoes. Direct distribution of MEG in soil or water resulted in rapid dissipation. When tested at 32°C, MEG had a half-life of 6 h in soil and water, whereas at 22°C, half-life values for MEG were 16 h in soil and 34 h in water. Because approximately 95% of a material is accounted for in 4 to 5 half-lives, dissipation from the soil and water would occur in approximately 30 h at 32°C (soil or water) and 80 h (soil) and 170 h (water) at 22°C. Based on these results, MEG would not persist in soil or water beyond approximately 7 d. (As a note of interest, MEG-treated cigarette filters placed on top of soil, embedded in soil, or suspended in the air had half-life values that ranged from 3.5 to 4.5 d.) Soil-leaching studies showed that MEG did not migrate to any appreciable extent in sand, Houston clay, or Lufkin (Texas) fine sandy loam, although there was some streaking found in the sand samples. Application of MEG to field-grown tomatoes indicated that MEG disappeared rapidly from the surface of the fruit. After 24 h, only 3.4% of the applied dose was found, and no MEG was detected on the tomato surface after 3 d. Total internal accumulation of MEG-related residues in the fruit leveled off to concentrations of 4 to 5 ppm after 14 d. The low internal concentrations were attributed to evaporation from the surface prior to absorption.

MEG in aquatic or atmospheric environments dissipates rapidly (Hazardous Substance Data Base [HSDB], 1996). MEG volatilizes from water and has been estimated to have a half-life of 9 and 68 d in a model river and lake, respectively. The bioconcentration factor of MEG in aquatic organisms is estimated to be 120, where any value less than 1000 is generally considered too water soluble for bioaccumulation to occur to any significant extent. In the ambient atmosphere, MEG exists as a vapor. In the gaseous state, MEG can react with hydroxyl radicals that are the product of photochemical reactions. The half-life for MEG in the air is estimated to be 5 h.

Thus, the environmental research reported to date indicates that, while present in the environment, MEG concentrations are extremely low and, in those instances when concentrated quantities are introduced into the environment, MEG quickly dissipates and does not persist in the soil or water. The rapid loss of MEG from the environment can be attributed to its physical and chemical properties, which also help explain its pharmacological and toxicological effects in mammalian organisms. For these reasons, a general summary of chemical and physical properties and a brief description of relevant chemical nomenclature are presented in the following section.

Structure, Chemical Identification, and Physical–Chemical Properties

MEG is classified as an allylbenzene, or more precisely, an allylalkoxybenzene compound. Its chemical structure is shown in Figure 1.1. MEG is composed of a benzene ring with a 2-propene substituent at position 1 and methoxy substituents at positions 3 and 4. The propene chain (referred to in the literature as an allyl- or alkenyl-substituent or propenoid) is unsaturated. An understanding of these substituent groups is important because they are responsible for the high reactivity of MEG with macromolecules found in biological systems, as described in the toxicology sections.

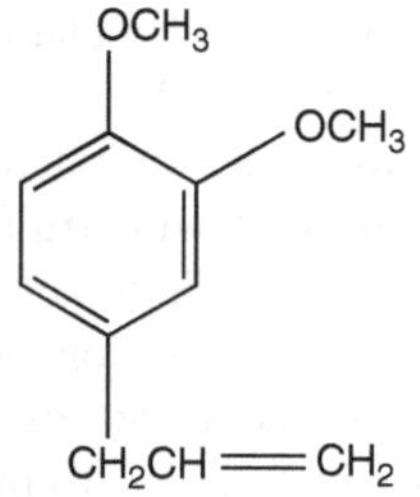

OCH_3 OCH_3 $CH_2CH{=}CH_2$

FIGURE 1.1 Structure of MEG.

Like most chemicals, MEG goes by a variety of different names. Other names given to MEG are listed in Table 1.4. The most widely used name and spelling for MEG is "methyeugenol." However, early manuscripts, as well as a recent review article, used a two-word form — "methyl eugenol" (Steiner, 1952; Smith et al., 2002). Earlier references commonly cited MEG as "eugenol methyl ether" (or similar derivation — eugenyl methyl ether or methyl eugenyl ether). MEG has a number of structurally similar analogs. Much of what has been learned about MEG came about as a result of earlier investigations with these analogs, especially eugenol, isoeugenol, estragole, and safrole. Table 1.5 summarizes some of the structural analogs of MEG.

Finally, the physical–chemical properties for MEG are summarized in Table 1.6. MEG has limited solubility in water but can be solubilized in semipolar solvents. It is readily soluble in organic solvents and can form azeotropic mixtures with some

TABLE 1.4
Synonyms for Methyleugenol (MEG)

1,2-Dimethoxy-4-(2-propenyl)benzene	4-Allylveratrole
1-(3,4-Dimethoxyphenyl)-2-propene	Veratrole methyl ether
Dimethoxy-4-(2-propenyl)benzene	Eugenol methyl ether
1-Allyl-3,4-dimethoxybenzene	*o*-Methyl eugenol ether
1,2-Dimethoxy-4-allylbenzene	Eugenyl methyl ether
4-Allyl-1,2-dimethoxybenzene	Methyl eugenyl ether
3,4-Dimethoxyallyl benzene	1,3,4-Eugenol methyl ether
2-Methoxy-4-propenylphenol methyl ether	ENT 21040
Allyl veratrole	

Source: Furia, T.E. and Bellanca, N. (Eds.) (1975), *Fenaroli's Handbook of Flavor Ingredients (1975)*, Cleveland, OH: The Chemical Rubber Co., 2nd ed., Vol. 2, p. 200; de Vincenzi, M., Silano, M., Stacchini, P., and Scazzocchio, B. (2000), Constituents of aromatic plants: I. Methyleugenol, *Fitoterapia*, 71: 216–221; Lewis, R.J., Jr. (Ed.) (2001), *Hawley's Condensed Chemical Dictionary*, New York: John Wiley & Sons, p. 735; ChemFinder (2003), Methyleugenol, available at http://www.chemfinder.camsoft.com/,CambridgeSoft (accessed December 2002).

solvents, i.e., ethylene glycol, benzoic acid, etc. MEG is volatile and, if left exposed to air at room temperature, will soon completely evaporate. Also, exposure to air changes MEG from a colorless to pale yellow oily liquid to a dark yellow to orange color that has a more viscous consistency, e.g., syrup-like (Lide, 1998). Physical–chemical properties that play a major role in leading to the toxicological effects produced by MEG include its size, i.e., molecular weight, and lipid solubility, i.e., high lipid–water partition coefficient. This latter property is especially important and contributes to its rapid and extensive absorption and widespread distribution, thereby accounting for several different sites of action at pharmacological doses and a variety of target organs at toxicological doses, which are the focus of the remainder of this review.

PHARMACOLOGICAL EFFECTS

If there is a single quotation that would be recognized by all who study toxicology, it would have to be this statement by Paracelsus, "All substances are poisons; there is none which is not a poison. The right dose differentiates a poison from a remedy" (Gallo, 1996). MEG is no exception. Prior to a description about the toxicology of MEG, this section reviews what is known about the pharmacological effects of MEG. In addition to enhancing the taste of foods and drinks and the fragrance of perfumes, MEG has beneficial properties that have been investigated for possible treatment of various conditions and diseases. Table 1.7 summarizes MEG's potentially beneficial and therapeutic pharmacological effects.

TABLE 1.5
Selected Analogs of MEG

Chemical Name	CAS	Structure
Methyleugenol	93-15-2	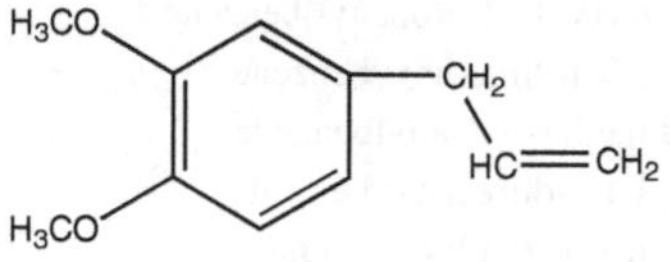 H_3CO, H_3CO, CH_2, $HC{=}CH_2$
Eugenol	97-53-0	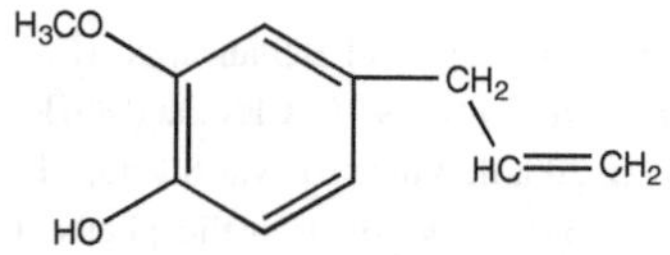 H_3CO, HO, CH_2, $HC{=}CH_2$
Isoeugenol	97-54-1	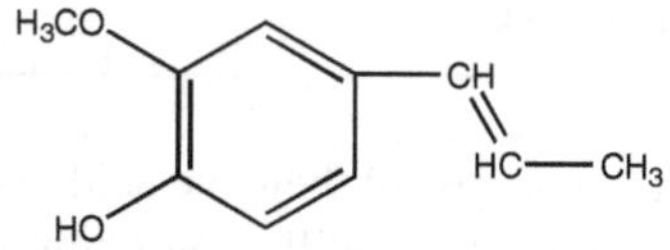 H_3CO, HO, CH, $HC—CH_3$
Estragole	140-67-0	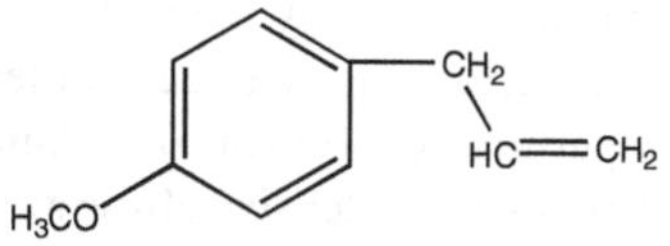 H_3CO, CH_2, $HC{=}CH_2$
Safrole	94-59-7	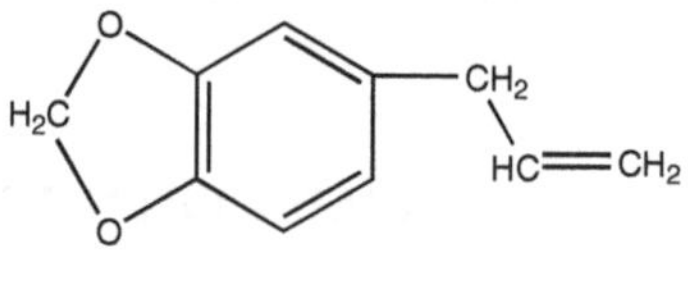 O, H_2C, O, CH_2, $HC{=}CH_2$

Biochemical Effects

MEG may be useful in the prevention and treatment of anaphylactic reactions. Histamine is produced when the substrate histidine is catalyzed by L-histidine decarboxylase (HDC). Mast cell degranulation leads to histamine release and produces vasodilation, which can progress to anaphylactic shock. Shin et al. (1997) used a variety of *in vivo* and *in vitro* assays to determine the effects of MEG on histamine release. First, they investigated systemic anaphylaxis by giving BALB/c mice a single intraperitoneal (i.p.) injection of 48/80, a mast-cell degranulator. Other

TABLE 1.6
Summary of MEG Physical–Chemical Properties

Property	Value/Description	Reference
CAS No.	93-15-2	ChemFinder, 2003
RTECS Code	CY2450000	ChemFinder, 2003
Chemical formula	$C_{11}H_{14}O_2$	ChemFinder, 2003
Molecular weight	178.2304 g/mol	ChemFinder, 2003
Color	Colorless to pale yellow	Furia and Bellanca, 1975; ChemFinder, 2003
Odor	Clover-carnation	Furia and Bellanca, 1975; Lide, 1998; HSDB, 1996
Taste	Bitter, burning	Furia and Bellanca, 1975; HSDB, 1996
Physical state (@ RT)	Liquid	ChemFinder, 2003
Refractive index (@ 20°C)	1.5320–1.5360	Furia and Bellanca, 1975
Melting point	4°C	ChemFinder, 2003
Boiling point	244–245°C; 254.7°C	Furia and Bellanca, 1975; Lide, 1998; HSDB, 1996
Specific gravity		HSDB, 1996; Lewis, R.J., Jr., 2001; Furia and Bellanca, 1975
(@ 20°C/4°C)	1.0369	
(@15 °C)	1.055	
Vapor pressure (@ 85°C)	1 mm Hg	HSDB, 1996
Flash point	117°C	ChemFinder, 2003
Solubility: Water	<0.1 g/100 l	ChemFinder, 2003
Ethanol	Soluble; 1:4 in 60% alcohol; 1:2 in 70% alcohol	HSDB, 1996; Furia and Bellanca, 1975
Ether	Soluble	HSDB, 1996
Chloroform	Soluble	HSDB, 1996
Glycol	Insoluble	HSDB, 1996
Propylene glycol	Insoluble	HSDB, 1996
Ethylene glycol	Forms azeotropic mixture (31.5%)	Furia and Bellanca, 1975
Eugenol	Forms azeotropic mixture (55%)	Furia and Bellanca, 1975
Benzoic acid	Forms azeotropic mixture (11%)	Furia and Bellanca, 1975

mice were pretreated (1 h) with a single i.p. injection of saline or MEG (0.01 to 100 mg/kg). The status of the mice was evaluated 1 h after the compound 48/80 injection. Fatal shock was observed in the saline group and the 0.1 and 1.0 mg/kg MEG groups. However, at 1 mg/kg of MEG, mortality decreased to 80%, and at 10 and 100 mg/kg of MEG, all animals survived. Analysis of serum samples taken from the animals showed that MEG treatment reduced serum histamine concentrations. Figure 1.2 shows the dose-related inhibitory effect of MEG treatment on compound 48/80-induced histamine release. Thus, MEG inhibited 48/80-induced fatal anaphylaxis in 100% of the animals at doses ≥10 mg/kg. It is from this research that MEG has been cited as an inhibitor of histamine release from mast cells after degranulation by specific antigens.

TABLE 1.7
Pharmacological Effects of MEG in Animals

Species	Pharmacological Effect	Reference
Rats/mice	Inhibits histamine release from mast cells after degranualation by specific antigens. May be effective in the prevention/treatment of anaphylaxis	Shin et al., 1997
Guinea pig	Relaxant and antispasmodic effects on gastrointestinal smooth muscle. The effect is direct, does not involve alteration of the membrane potential, and is reversible	Lima et al., 2000
Rats/mice	Analgesic and anesthetic activity. Causes loss of righting reflex and insensitivity to tail pinch. Used to obtain a surgical plane of anesthesia without producing excitation during the induction phase, no bronchial secretions, and produces muscle flaccidity	Sell and Carlini, 1976; Carlini et al., 1981
Mice	Anticonvulsant activity. An essential oil containing MEG and related compounds produced a dose-related decrease in MES and PTZ-induced seizures	Sayyah et al., 2002

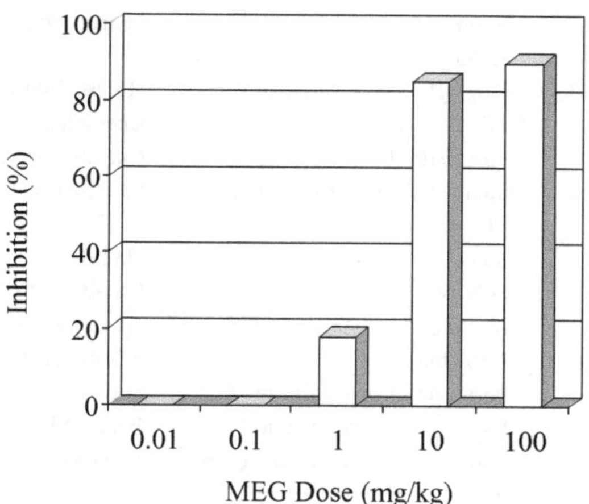

FIGURE 1.2 MEG inhibits compound 48/80-induced histamine release.

The mechanism of this antianaphylactic effect of MEG was further evaluated using mastocytoma P-815 cells. These cells synthesize HDC in response to various stimuli, e.g., substance P. Shin et al. (1997) incubated the P-815 cells in the absence or presence of 100 n*M* of substance P, or substance P with 100 n*M* of MEG for 6 h. The level of messenger-ribonucleic acid (mRNA) that codes for the enzyme HDC was determined using Northern blot analysis. MEG suppressed the increase in substance P-induced accumulation of HDC mRNA. This effect was not due to cytotoxicity based on trypan blue uptake. Thus, MEG appears to suppress stimuli-induced expression of the gene for HDC. As for a mechanism, because substance P binds to mast cells to trigger release of histamine, MEG may act as an antagonist at the same receptor, thereby reducing the signal transduction that would lead to increased expression of HDC. The outcome is a decrease in histamine production and release, suggesting that MEG may be effective in the prevention or treatment of conditions that can lead to anaphylactic reactions.

Gastrointestinal System Effects

MEG has relaxant and antispasmodic effects on the gastrointestinal tract. Lima et al. (2000), using the isolated guinea pig ileum, investigated MEG for its effects on contractility and the transmembrane potential. MEG progressively increased relaxation of the spontaneous tonus of the ileum at concentrations ranging from 0.5 to 672 μ*M*. The effective concentration that produced 50% of the maximal response (EC_{50}) was 52.2 ± 18.3 μ*M*. The MEG-induced relaxant effect was completely reversible by removing the MEG. Also, MEG inhibited contractions induced by treatment with potassium, acetylcholine, or histamine. Possible mechanisms for the MEG relaxant effects were investigated by adding other agonists and antagonists to the preparation. When MEG-induced relaxation was compared with the activation of intestinal adrenergic receptors using adrenaline, MEG reduced tone to a significantly greater extent than the adrenergic agonist. Based on this result, it is unlikely that MEG-induced relaxant activity is mediated in a similar manner because potassium chloride (KCl)-induced contractions were reduced only 16% by adrenaline, whereas the contractions were completely reversed by MEG. Pretreatment with tetrodotoxin (a sodium channel blocker) or hexamethonium (a neuronal ganglionic blocker) did not alter the MEG-induced relaxation. Therefore, it is unlikely that MEG produces relaxation indirectly. Finally, MEG significantly hyperpolarized the transmembrane potential of the guinea pig ileum, although in the presence of potassium, no effect on the transmembrane potential was observed. Although hyperpolarization may partially explain the relaxant effect of MEG, the authors suggest it is not the primary mechanism because the effective concentration for maximal relaxation produced a minimal hyperpolarization response. Furthermore, hyperpolarization does not appear to be significant because MEG completely relaxed the K^+-induced ileum contractions without altering the membrane potential. Thus, MEG has smooth muscle relaxant activity, and this effect is believed to be independent of membrane potential alterations and dependent on a direct effect of the smooth muscle.

Central Nervous System Effects

MEG has analgesic and anesthetic activity. Carlini and coworkers (Sell and Carlini, 1976; Carlini et al., 1981) reported that MEG caused in rats and mice a loss of righting reflex, changed electroencephalogram (EEG) patterns in a similar way as pentobarbital (rats only), and induced a surgical plane of anesthesia. Endpoints used to evaluate the level of anesthesia were the loss of righting reflex, tail pinch insensitivity, and loss of consciousness. Ten rats and mice per group (age, sex, and strain not specified) were given a single i.p. injection of MEG suspended in Tween-80

(1%, w/v) at doses of 200, 250, or 300 mg/kg. A loss of righting reflex was observed within approximately 1 to 2 min for both species, and insensitivity to the tail pinch was observed 5 to 6 min (rats) and 2 to 3 min (mice) after dosing. In regard to loss of consciousness, MEG caused a slight dose-related effect. Rats exhibited a loss of righting reflex for 62 and 144 min, and insensitivity to the tail pinch for 15 and 63 min at doses of 200 and 250 mg/kg, respectively (no data reported at 300 mg/kg). A similar trend was observed for the mice, but the duration of the endpoints was less than 40 min at the highest dose tested (Engelbrecht et al., 1972; Carlini et al., 1981). Although a clear dose–response trend was not produced over the narrow range of doses tested, use of lower doses enabled Engelbrecht and coworkers (1972) to determine a well-defined dose–response relationship between the loss of righting reflex and MEG. Using Swiss-Webster (SW) mice, they reported the MEG dose that causes a loss of righting reflex in 50% of the animals to be 26 and 88 mg/kg after a single i.v. or i.p. injection. Carlini and coworkers (1981) evaluated MEG as a surgical anesthetic by comparing it with sodium pentobarbital when performing a central nervous system (CNS) operation, i.e., electrolytic lesion of the substantia nigra. No deaths during or shortly after the operations were observed in 15 animals treated with MEG, compared with 7/15 deaths in animals treated with pentobarbital. In addition, no animals given MEG developed bronchial secretions, whereas all of the pentobarbital-treated rats had bronchial secretions. Finally, MEG did not produce an excitation phase like the one produced by pentobarbital, and the MEG-treated animals exhibited marked abdominal muscle flaccidity. Thus, MEG has an effect on CNS, and although relatively high doses are needed, a surgical plane of anesthesia can be obtained, in which induction does not involve an excitation phase, the duration is long, and muscles become flaccid.

MEG is included with the class of general depressants, more specifically, as a sedative-hypnotic. Engelbrecht et al. (1972) described the signs of MEG-induced generalized depression in SW mice, following a single i.p. injection at a dose of 200, 400, or 800 mg/kg. Within 3 minutes of dosing, ataxia was observed, followed by sedation. In this sedated state, introduction of external stimuli could induce hyperactivity in the mice. Later, or as the dose was increased, the animals lost their righting reflex and exhibited an ascending paralysis that first involved the hind limbs, next the fore limbs, and ultimately the neck and head area. Eventually, the muscles became flaccid and the respirations were slow and irregular. Even higher doses led to apnea and death, apparently due to respiratory failure. No tremors or convulsions were observed, although at the high dose, animals exhibited some writhing. In addition, evidence of MEG-induced neurotoxicity was apparent when experiments using a rotarod apparatus showed animals to have impaired motor functions.

MEG may have anticonvulsant activity against experimentally induced seizures in animals. Engelbrecht et al. (1972) administered single i.p. injections of MEG to male SW mice and determined the median protective dose (PD_{50}) that inhibited seizures induced by maximum electroshock (MES), strychnine, or pentylenetetrazole (PTZ). The PD_{50} values for MEG were 68, 258, and 78 mg/kg following MES, strychnine, and PTZ, respectively. Sayyah et al. (2002) extracted the leaf essential oil and examined its effectiveness in inhibiting seizures induced by MES or PTZ in

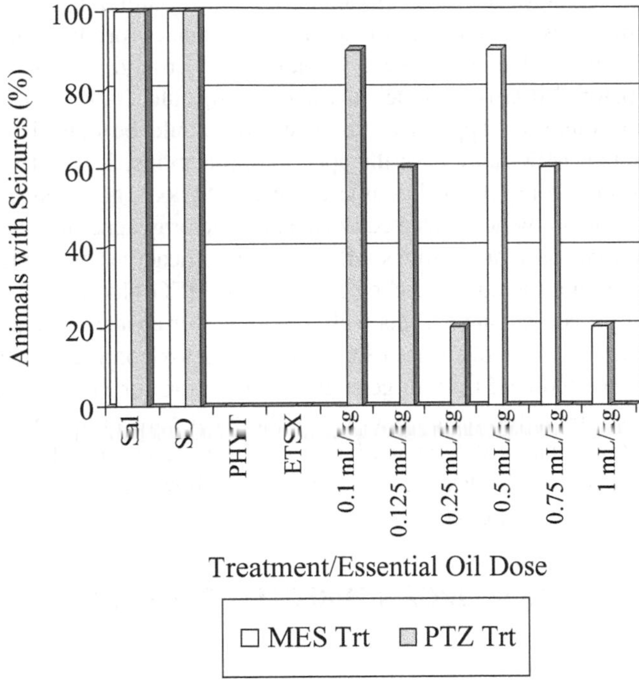

FIGURE 1.3 Effect of essential oil treatment on seizure induction.

male NMRI mice. The essential oil from *Laurus nobilis*, an evergreen shrub indigenous to parts of Europe and Mediterranean areas, contains MEG (2.5%), as well as other related substances such as cineol (44%), eugenol (15%), sabinene (6%), 4-terpineol (3.6%), α-pinene (2.7%), α-terpineol (2.2%), and β-pinene (2.1%) (Riaz et al., 1989). For the MES test, groups of ten mice each were pretreated (30 min) i.p. with graded doses of the essential oil (0.5 to 1 ml/kg), phenytoin (25 mg/kg, positive control), or sesame oil or saline (vehicle and negative controls). Following a transauricular electroshock, they determined the percentage of animals that did or did not exhibit hind limb tonic extensions within 10 sec after the shock. For the PTZ test, the groups were pretreated (30 min) i.p. with the essential oil (0.1 to 1 ml/kg) and vehicle and negative controls as described earlier, or ethosuximide (150 mg/kg, positive control) and PTZ (110 mg/kg). They observed the animals for 30 min after PTZ administration for the absence or presence of hind limb tonic extensions. The essential oil produced a dose-dependent decrease in the percentage of tonic seizures of MES- or PTZ-treated animals (Figure 1.3). All animals treated with the saline (Sal) and sesame oil (SO) exhibited seizures following treatment with MES or PTZ, whereas phenytoin (PHYT) completely protected against MES-induced seizures and ethosuximide (ETSX) against PTZ-induced seizures. The essential oil of *L. nobilis* protected against PTZ-induced seizures more selectively than against seizures produced by MES treatment. Although these investigators did not conduct any studies to further characterize the mechanism, they theorized that the active ingredients of the essential oil may modulate glutamatergic and gamma amino butyric acid (GABA)-ergic transmission because similar substances produce anticonvulsant activity in this way (Wie et al., 1997; Szabadics and Erdelyi, 2000).

In conclusion, MEG is a biologically active molecule. It possesses activity that under certain conditions appears to produce what could be considered beneficial effects. Although MEG may have therapeutic applications, it is not recognized as the agent of choice for such conditions as acute anaphylaxis, gastrointestinal spasms, and seizures, nor is it widely accepted as an analgesic or general anesthetic. Relative to more commonly used medications, MEG is not as efficacious, which is the primary reason that it is not the agent of choice. Although MEG has structural and physical–chemical properties that are relatively favorable for absorption and distribution to the site of action, metabolism and excretion result in toxicological and pathological findings that preclude MEG from general consideration for general drug development. This may be unfortunate because, for the most part, both the beneficial and hazardous effects can be determined based on the dose and duration of exposure. Having reviewed MEG's beneficial effects, the following sections focus on its toxicity and pathological effects.

NONORGAN-DIRECTED TOXICITY

SURVIVAL

Exposure levels sufficient to have an effect on survival have been determined for MEG in some types of animals, using specified routes of administration. With regard to acute exposure, MEG is relatively harmless because LD_{50} values (the lethal dose that produces death in 50% of the treated animals) are in the g/kg range (Table 1.8).

TABLE 1.8
LD_{50} Values for MEG in Animals

Species	Route of Administration	LD_{50} (g/kg)	Reference
Rat	Oral	1.56	Jenner et al., 1964
		0.81	Keating, 1972
		1.179	Beroza et al., 1975
Mouse	i.p.	0.54	Engelbrecht et al., 1972
	i.v.	0.112	
Rabbit	Dermal	>5	Keating, 1972
		2.025	Beroza et al., 1975

Note: For comparison, the LD_{50} value for morphine is approximately 1 mg/kg, and for dioxin (TCDD) it is 1 μg/kg. (Eaton, E.L. and Klaassen, C.D. (1996), Principles of toxicology, in C.K. Klaassen, M.O. Amdur, and J. Doull (Eds.), *Toxicology: The Basic Science of Poisons*, New York: McGraw-Hill, p. 14.)

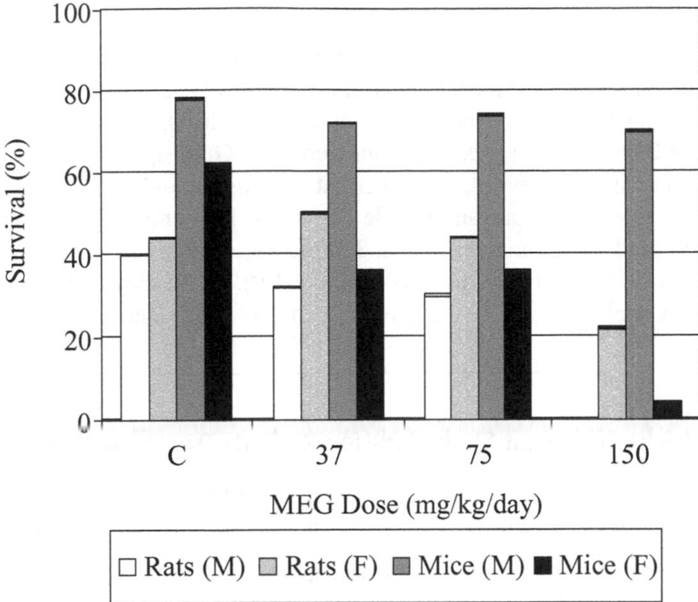

FIGURE 1.4 Survival of rats and mice given MEG by gavage for 14 weeks.

However, acute lethality does not provide a complete picture of the toxicity spectrum of MEG, because although low doses are not acutely toxic, prolonged exposure to MEG at low doses can produce deleterious effects, e.g., carcinogenicity, which can decrease survival. Subchronic rodent studies (Abdo et al., 2001), used to evaluate toxicity and set dose levels for chronic studies, showed that male and female F344 rats could be administered up to 1000 mg MEG/kg, 5 d a week for 14 weeks, with no effect on survival. For the B6C3F$_1$ mice, all but one male survived a dose of 1000 mg/kg, whereas survival was unaffected in female mice given up to 300 mg/kg. In a chronic study (Johnson et al., 2000), survival was affected by MEG treatment at doses greater than 75 mg/kg for male F344 rats, and 150 mg/kg for female F344 rats and male B6C3F$_1$ mice, and 37 mg/kg for female B6C3F$_1$ mice (Figure 1.4). Survival was affected at these relatively lower-dose and multiple exposures because MEG caused cancer. The organs affected and the types of cancer found in these organs are described in greater detail in the section on target organs.

In the chronic rat and mouse studies, the survival responses differed based on gender. In general, survival was increased at a given dose level for female rats when compared with male rats. The basis for the gender effects on survival is apparent from the toxicological findings at necropsy. In contrast, survival was unaffected by MEG treatment in male mice, whereas in the female mice survival was affected at the lowest dose tested (37 mg/kg). However, the incidences of severe pathology findings were higher in the males, thereby making the survival results for the male and female mice difficult to explain. At the present time there is no clear explanation for the apparent gender effect on survival for the mice.

Body Weight

Body weight (BW) is commonly used as a toxicological endpoint to assess the effect of a toxicant on growth, nutritional status, and the general health of a test system. MEG affects BW and BW gain in rodents (Johnson et al., 2000; Abdo et al., 2001). The group mean BW values for rodents administered MEG for up to 14 weeks decreased slightly less than 10% (relative to control) at 300 mg/kg and approximately 30 and 15% at 1000 mg/kg for male and female F344 rats. For female B6C3F$_1$ mice, these values decreased approximately 10% at 300 mg/kg, whereas doses up to 300 mg/kg had no effect on male B6C3F$_1$ mice BW (Table 1.9). When the duration of exposure to MEG was prolonged for 2 yr, a greater than 10% BW decrease was observed at

TABLE 1.9
Body Weight Effects (at Study End) for Rodents Treated with MEG

Species	Sex	Parameter	Dosage (mg/kg)					
			C	10	30	100	300	1000
			14-Week Study					
F344 rat	M	BW (g)	331	331	332	322	304**	231**
		BW gain (g)	214	218	216	203	188**	119**
	F	BW (g)	196	184*	187*	183**	180**	164**
		BW gain (g)	99	89*	89*	86**	83**	69**
B6C3F$_1$ mice	M	BW (g)	34.0	34.1	33.7	34.2	32.0	—
		BW gain (g)	11.1	10.5	10.3	10.5	8.4**	—
	F	BW (g)	30.3	30.7	30.7	29.6	27.6**	—
		BW gain (g)	10.7	10.7	9.9	9.6	8.0**	—

2-Yr Study

Species	Sex	Parameter	Dosage (mg/kg)				
			C	37	75	150	300[a]
F344 rat	M	BW (g)	414	387**	397**	—[b]	—[b]
	F		344	316**	277**	257**	256**
B6C3F$_1$ mice	M		52.2	47.1**	44.0**	45.4**	—
	F		60.9	37.4**	34.2**	33.0**	—

Note: Different from vehicle control (C) at *($p \leq .05$) or **($p \leq .01$).

[a] Group dosed for 52 weeks with MEG and then with the vehicle alone for the remainder of the study (rats only).

[b] No survivors.

Source: Data from Johnson, J.D., Ryan, M.J., Toft, J.D., II, Graves, S.W., Hejtmancik, M.R., Cunningham, M.L., Herbert, R. and Abdo, K.M. (2000), Two-year toxicity and carcinogenicity study of methyleugenol in F344/N rats and B6C3F$_1$ mice, *Journal of Agricultural and Food Chemistry*, 48: 3620–3632 and Abdo, K.M., Cunningham, M.L., Snell, M.L., Herbert, R.A., Travlos, G.S., Eldridge, S.R., and Bucher, J.R. (2001), 14-Week toxicity and cell proliferation of methyleugenol administered by gavage to F344 rats and B6C3F$_1$ mice, *Food and Chemical Toxicology*, 39: 303–316.

a dose ≥75 mg/kg for F344 rats and 37 mg/kg for B6C3F$_1$ mice. Thus, although relatively large and repeated doses of MEG are needed to affect the BW of rodents, it appears that the effective dose decreases as the duration of exposure increases.

The MEG dose-related decreases in BW could be explained by structural and functional impairment of various target organs affected by MEG exposure. As described in the following section on target organs, the liver and stomach can be severely affected by MEG, which probably explains the effect on BW. Briefly, MEG-induced hepatic toxicity may alter protein, fat, and carbohydrate anabolism and catabolism, as suggested by such clinical findings as hypoproteinemia and hypoalbuminemia, thereby indicating that the animals had a reduced nutritional status. Furthermore, MEG-induced glandular stomach atrophy (parietal and chief cell loss) contributes to the poor nutritional status and subsequent decreased BW. Destruction of these cells leads to hypochlorhydria, increased pH, reduced release of digestive enzyme precursors, and decreased activity of acid-activated digestive enzymes, e.g., pepsin, trypsin, and chymotrypsin. The toxicity of MEG on the stomach and liver appears to sufficiently lower efficient utilization of food, thereby leading to depressed BW and decreased BW gain, as well as conditions associated with poor nutrition, such as anemia (Abdo et al., 2001).

Food and Water Consumption

Little information pertaining to an inhibitory effect or enhancement by MEG on food and water consumption is available. Osborne et al. (1981) conducted a 91-d feeding study using Sprague–Dawley (SD) rats, which resulted in an average daily intake of 18 mg MEG/kg. Weekly food consumption measurements for the animals with MEG in the feed were not significantly different from those for the animals fed untreated feed. There were no other food consumption results available from the Osborne citation, and no additional results were found in the literature regarding further effects on food consumption or whether MEG affects water consumption.

TARGET ORGAN TOXICITY

MEG exposure has been shown to damage cells of certain organs more than others. The organs that display toxicity and altered pathological findings as a result of chemical insult are referred to as target organs, which are often identified by conducting subchronic and chronic animal toxicology studies. A summary of the toxicology studies conducted and a description of the experimental designs used are presented in Table 1.10. From such research, the target organs for MEG include the liver, stomach, blood, testes and uterus (reproductive system), kidney, peritoneum, and skin. The MEG-related toxicological and pathological findings in these organs, as well as the etiology, are reviewed in the sections that follow.

Liver

MEG causes morphological damage to the liver, based on microscopic examination of the hepatocytes (Table 1.11), and liver dysfunction as assessed by evaluating

TABLE 1.10
Experimental Design Summaries for Referenced Toxicology Studies of MEG

Duration	Route of Administration	Species	Strain	Sex	Number of Animals/ Group	Exposure Levels	Reference
91 Days	Dosed feed	Rat	Sprague–Dawley	M/F	24/sex	18 mg/kg/d (average daily intake)	Osborne et al., 1981
18 Months	Intraperitoneal injection	Mice	B6C3F$_1$	M	58	Injected on days 1, 8, 15, and 22 of age. Total dosage: 4.75 μmol (0.85 mg).	Miller et al., 1983
14 Weeks	Gavage[a]	Rat	Fischer 344	M/F	10/sex	0 (vehicle control), 10, 30, 100, 300, or 1000 mg/kg/d	Abdo et al., 2001
14 Weeks	Gavage[a]	Mice	B6C3F$_1$	M/F	10/sex	0 (vehicle control), 10, 30, 100, 300, or 1000 mg/kg/d	Abdo et al., 2001
2 Yr	Gavage[a]	Rat	Fischer 344	M/F	50/sex	0 (vehicle control), 37, 75, or 150 mg/kg/d; and 300 mg/kg/d for the first 53 weeks	Johnson et al., 2000
2 Yr	Gavage[a]	Mice	B6C3F$_1$	M/F	50/sex	0 (vehicle control), 37, 75, or 150 mg/kg/d	Johnson et al., 2000

[a] Formulated as a suspension using 0.5% aqueous methylcellulose as the vehicle.

clinical pathology end points (Table 1.12). The toxicological and pathological findings attributed to MEG exposure are divided below into hepatic effects, i.e., damage to the hepatocytes, and intrahepatic effects, i.e., damage to the biliary system.

Hepatocellular Pathology

Subchronic Exposure

The pathological changes found in the liver are dependent on the dose and duration of exposure. In rats and mice exposed to MEG for 14 weeks by the gavage route of

TABLE 1.11
Summary of MEG-Related Microscopic Findings in the Liver

Rats	Mice
Subchronic Exposure	
Cytologic alterations	Cytologic alterations
Cytomegaly	Necrosis
Kupffer cell pigment	Bile duct hyperplasia
Basophilic (males) and mixed cell foci	Subacute inflammation
Bile duct hyperplasia	
Adenoma (males)	
Chronic Exposure	
Rats	**Mice**
Hypertrophy	Hypertrophy
Bile duct and oval cell hyperplasia	Bile duct and oval cell hyperplasia
Cystic degeneration	Eosinophilic foci
Basophilic, eosinophilic, and mixed cell foci	Hematopoietic cell proliferation
	Hemosiderin pigmentation
Adenoma	
	Chronic active inflammation
Carcinoma	
	Necrosis
Cholangioma	
	Adenoma
Cholangioma and cholangiocarcinoma	
	Carcinoma
	Hepatoblastoma

Source: From Johnson, J.D., Ryan, M.J., Toft, J.D., II, Graves, S.W., Hejtmancik, M.R., Cunningham, M.L., Herbert, R. and Abdo, K.M. (2000), Two-year toxicity and carcinogenicity study of methyleugenol in F344/N rats and B6C3F$_1$ mice, *Journal of Agricultural and Food Chemistry*, 48: 3620–3632; Abdo, K.M., Cunningham, M.L., Snell, M.L., Herbert, R.A., Travlos, G.S., Eldridge, S.R., and Bucher, J.R. (2001), 14-Week toxicity and cell proliferation of methyleugenol administered by gavage to F344 rats and B6C3F$_1$ mice, *Food and Chemical Toxicology*, 39: 303–316.

administration, doses of 300 mg/kg/d or greater resulted in microscopic findings, whereas 100 mg/kg/d or less produced no observable effects (Abdo et al., 2001). Likewise, no microscopic findings were observed in the liver (or 27 other major organs and tissues) when SD rats were fed MEG at an average daily intake of 18 mg/kg for 91 d (Osborne et al., 1981).

Microscopic findings in the hepatocytes of rats and mice were somewhat similar, although the latter species exhibited additional toxicity not manifested by the former (Table 1.11). Cytologic alterations were observed in the hepatocytes of both species. In the F344 rat, the cytoplasm of individual hepatocytes appeared to have altered tinctorial staining qualities (see the subsection intrahepatic biliary pathology), and

TABLE 1.12
Summary of Clinical Pathology Findings in Rats Given MEG by Gavage for 14 Weeks

Hematology	Serum Chemistry
↓Hemoglobin	↓Total protein
↓Hematocrit	↓Albumin
↓MCV, MCH	↑Alanine aminotransferase
↑Platelets	↑Sorbitol dehydrogenase
	↓Alkaline phosphatase
	↑Bile acids

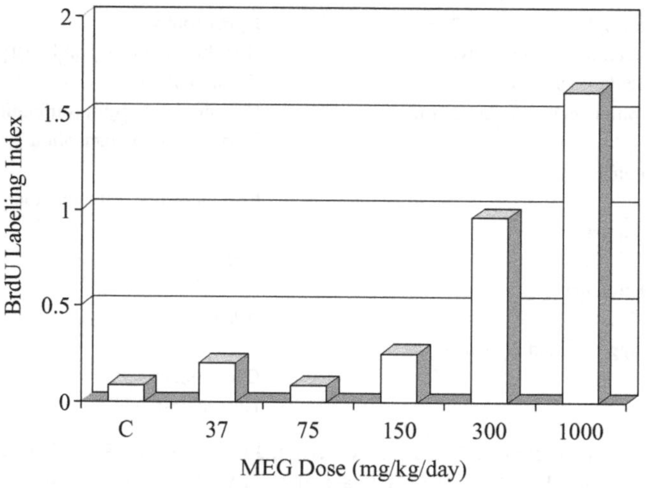

FIGURE 1.5 Effect of MEG on hepatocellular proliferation.

periportal hepatocytes were enlarged (cytomegaly). In B6C3F$_1$ mice, hepatocytes exhibited nuclear and cytoplasmic enlargement (hypertrophy) and increased cytoplasmic eosinophilia. MEG appeared to increase hepatocellular mitotic activity, which supported the results obtained when the effect of MEG on cellular proliferation was measured. Proliferation was determined by measuring the incorporation of bromodeoxyuridine (BrdU) into the cell nucleus. The hepatocellular proliferation caused by MEG treatment was extensive. The livers of female F344 rats exhibited increases of 195, 1020, and 1780% relative to control in the incorporation of BrdU at doses of 150, 300, or 1000 mg MEG/kg/d, respectively, after 30 d of dosing (Figure 1.5).

In addition to morphological changes in the hepatocyte, MEG-related necrosis was observed for individual hepatocytes (Table 1.11). Necrotic cells were found scattered throughout the liver. Additional evidence for hepatocellular damage was obtained by assaying the blood (serum) for liver cytosol-derived enzymes. Alanine aminotransferase (ALT) and sorbitol dehydrogenase (SDH), two enzymes found in the cytoplasm of hepatocytes, were increased in serum from male and female F344 rats given 100 mg MEG/kg/d or more. The percentage increases (relative to control) ranged from 28 to 74% for ALT and were approximately 50% for SDH.

MEG impairs hepatic protein synthesis, but this appears to be a secondary effect of glandular stomach lesions. The effect on protein synthesis is based on changes in serum protein and albumin concentrations. Decreases in total protein (TP) and albumin ranged from 11 to 13% and from 10 to 14% (relative to control), respectively, when F344 rats were given at least 100 mg/kg/d (TP) or 300 mg/kg/d (albumin). These results indicated that exposure to MEG could lead to the development of hypoproteinemia and hypoalbuminemia. The mechanism of this effect on hepatic protein synthesis is described in further detail under the subsection that describes glandular stomach toxicity.

Chronic Exposure

Lifetime exposure to MEG resulted in nonneoplastic and neoplastic lesions when tested in F344 rats and B6C3F$_1$ mice (Johnson et al., 2000). The types of lesions found in this 2-yr study are summarized in Table 1.11. Nonneoplastic and neoplastic lesions were observed at doses as low as 37 mg/kg/d. These lesions increased with increasing dose. In addition, the experimental design for this study included a stop exposure group (Table 1.10). The lesion incidence did not decrease when MEG treatment was discontinued. Rather, there was an increase in nonneoplastic lesions and some types of neoplastic lesions.

The nonneoplastic lesions that developed from chronic exposure to MEG were more severe than those observed after subchronic exposure. Use of interim terminations during the chronic study, together with the results of the terminal necropsy, provide insight into the time course of these lesions and the progressive damage that developed with increasing duration of exposure (Johnson et al., 2000). After 6 months, there were no control group animals, but all 5 male and 5 female F344 rats receiving 300 mg/kg/d had oval cell hyperplasia and hypertrophy, and the males also had mixed cell foci. The incidence of these lesions was the same after 12 months, but the severity of the oval cell hyperplasia and hypertrophy increased for the females. Additional lesions found in the rats at 12 months included eosinophilic and mixed cell foci (males and females). These nonneoplastic lesions found at 6 and 12 months were also observed after 2 yr of exposure in the F344 rat. For rats and mice, the incidences and severities of oval cell hyperplasia and hepatocellular hypertrophy were increased significantly, compared with the control at doses equal to or greater than 37 mg/kg/d. A positive correlation with dose was also evident for the eosinophilic and mixed cell foci (at least 37 mg/kg/d). For rats, the incidences of the basophilic foci tended to decrease for the male 150 and 300 mg/kg/d groups but were statistically decreased for the female 150 and 300 mg/kg/d groups. Mice also exhibited hematopoietic proliferation and pigmentation hemosiderin (at least 37 mg/kg/d), as well as chronic active inflammation and necrosis (at least 75 mg/kg/d). Thus, MEG caused the development of nonneoplastic lesions, and the type of lesion and the incidence and severity of the lesions clearly increased as the exposure to MEG was extended.

Chronic exposure to MEG resulted in the development of neoplastic lesions. Liver tumors attributed to MEG exposure were initially demonstrated using a preweaning exposure bioassay (Miller et al., 1983). Prior to weaning, male B6C3F_1 mice were injected (i.p.) with MEG, total dose 4.75 μmol, on days 1, 8, 15, and 22. Two groups of controls were used: uninjected or vehicle (trioctanoin)-injected. Mice were examined for liver tumors at 13 months (≥20 animals/group) and at 13 to 18 months (≥58 animals/group). The rates of hepatoma-bearing mice were 12, 5, and 70% at 13 months, and 28, 41, and 96% at 13 to 18 months for the uninjected controls, vehicle controls, and MEG-treated mice, respectively. In addition, the average number of hepatomas per mouse for both control groups at either time point was 0.5 or less, whereas the MEG-treated mice had 1.5 tumors per mouse at 13 months and 3.2 tumors per mouse at 13 to 18 months.

In a more recent study, neoplastic lesions were found in F344 rats and B6C3F_1 mice given MEG for 2 yr (Johnson et al., 2000). These lesions are summarized in Table 1.11. For the F344 rats, the presence of an adenoma was found as early as 14 weeks (1 of 10 rats at 1000 mg/kg/d), and in the 2-yr study, an interim necropsy at 12 months showed that 4 of 5 male F344 rats given 300 mg/kg/d had adenomas. At termination of the 2-yr study, the incidence of liver neoplasms was quite dramatic (Figure 1.6). MEG caused a dose-related increase in the incidence of adenomas. Whereas only 10% of the male and 2% of the female controls exhibited hepatic adenomas, 76% of the males and 66% of the females receiving 150 mg/kg/d had adenomas. In B6C3F_1 mice, MEG induced adenoma formation in most treated male and female animals. The rates of animals bearing adenomas for dosages of 0 (control), 37, 75, or 150 mg/kg/d for 2 yr were 53, 86, 76, and 78% for the males, and 40, 96, 94, and 82% for the females, respectively.

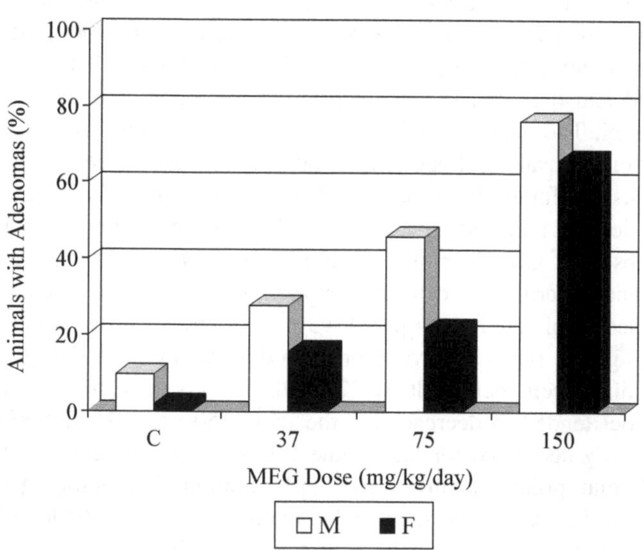

FIGURE 1.6 Hepatic adenomas in F344 rats given MEG by gavage for 2 yr.

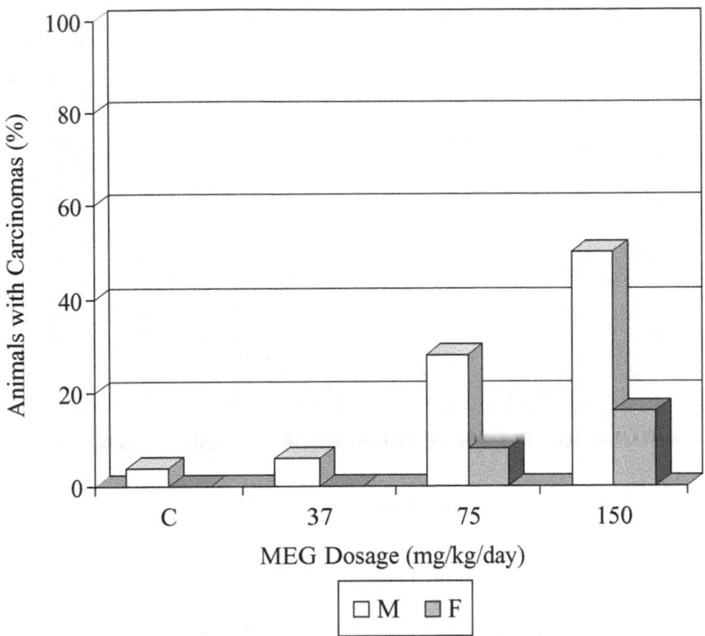

FIGURE 1.7 Hepatic carcinomas in F344 rats given MEG by gavage for 2 yr.

In addition, carcinomas were associated with MEG exposure. Figure 1.7 illustrates the percentages of F344 rats that developed carcinomas after a 2-yr exposure. In $B6C3F_1$ mice, MEG induced carcinoma formation in most treated male and female $B6C3F_1$ animals. Rates of animals bearing carcinomas following doses of 0 (control), 37, 75, or 150 mg/kg/d for 2 yr were 20, 40, 38, and 18% for the males and 14, 74, 94, and 94% for the females, respectively.

Finally, MEG caused the development of biliary system neoplasms in F344 rats and $B6C3F_1$ mice, and in the mice, also resulted in the generation of hepatoblastomas. Further details about the biliary system neoplasms are given in the following section. As for the hepatoblastomas, the incidences of this tumor in female $B6C3F_1$ mice were 0, 6, 11, and 15 out of 50 animals in the 0 (control), 37, 75, and 150 mg MEG/kg/d dose groups, respectively. A hepatoblastoma is a malignant neoplasm, similar in metastatic potential to that of a hepatocellular carcinoma. The presence of these types of neoplasms, i.e., adenoma, carcinoma, and hepatoblastoma, is considered very important in the analyses of evaluating the carcinogenic potential of a chemical on the liver (Johnson et al., 2000).

Intrahepatic Biliary Pathology

Subchronic Exposure

Intrahepatic biliary processes were altered when F344 rats and $B6C3F_1$ mice were treated with MEG (Abdo et al., 2001). Periportal hepatocytes accumulated yellow-gold to green pigment within the cytoplasm (Kupffer cell pigmentation; Table 1.11).

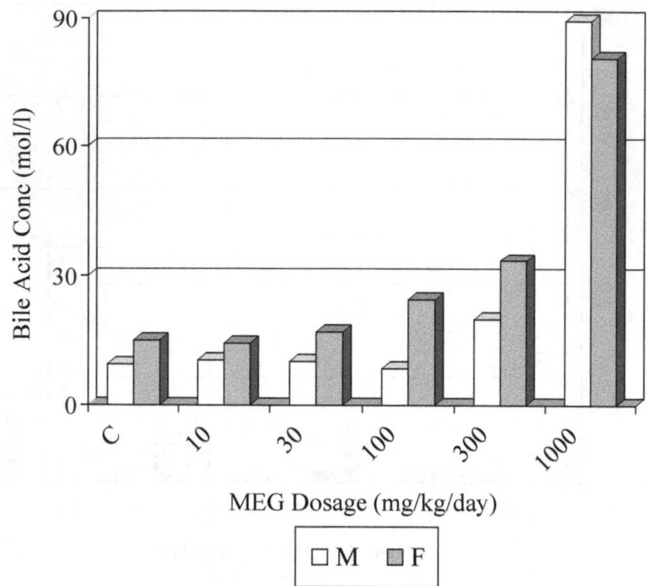

FIGURE 1.8 Serum bile acid levels in F344 rats given MEG by gavage for 14 weeks.

Based on the type of cell involved (macrophagic), location in the lobule (nearest the portal triad — Zone 1), and the characteristic color of the pigment (Cheville, 1994; Kanel and Korula, 2000), this finding suggested that an accumulation of bilirubin-derived components and bile occurred. This finding is consistent with canalicular cholestasis, which describes an impaired secretion of bile, as well as compromised secretion of solutes normally found in the bile. Other microscopic findings within the portal areas included small bile ductules and immature biliary cells. Repeated dosing and chronic exposure are known to cause bile duct destruction, which leads to biliary duct proliferation (bile duct hyperplasia) and even fibrosis (Moslen, 1996). These results suggest that MEG can damage intrahepatic bile ducts, thereby producing a cholangiodestructive cholestasis.

Clinical chemistry endpoints indicated that MEG altered normal biliary system function. In the Abdo et al. (2001) study, doses equal to or greater than 300 mg MEG/kg/d for 14 weeks increased bile acid concentrations in the serum (Figure 1.8). Relative to control, bile acid levels for male and female F344 rats increased 107 and 122% at a dose of 300 mg MEG/kg/d and 841 and 438% at a dose of 1000 mg MEG/kg/d, respectively.

Chronic Exposure

A longer duration of exposure to MEG increased the extent of damage to the biliary system (Johnson et al., 2000). After 6 months of exposure to 300 mg MEG/kg/d, there were no MEG-related microscopic findings in the biliary tract of male F344 rats. However, at 12 months, 5 of 5 male F344 rats had cystic (focal) degeneration. After 2 yr of exposure, cystic degeneration was observed in 4/50, 2/50, 25/50, or 38/50 male F344 rats given doses of 0, 37, 75, or 150 mg/kg/d, respectively. For female F344 rats, cystic degeneration significantly increased for the 150 mg/kg/d group (29/50), as was bile duct hyperplasia for the 150 and 300 mg/kg/d groups (22/49 and 30/50). Female B6C3F$_1$ mice exhibited a similar response, but the incidence was approximately half. Male F344 rats in the 37 to 150 mg/kg/d groups actually exhibited a decreased incidence of bile duct hyperplasia. In addition to these nonneoplastic findings, MEG also produced neoplastic changes in the bile ducts. Cholangioma and cholangiocarcinoma were present in F344 rats given 150 mg/kg/d. Although the incidence of these neoplasms was not significant, the incidence in the "stop exposure" group was statistically significant (p .01). The stop exposure group was composed of animals given 300 mg MEG/kg/d for 52 weeks, after which the animals were given only the vehicle for the remainder of the 2-yr study. There were no control animals that exhibited these neoplasms, but the rates of 300 mg/kg/d stop exposure group animals with cholangioma were 12 and 16%, and with cholangiocarcinoma were 14 and 18% for males and females, respectively. Thus, long-term exposure to MEG at the dose levels tested in these rodent species indicated that MEG is carcinogenic to intrahepatic biliary tissue.

Metabolism and Bioactivation

The toxicity of MEG occurs because of reactive metabolite formation. Rather than detoxification, the liver metabolizes MEG to a biologically active metabolite, i.e., 1-hydroxymethyleugenol. The metabolic pathway for MEG is illustrated in Figure 1.9. The metabolism of MEG and other alkenebenzene derivatives was initially described by Solheim and Scheline (1973, 1976). Since then, the metabolism and bioactivation of MEG have been extensively studied and reviewed (Drinkwater et al., 1976; Rompelberg et al., 1996; Gardner et al., 1997; Burkey et al., 2000; Guenthner and Luo, 2001; Smith et al., 2002). The following sections describe this bioactivation process, the candidate reactive metabolites, macromolecular adduct formation, and oncogene alterations that lead to nonneoplastic and neoplastic liver lesions.

The metabolic pathway for MEG was first reported by Solheim and Scheline (1976). The identification of the components of this pathway was aided by the work done previously on structurally similar alkenebenzene derivatives, e.g., safrole, estragole, and eugenol (Borchert et al., 1973a, 1973b; Stillwell et al., 1974). The structure of MEG lends itself to a variety of possible metabolic pathways. First, hydroxylation of the benzene ring or the methoxy or allyl substituents can occur. In regard to the benzene ring, hydroxy substituents can be added at the 2nd or 6th positions. However, the 2-hydroxy metabolite is more likely due to steric hindrance at the number 6 position. In addition, after removal of either of the two or both methoxy substituents, the benzene ring can be hydroxylated at positions 3 or 4. Finally, the allyl substituent can be hydroxylated at the 1 position. The formation of this 1-hydroxy metabolite is considered responsible for MEG-related toxicity, and the bioactivation of this molecule is described in further detail in the following text. A second metabolic pathway, also an oxidative process, involves epoxidation. An epoxide can form at the 2–3 double bond on the allyl substituent. Solheim and Scheline reported that the epoxide is oxidized to the diol and, ultimately, to the keto

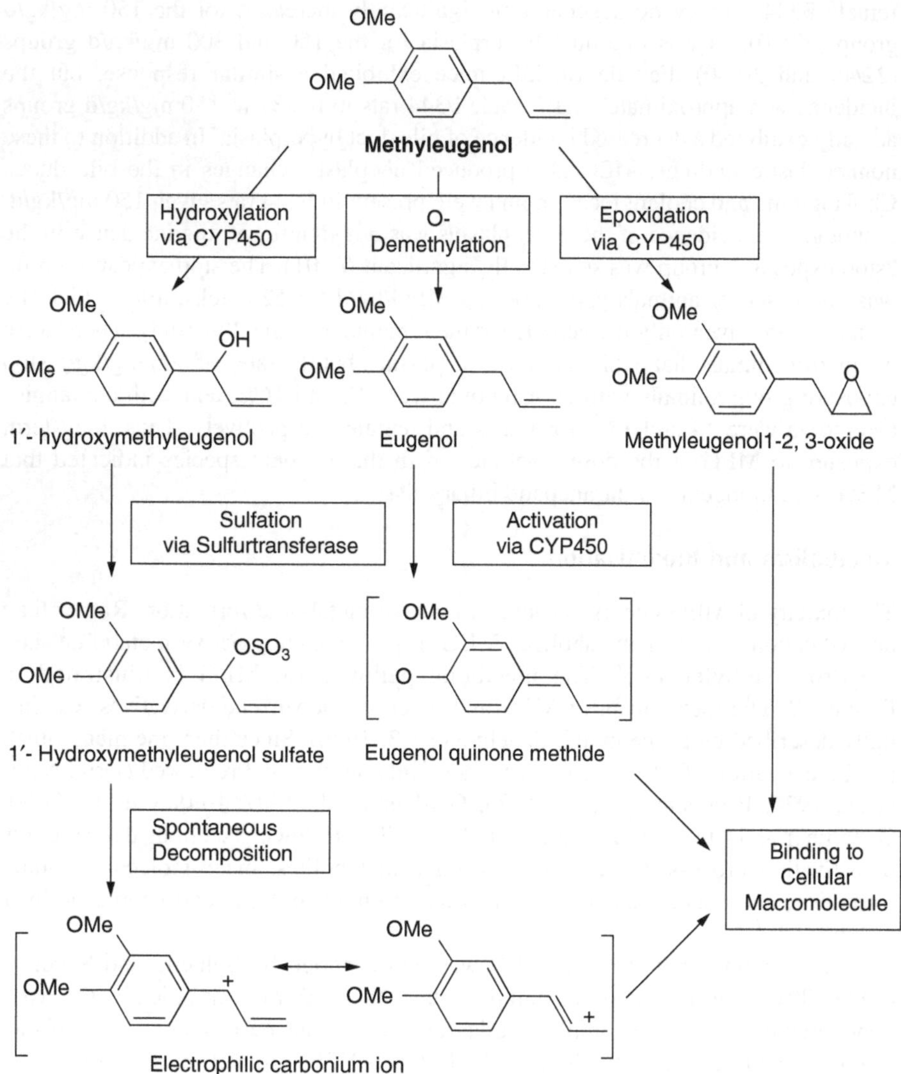

FIGURE 1.9 Metabolic pathway for MEG. (Adapted from Gardner, I., Bergin, P., Stening, P., Kenna, J.G., and Caldwell, J., 1996, Immunochemical detection of covalently modified protein adducts in livers of rats treated with methyleugenol, CRC, *Critical Reviews in Toxicology*, 9: 713–721.)

acid before being excreted. A third pathway for the biotransformation of MEG is also an oxidation reaction and involves cleavage of the 2nd or 3rd carbons of the allyl substituent, thereby producing a 1 or 2 aldehyde, respectively. Fourth, the allyl substituent of MEG can undergo a simultaneous ω-oxidation and migration of the double bond that produces an alcohol group at the 3 position and shifts the double bond from the 2–3 to the 1–2 position. Finally, MEG and related congeners may be metabolized to a quinone methide complex, an extremely reactive electrophile (Burkey et al., 2000). For this metabolite to form, the para-methoxy group must undergo O-demethylation, and a methylene or substituted methylene group must be put in place for one of the carbonyl oxygens. Thus, biotransformation of MEG occurs by oxidation, which results in the formation of several metabolites.

The work of Miller and coworkers (1983) also showed the significance of an MEG metabolite in tumor formation. Using the preweaning bioassay described above, they also treated male $B6C3F_1$ mice with 1-hydroxyMEG (total dose 2.85 μmol). At 13 months, 68% of the mice had hepatomas with an average of 2.3 liver tumors per mouse. These endpoints increased to 93% and 3.5 at 13 to 18 months. Although the parent and metabolite caused a similar percentage of animals to develop tumors, the number of tumors per animal was greater for 1-hydroxyMEG than for MEG. These results showed that there was good reason to believe that bioactivation played a major role in MEG-induced tumor formation.

1-HydroxyMEG Metabolite

The 1-hydroxyMEG metabolite is considered to be the most biologically active of the MEG metabolites. The first step in its formation involves hydroxylation at the 1 position of the allyl substituent. The second step involves sulfation, thereby producing a 1-sulfoxyMEG metabolite. This sulfate ester is unstable, and it readily hydrolyzes. Hydrolysis results in a carbonium ion intermediate, which is a highly reactive electrophile that has been shown to covalently bind with DNA and proteins (Randerath et al., 1984; Chan and Caldwell 1992; Gardner et al., 1996).

Gardner and coworkers (1997) described the kinetics of the 1-hydroxylation of MEG. MEG (1 to 2000 *μM*) was incubated with liver microsomes collected from untreated F344 male rats, and the formation of 1-hydroxyMEG was measured over time using a GC-MS analytical method. A velocity vs. velocity/substrate concentration plot was nonlinear, which indicated that 1-hydroxylation of MEG was catalyzed by high- and low-affinity enzyme components. The K_m (binding constant) and V_{max} (maximum reaction velocity) values for the high-affinity activity was 74.9 ± 9.0 μ*M* and 1.42 ± 0.17 nmol/min/nmol, respectively. However, the kinetic estimates for the low-affinity activity could not be defined accurately due to the low solubility of MEG in the reaction mixture. An approximation of the K_m for the low-affinity activity was >2 m*M*. In addition, these investigators showed that the kinetics of this reaction may be sex dependent. When microsomes were collected from female F344 rats and incubated with a low concentration of MEG (20 μ*M*), there was a statistically significant increase in 1-hydroxyMEG formation when compared with the males. However, at higher concentrations there were no apparent differences noted between the sexes.

The kinetic values of this reaction are useful for estimating formation of the 1-hydroxy metabolite for a given level of exposure. Using the apparent K_m values of these enzymes and the plasma concentrations that result from various doses, it is possible to estimate whether bioactivation of the high-affinity enzymic components will occur. For example, following a single oral administration of MEG to F344 rats, using a dose of 37 to 150 mg/kg resulted in plasma concentrations of approximately 1 to 7 μg/ml (~6 to 39 μ*M*) (Graves and Runyon, 1995), which approaches the K_m of the high-affinity enzymic components. Consequently, MEG would be metabolized to the 1-hydroxy metabolite when the internal dose reached such levels.

Gardner and coworkers (1997) also investigated the nature of the cytochrome P450 (CYP) isozymes that catalyze the formation of the 1-hydroxy metabolite, as well as characterized the changes associated with autoinduction of the microsomal enzymes following repeated exposure with MEG. In these experiments, induction and inhibition of liver microsomal 1-hydroxylation of MEG was determined after pretreatment with specific microsomal enzyme inducers and inhibitors. The results were different for the high- and low-affinity reactions. There are several CYPs that can catalyze the high-affinity reaction, e.g., CYP 2B1/2 and CYP 1A2. In contrast, at low substrate concentrations the 1-hydroxylation is catalyzed primarily by CYP 2E1 as well as at least one other enzyme, which Gardner and coworkers attributed to CYP 2C6. Because the microsomal enzymes are inducible, any treatment that induces their activity increases the fraction of MEG catalyzed to the 1-hydroxy metabolite. Gardner et al. (1997) showed that autoinduction of this conversion occurred in the liver when rats were given 30 to 300 mg/kg/d for 5 d. However, it did not occur at a dose of 10 mg/kg/d. They suggest that because human exposure to MEG is on the order of μg/kg/day, autoinduction will not occur, thereby lowering the possibility that the 1-hydroxy metabolite is produced.

2,3-Epoxide Metabolite

The allyl side chain of MEG is metabolized to an epoxide. Epoxidation yields the 2, 3-epoxide metabolite. Although no actual detection of an MEG epoxide has been reported, the presence of epoxide-related metabolites would suggest that epoxidation of MEG does occur. Solheim and Scheline (1976) inferred the presence of an MEG epoxide based on detection of the 2, 3-diol and keto acid, both products of further metabolism of an epoxide. Also, the epoxide and diol metabolites of chemically related structures (e.g., safrole and estragole) have been measured, and epoxidation has been estimated to occur to the same extent as 1-hydroxylation, i.e., 20 to 30% (Luo and Guenthner, 1995). Thus, it is generally accepted that MEG is metabolized to an epoxide.

Recently, the importance of the epoxide metabolite as a principal intoxicant of MEG-induced toxicity has been studied. For the epoxide to cause toxicity, it must be sufficiently stable until it can react with macromolecules located in other parts of the cell or other cells from where it was formed. A further basis for evaluating the importance of the epoxide is that no evidence has been produced to implicate it in the formation of deoxyribonucleic acid (DNA) adducts by allylbenzene analogs. Although the epoxide appears to be formed and, when present, readily forms covalent bonds with macromolecules, detoxification prevents it from being the principal intoxicant of MEG. Guenthner and Luo (2001) reported that the observed lack of *in vivo* epoxide-related genotoxicity can be attributed to two pathways that metabolize the epoxide metabolite. Using a liver perfusion system, epoxides were almost simultaneously converted to the dihydrodiol by epoxide hydrolases and then conjugated to glutathione by glutathione transferase. The K_m for the epoxides was in the low-millimolar range, and the V_{max} was approximately 1 μmol of epoxide hydrolyzed/min/g liver. In essence, the epoxide is metabolized more rapidly than it is created, as well as conjugated as fast as the epoxide is hydrolyzed. Even if saturation were to occur, e.g., tenfold higher concentrations, efficient detoxification of the epoxide would still be observed, according to these researchers. The lack of genotoxicity from the epoxide metabolite has also been reported for a molecule structurally similar to MEG, i.e., estragole (Phillips et al., 1981). Thus, the epoxide metabolite does not appear to be responsible for MEG-induced genotoxicity.

The epoxide metabolite has also been studied for its cytotoxicity. An isolated liver perfusion model was used to measure cytotoxicity (release of alanyl aminotransferase into the perfusate) when perfused with allylbenzene, estragole, or their respective epoxide metabolites (Guenthner and Luo, 2001). No cytotoxicity was observed following perfusion with the epoxides. The parent molecules were cytosolic but not the epoxide metabolites, even at very high concentrations. The epoxide metabolites formed small amounts of covalent adducts to proteins but not with DNA. Using a different method, the cytotoxicity of the MEG epoxide has also been evaluated by measuring lactate dehydrogenase (LDH) release from cultured primary hepatocytes isolated from male F344 rats and female B6C3F$_1$ mice (Burkey et al., 2000). LDH leakage was approximately 50% from 600 to 3000 μM, using rat hepatocytes, and 1000 μM, using mouse hepatocytes (leakage increased to 80 to 90% at 3000 to 5000 μM). When cyclohexane oxide (CHOX) (2000 μM), a competitive inhibitor of epoxide hydrolase, was incubated with noncytotoxic concentrations of MEG (10 μM), then the leakage of LDH increased from approximately 50 to 80% for the rat and mouse hepatocytes. Thus, the epoxide, although formed, is rapidly detoxified and does not appear to be responsible for MEG-related cytotoxicity.

Quinone Methide Metabolite

One other highly reactive electrophilic species that can be formed by metabolism of MEG is a quinone methide complex. The formation of this metabolite involves more than one step. First, the para-methoxy group is removed by O-demethylation, which results in the production of eugenol (Solheim and Scheline, 1976). A methylene group is then substituted for a carbonyl oxygen, which can be catalyzed by a peroxidase or CYP enzyme (Thompson et al., 1990). MEG metabolism to the quinone methide metabolite is considered to be very limited, if formed at all. As noted by Burkey et al. (2000), because the extent to which MEG is metabolized to eugenol is minor — only 7% of the urinary metabolites 24 h after a dose of 200 mg/kg (Solheim and Scheline, 1976) — the presence of a quinone methide may occur but only to a minor extent for MEG. Thus, although the quinone methide species has been implicated in the toxicity of related allkylbenzene molecules (Thompson et al., 1991; Thompson et al., 1998), it does not appear to play a role in the mechanism of toxicity for MEG.

Etiology

MEG Forms DNA and Protein Adducts

The ultimately reactive metabolites of MEG can covalently bind to proteins and DNA. The formation of such adducts may lead to cellular damage and genotoxicity. Evidence of MEG adduct formation was first reported by Randerath and coworkers

(1984) using a technique of ^{32}phosphorous (P) postlabeling of DNA adducts. These results were confirmed when the technique was made more efficient using the high-performance liquid chromatography (HPLC) technique (Levy and Weber, 1988). In the Randerath et al. (1984) study, adult female CD-1 mice were given a single i.p. administration of MEG (approximately 80 or 400 mg/kg), and the livers were removed 24 h after dosing. The total numbers of DNA adducts per 10^7 nucleotides were 150 ± 15 and 646 ± 88 for the low- and high-dose groups, respectively. For the low-dose group this amounts to 1 adduct in 66,000 normal nucleotides. Based on these results, if there are approximately 6×10^9 base pairs in a diploid genome of a single liver cell, then the total number of DNA adducts that would be formed from a dose of 80 mg/kg would be approximately 9×10^4. Thus, the amount of genetic damage that occurred as a result of generating MEG adducts in mouse liver DNA was extensive. In addition to estimating the number of adducts, these investigators identified the adducts formed, which was accomplished by taking the adduct patterns for MEG and comparing them to the chromatographic evidence obtained by administering [^{3}H]-1-hydroxysafrole to mice. The nucleotide adducts attributed to MEG metabolism were 3, 5-bisphosphate of an N^2-(*trans*-propenylbenzene-3-yl) deoxyguanosine adduct, which were approximately 80% of the adducts formed. In addition, N^2-(allylbenzen-1-yl)deoxyguanosine and N^6-adenine derivative adducts were found. Finally, a measure of the DNA-binding activity of MEG was determined using a covalent binding index (CBI). Of the ten different allyl- or propenylbenzenes tested, MEG had the highest CBI, 31.1, which exceeded estragole and safrole by about twofold and fivefold, respectively. Thus, the outcome of this study prompted these investigators to evaluate MEG further for evidence of carcinogenic activity.

In a follow-on study (Randerath et al., 1984), newborn male B6C3F$_1$ mice were injected with MEG (0.25, 0.5, 1, or 3 μmol on days 1, 8, 15, or 22 after birth, respectively), and different groups of mice were used on days 23, 29, or 43 for analysis of liver DNA. On day 23, the amount of covalently bound MEG was 72.7 ± 10.7 pmol/mg DNA. The adducts decreased to 37.1 ± 9.7 and 25.6 ± 6.2 pmol/mg DNA by days 29 and 43. The investigators noted that the decrease in adducts over time was much slower in newborns than adults, where >80% of the adducts initially formed were lost within the first week of treatment. They theorized that this may explain the higher incidence of cancer in newborn mice relative to adult mice because the former had a greater proportion of DNA adducts, which also persisted for a longer period of time. Thus, this series of studies by Randerath and coworkers provided convincing evidence that MEG had carcinogenic activity.

MEG also forms liver protein adducts. Gardner et al. (1996) prepared specific polyclonal antisera by immunizing rabbits with MEG coupled to rabbit serum albumin, which allowed them to identify protein adducts formed in rats given MEG. They treated male F344 rats with a single i.p. injection or daily i.p. injections for 5 d at doses of MEG ranging from 10 to 300 mg/kg. Analysis by the enzyme-linked immunosorbant assay (ELISA) and immunoblotting techniques showed that MEG produced a dose-dependent expression of many protein antigens in rat livers. Protein adducts were found primarily in the microsomal fraction, but the nuclear and mitochondrial fractions also had extensive protein adducts. Of the protein adducts formed, the most prevalent was a 44-kDa protein. At MEG doses of 10 or 30 mg/kg, the 44-kDa protein was the only novel protein adduct formed. At higher doses (>100 mg/kg), the 44-kDa adduct was identified more frequently, but there was also a range of other adducts formed. The solubility characteristics of the 44-kDa protein provided some evidence as to its location in the cell. This protein adduct was soluble in sodium carbonate, which is known for solubilizing peripheral, but not integral, membrane proteins. These results suggested to the investigators that the protein adduct may be a peripheral membrane protein.

Gardner and coworkers (1996) extended what is known about the mechanism of the protein adduct formation by using an *in vitro* rat-isolated hepatocyte assay. They examined the adduct pattern formed following incubation with the bioactive metabolite 1-hydroxyMEG. The 44-kDa protein adduct, as well as a range of other adducts, were produced when 1-hydroxyMEG was incubated with the hepatocytes. In fact, the adduct pattern produced by the *in vitro* assay was similar to the one obtained following the *in vivo* study. Additional evidence for the role that metabolism plays in adduct formation was demonstrated by incubating MEG with microsomes or postmitochondrial supernatants. The metabolizing capacity of these fractions with, but not without, nicotinamide adenine dinucleotide phosphate (NADPH) resulted in a range of protein adducts being formed. In a follow-on study, Gardner and coworkers (1997) pretreated rats for 3 d with dexamethasone (80 mg/kg/d) and then gave them a single i.p. injection of 100 mg/kg of MEG. Dexamethasone induces the microsomal enzymes that catalyze the formation of the 1-hydroxy metabolite. The livers had a markedly higher level of the protein adducts. These results demonstrated the crucial role of metabolism and the 1-hydroxyMEG metabolite in the formation of the protein adducts.

A Mechanism for MEG-Related Liver Cancer

The formation of the 1-hydroxy metabolite appears to play an important role in the genotoxicity of MEG. The unscheduled DNA synthesis (UDS) assay was first used to show that MEG and related compounds damage DNA (Chan and Caldwell, 1992; Howes et al., 1990). In this assay, an increase in UDS, as measured by incorporation of [^{3}H]-thymidine into the DNA, indicates that the DNA has been damaged and repair processes have been initiated. The genotoxic effect of MEG using the UDS assay was confirmed and extended by Burkey et al. (2000) using F344 rat and B6C3F$_1$ mouse hepatocytes. In both rat and mouse hepatocytes, MEG (10 μM) was incubated with or without CHOX, an inhibitor of expoxide hydrolase, or pentachlorophenol (PCP), an inhibitor of sulfation. MEG increased UDS, thereby indicating that damage to DNA had occurred. However, CHOX had no effect on MEG-induced UDS, which suggested that the epoxide metabolite was not involved in the DNA damage. In contrast, PCP completely reversed the MEG-induced UDS. Thus, PCP-mediated inhibition of the sulfated metabolite supports the conclusion that the 1-hydroxy metabolite is important in MEG-induced genotoxicity.

Gene function may be compromised when chemicals modify the DNA sequence by adduct formation or mutations. Oncogenes are those genes that, when altered by

TABLE 1.13
β-Catenin Mutations in B6C3F$_1$ Mouse Hepatocellular Neoplasms

Group	Frequency	Codon	Mutation	Bases	N	Amino Acid
Control	2/22 (9%)	32	GAT to GCT	A to C	1	Asp to Ala
		33	TCT to TTT	C to T	1	Ser to Phe
MEG	20/29 (69%)	32	GAT to GTT	A to T	5	Asp to Val
			GAT to TAT	G to T	1	Asp to Tyr
			GAT to GGT	A to G	3	Asp to Gly
			GAT to CAT	G to C	1	Asp to His
		33	TCT to TAT	C to A	1	Ser to Tyr
			TCT to TTT	C to T	1	Ser to Phe
		34	GGA to AGA	G to A	1	Gly to Arg
			GGA to CGA	G to C	1	Gly to Arg
			GGA to GTA	G to T	2	Gly to Val
		41	ACC to ATC	C to T	1	Thr to Ile
			ACC to GCC	A to G	1	Thr to Ale
		5–10	Deletions	—	2	—

Source: From Devereux, T.R., Anna, C.H., Foley, J.F., White, C.M., Sills, R.C., and Barrett, J.C. (1999), Mutation of β-catenin is an early event in chemically induced mouse hepatocellular carcinogenesis, *Oncogene*, 18: 4726–4733.

mutagenic activity of a chemical, cause neoplastic transformation of the cell. Oncogenes are involved in such processes as regulation of cellular growth, signal transduction, and nuclear transcription (Pitot and Dragon, 1996). Consequently, it is obvious how modifications to these genes can be detrimental to the organism.

Devereux et al. (1999) examined the MEG-induced liver tumors from B6C3F$_1$ mice for oncogene activation. They obtained liver specimens from the NTP 2-yr mouse gavage toxicity study (NTP, 1998; Table 1.10) and evaluated the effect of MEG on the β-catenin gene. This gene is classified as an oncogene because mutations produce β-catenin accumulation and, together with upregulation of the Wnt-signaling pathway, leads to cancer as a result of increased cell proliferation and decreased apoptosis (Gumbiner, 1997). An examination of the β-catenin gene for molecular alterations in spontaneous hepatocellular neoplasms (control group) resulted in 2 of 22 (9%) mice with a mutation, whereas 20 of 29 (69%) of MEG-treated mice exhibited β-catenin gene mutations. There were no measurable H-*ras* mutations (C→A) at codon 61 detected in the 29 MEG-induced hepatocellular neoplasms that were tested. A summary of the β-catenin mutations is summarized in Table 1.13.

Based on these results, MEG produced both deletions and point mutations to the β-catenin gene. All point mutations occurred at codons 32, 33, 34, and 41. In addition, apparently, any of the bases for a given codon was susceptible to mutation as there were 11 different base substitutions in the four affected codons. These investigators noted that there was an absence of a specific β-catenin mutation pattern for MEG even though a clear relationship existed between exposure to MEG and mutation of the β-catenin gene. Examination of the hepatocellular tumors by dose group showed a similar incidence of mutations (63 to 75%) for all groups (37 to 150 mg/kg). The absence of a dose–response effect suggests that maximum effect is observable at a dose of at least 37 mg/kg when administered for up to 2 yr. Lower doses would likely produce mutation frequencies that would allow a dose response to be observed. These researchers also noted that β-catenin mutations were observed in almost equal frequency when adenomas and carcinomas were examined, which led them to suggest that mutation to the β-catenin gene may be an early event in hepatocellular tumor formation.

The findings by Devereux et al. (1999) have important implications in understanding carcinogenesis in humans. The β-catenin mutations they observed on these specific DNA codons affect the production of molecules that control β-catenin synthesis and degradation. As a consequence of altering the regulation of β-catenin, these investigators demonstrated that an accumulation of β-catenin occurs, and the pattern of its distribution is altered in the affected organ. Because alteration of the β-catenin gene and accumulation of β-catenin were predominant in the MEG-induced liver tumors, β-catenin imbalance is implicated in carcinogenesis. The β-catenin alterations described by Devereux et al. (1999) are the same mutations as those found in human hepatocellular carcinomas. In addition, the regulation of β-catenin is believed to play a role in the formation of human colon cancer and melanoma.

Stomach

MEG, when administered orally, alters gastric secretions, damages stomach cells, and causes neoplastic cell growth. The types of lesions found are summarized in Table 1.14. The incidence and severity of the effects are related to dose and duration of exposure, as described in the following text.

Subchronic Exposure

Abdo et al. (2001) reported that daily gavage administrations of MEG for 14 weeks to F344 rats and B6C3F$_1$ mice (Table 1.10) produced significant effects on serum gastrin concentrations, stomach pH, cell proliferation, and glandular stomach lesions.

In regard to serum gastrin, MEG produced a biphasic response. The response was dose dependent and apparently related to the duration of exposure. As illustrated in Figure 1.10, serum gastrin decreased approximately 65% relative to controls when F344 female rats were given single daily gavage administrations of MEG for 30 d at doses of 37, 75, and 300 mg/kg/d. A similar effect was observed after 90 d but only for the 150 mg/kg/d group. However, serum gastrin was significantly elevated (approximately 875%) after 30 d when the dose was 1000 mg/kg/d and, after 90 d at doses of 300 or 1000 mg/kg/d, reached increases of approximately 900 and 1100%, respectively. In male B6C3F$_1$ mice, the serum gastrin increased 3- to 4-fold after 30 d of dosing for the 150 and 300 mg/kg/d groups but was similar to control at doses from 9 to 75 mg/kg/d. Likewise, there was no effect exhibited by the mice treated groups after 90 d of dosing.

TABLE 1.14
Summary of MEG-Related Microscopic Findings in the Stomach

Subchronic Exposure	
Rats	**Mice**
Atrophy	Atrophy
Chronic inflammation	Degeneration
	Necrosis
	Edema
	Mitotic alteration
	Cystic glands
Chronic Exposure	
Rats	**Mice**
Atrophy	Atrophy
Neuroendocrine cell hyperplasia	Ectasia
Benign neuroendocrine tumor	Hyperplasia
Malignant neuroendocrine tumor	Chronic active inflammation
	Malignant neuroendocrine tumor

Source: From Johnson, J.D., Ryan, M.J., Toft, J.D., II, Graves, S.W., Hejtmancik, M.R., Cunningham, M.L., Herbert, R. and Abdo, K.M. (2000), Two-year toxicity and carcinogenicity study of methyleugenol in F344/N rats and B6C3F$_1$ mice, *Journal of Agricultural and Food Chemistry*, 48: 3620–3632; Abdo, K.M., Cunningham, M.L., Snell, M.L., Herbert, R.A., Travlos, G.S., Eldridge, S.R., and Bucher, J.R. (2001), 14-Week toxicity and cell proliferation of methyleugenol administered by gavage to F344 rats and B6C3F$_1$ mice, *Food and Chemical Toxicology*, 39: 303–316.

Stomach pH was affected by MEG treatment in the F344 rats and B6C3F$_1$ mice. The rat control groups had stomach pH values of 2.2 (±0.3) at 30 or 90 d after dosing, respectively. Stomach pH became more alkaline following treatment with 1000 mg/kg/d after 30 d of dosing (6.4 ± 0.1) and at 300 and 1000 mg/kg/d after 90 d of dosing (3.1 ± 0.5 and 3.8 ± 0.3).

Cell proliferation was affected by MEG treatment, based on changes in the BrdU labeling index (number of BrdU-positive cells). Although there were no overt MEG-related effects observed in the forestomach, gastric pits, and pylorus, there were substantial dose-related changes observed in the fundus. After 30 d of dosing, the labeling index values for the F344 rat female control and 37, 75, 150, 300, and 1000 mg/kg/d groups were 1, 2, 1, 11, 54, and 61, respectively. After 90 d of dosing, statistically and biologically significant increases were observed at doses of at least 75 mg/kg/d. Likewise, for the male B6C3F$_1$ mice, the labeling index values for the control and

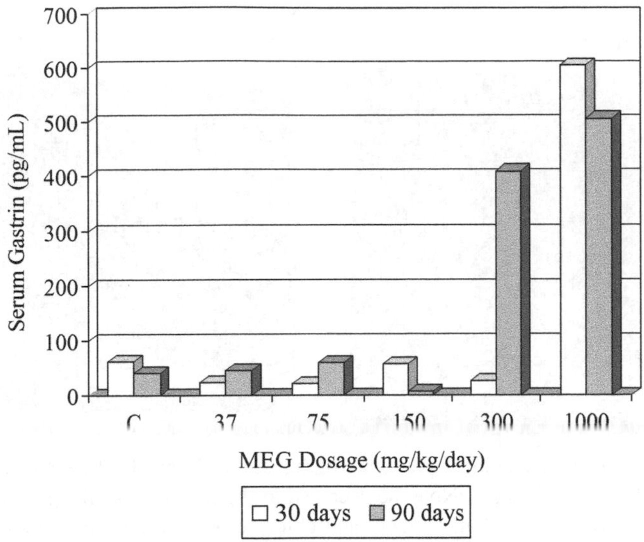

FIGURE 1.10 Effect of MEG on female F344 rat serum gastrin concentration.

9, 18.5, 37, 75, 150, and 300 mg/kg/d groups were 3, 1, 1, 1, 3, 57, and 18, respectively, after 30 d of dosing. Statistically and biologically significant increases were observed after 90 d of dosing at doses of at least 18.5 mg/kg/d.

Microscopic examination of the stomach indicated that MEG caused cellular damage. For the F344 rats, the incidence of atrophy and chronic inflammation of the mucosa of the glandular stomach were significantly increased in the male and female rats given 300 or 1000 mg/kg/d. The severity of these findings was considered minimal to mild at 300 mg/kg/d and mild to moderate at 1000 mg/kg/d. Photomicrographs of the anatomical effects on the glandular stomach are shown in Figure 1.11. The atrophy appeared as decreased gastric mucosal thickness, which was attributed to a loss of glandular epithelial, parietal, and chief cells. There was also condensation of the lamina propria. The inflammation consisted of fibrosis and a diffuse infiltration of the lamina propria by lymphocytes, neutrophils, and macrophages. Other findings included increased mitotic activity and glandular dilatation. For the B6C3F$_1$ mice, the stomach findings were more extensive and included atrophy, degeneration, necrosis, edema, mitotic alteration, and an increased incidence of cystic glands of the fundic region at doses of at least 30 mg/kg/d. For the rats, the atrophy appeared as described in the preceding text. The necrosis involved the parietal cells and, to a lesser extent, the chief cells. The degeneration appeared as dilated glands lined by dysplastic atypical epithelial cells and cellular detritus. The cystic glands were dilated and lined by flattened epithelium. The regenerative areas of the glandular epithelium contained increased numbers of morphologically normal mitotic figures. The edema occurred in the lamina propria (severity was minimal to mild).

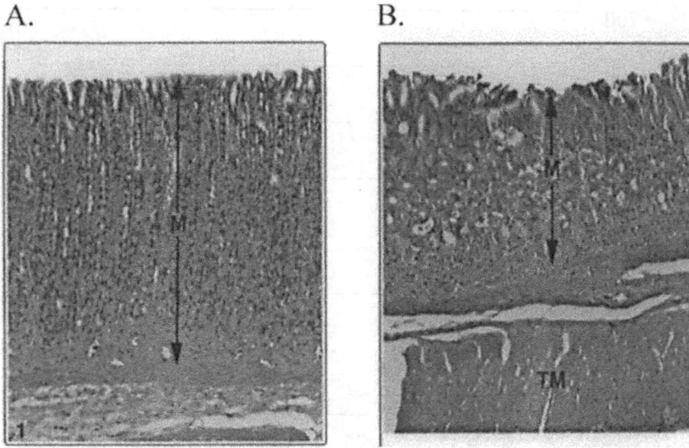

FIGURE 1.11 Glandular stomach photomicrographs. (A). Control. (B). Rat treated with 1000 mg/kg/d of MEG for 14 weeks. Note the decrease in the thickness of the gastric mucosa due to generalized loss of epithelial, parietal, and chief cells, accompanied by condensation of the lamina propria. (From Abdo, K.M., Cunningham, M.L., Snell, M.L., Herbert, R.A., Travlos, G.S., Eldridge, S.R., and Bucher, J.R. (2001), 14-week toxicity and cell proliferation of methyleugenol administered by gavage to F344 rats and $B6C3F_1$ mice, *Food and Chemical Toxicology*, 39: 303–316.)

Chronic Exposure

As was observed for the liver, lifetime exposure to MEG resulted in nonneoplastic and neoplastic lesions when tested in F344 rats and $B6C3F_1$ mice (Johnson et al., 2000). The types of lesions found are summarized in Table 1.14. These lesions increased with increasing dose. In addition, the incidence of most lesions tended to decrease when MEG treatment was discontinued. However, this was not the case for the malignant neuroendocrine tumor in the female rat. Additional details are provided in the following text.

The nonneoplastic lesions that were observed after subchronic exposure were substantiated from the chronic rat-study interim terminations. After 6 and 12 months of exposure, the F344 rat control groups had no evidence of glandular atrophy, but all 5 males and 5 females receiving 300 mg MEG/kg/d exhibited this finding. After 2 yr of exposure, atrophy was the most prevalent nonneoplastic finding. From 28 to 74% of male rats and 78 to 90% of female rats had this lesion following administration of doses at 37, 75, or 150 mg/kg/d. The severity was scored minimal to mild at 37 mg/kg/d but was upgraded to mild to moderate at doses of at least 75 mg/kg/d. Significant increases in the incidence of atrophy using $B6C3F_1$ mice were observed at doses of at least 75 mg/kg/d. Another nonneoplastic lesion common to both species was neuroendocrine cell hyperplasia. For the rats, the incidence of this lesion was statistically significant at doses of at least 150 mg/kg/d for the males and 37 mg/kg/d for the females, although the percentage of animals affected was low, ranging from 10 to 20%. For $B6C3F_1$ mice, the effective dose was similar (at least 75 mg/kg/d), but the percentage affected increased, ranging from 10 to 40%. The severity of

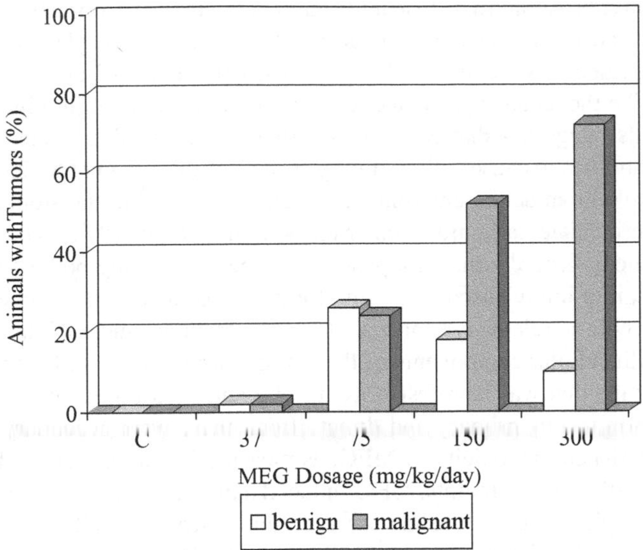

FIGURE 1.12 Neuroendocrine tumors in female F344 rats given MEG by gavage for 2 yr.

hyperplasia was minimal to mild. Although the F344 rats did not exhibit any other nonneoplastic lesions, the B6C3F$_1$ mice also presented with ectasia (extended stomach) and chronic active inflammation. Thus, lower doses of MEG for extended exposure periods led to similar types of lesions as was observed with higher doses for shorter periods of exposure. These types of lesions, such as atrophy and hyperplasia, often precede neoplastic lesions.

Chronic exposure to MEG resulted in the development of neoplastic lesions (Johnson et al., 2000). These lesions are summarized in Table 1.14. After 2 yr of exposure to MEG, there was a marked increase in the incidence of stomach neoplasms. F344 rats developed both benign and malignant neuroendocrine tumors. Although the tumor incidence for the males was statistically significant only for the malignant tumor at 150 mg/kg/d, the females exhibited a high incidence of both tumor types at doses of ≥75 mg/kg/d. Figure 1.12 shows the percentage of female F344 rats that developed benign or malignant neuroendocrine tumors as a result of chronic MEG treatment. Stopping exposure after 1 yr decreased the incidence of benign tumors but not the malignant tumors (300 mg/kg/d group). B6C3F$_1$ mice did not exhibit a statistically significant increase in the incidence of neuroendocrine tumors: two male 150 mg/kg/d group animals had malignant tumors, but there were no animals with this finding in the control and 37 and 75 mg/kg/d groups.

Etiology

MEG-related toxicity was observed in the glandular stomach, and the effects were observed within 30 d of gavage administration (Abdo et al., 2001). The toxicity was characterized by atrophy (decreased gastric mucosal thickness as a result of glandular epithelial, parietal, and chief cell loss), condensation of the lamina propria, chronic inflammation (fibrosis and a diffuse infiltration of the lamina propria by lymphocytes, neutrophils, and macrophages), increased mitotic activity and glandular dilatation, degeneration, necrosis, edema, and an increased incidence of cystic glands of the fundus. Unlike the damage attributed to MEG in the liver, the injury incurred by the stomach cells suggested that MEG was directly toxic to these cells and did not require metabolism to cause cellular injury. The allyl side chain of MEG can become highly reactive in an acidic environment, such as is found in the stomach, and can form an intermediate carbonium ion (Morrison and Boyd, 1974). The phenyl ring and the methoxy group in the para position on the phenyl ring both confer stability to the carbonium ion. Consequently, the formation of a carbonium intermediate of MEG represents a relatively stable and highly reactive nucleophile. Apparently, reactions with cellular components of the cell membrane were sufficient to produce cellular damage that was manifested as atrophy and, to a limited extent, necrosis.

In addition to this primary and direct effect, there were secondary effects that further exacerbated the toxicity of MEG, as measured by alterations in biochemical and clinical pathology endpoints. MEG-induced damage to the parietal cells provides a plausible explanation for the increased serum gastrin as well as decreased iron levels (see the following section on blood effects). The high-dose group F344 rats (at least 300 mg/kg/d) exhibited increases in serum gastrin of 800 to 1100% relative to control. Estragole-induced iron deficiency was also observed (at least 300 mg/kg/d) in the male and female F344 rats (NTP, 2002). Under normal conditions, the ingestion of food that contains proteins and amino acids stimulates G-cells to release gastrin, which acts indirectly [through stimulating enterochromaffin-like (ECL) cell release of histamine] and directly on parietal cells to secrete gastric acid. As the stomach pH declines below 3.5, the D-cells release somatostatin, which inhibits G-cell release of gastrin, thereby providing a negative feedback loop mechanism (Dockray, 1999; Hinkle and Samuelson, 1999). In the present study, repeated administration of MEG damaged the parietal cells, thereby reducing secretion of acid. Without the drop in acid pH to initiate a negative feedback response, the D-cells were not stimulated, somatostatin was not released, G-cells continued releasing gastrin, and gastrin levels became elevated. Gastric acid fosters the formation of chelates between inorganic iron and other food substances, e.g., sugars, amino acids, and bile. These macromolecular iron complexes allow iron to remain soluble when it passes into the more alkaline environment of the small intestine where iron absorption occurs (Greenberger and Isselbacher, 1994). Consequently, a reduction in gastric acid production prevents formation of soluble iron macromolecular complexes, thereby compromising iron absorption and resulting in the development of anemia.

The sustained release and elevation of serum gastrin levels have significant toxicological implications. In addition to stimulating gastric acid release, gastrin regulates epithelial cell proliferation, differentiation of parietal and ECL cells, and expression of genes associated with histamine synthesis and storage (Dockray, 1999). In addition to MEG, butachlor has been associated with similar findings. Long-term gastrin stimulation by butachlor resulted in atrophy of parietal cells, hypochlorhydria, and neuroendocrine tumors (Thake et al., 1995). ECL cells, unlike parietal cells, proliferate in response to normal and exaggerated gastrin levels (Dockray, 1999). Further support for this mechanism comes from genetically altered test systems. Transgenic mice that overexpress gastrin (INS-GAS mice) exhibit increased cell proliferation and thickness in both the stomach and colon (Wang et al., 1996). Recent reviews describe studies that demonstrate an association between ECL cell hyperplasia or tumor formation and elevated gastrin levels (Robinson, 1999; Rozengurt and Walsh, 2001). Gastrin may mediate these effects by altering a regulatory gene. Recently, gastrin was shown to increase the expression of the protein Reg1-α, a growth factor for gastric mucus cells (Fukui et al., 1998). Dockray (1999) suggests that Reg1-α may be an inhibitor of ECL cell growth, and a mutation to the gene that codes for Reg1-α could explain ECL proliferation and tumor formation. Just such a mutation has been found in several ECL cell carcinoid tumors (Higham et al., 1999).

Blood

Subchronic Exposure

MEG treatment alters hematology endpoints. Abdo et al. (2001) showed that after 14 weeks of exposure a slight effect on red blood cell indices is exhibited by F344 rats (Table 1.15). Briefly, MEG decreased the hematocrit (Hct), hemoglobin (Hgb), mean cell volume (MCV), and mean cell hemoglobin (MCH); had no effect on the red blood cell (RBC) count, mean cell hemoglobin concentrations (MCHC), and reticulocyte count; and increased the platelet and white blood cell (WBC) counts (females only). In general, these effects were observed at doses of 1000 mg/kg/d for the males and at least 300 mg/kg/d for the females, except for the change in platelet count that occurred at a dose of 100 mg/kg/d for both the male and female F344 rats. Doses of 75 mg/kg/d had no effect on hematology endpoints. In the Osborne (1981) study, SD rats maintained on diets with an average daily intake of 18 mg/kg/d exhibited hematological values similar to controls at study weeks 6 and 12.

TABLE 1.15
Effect of MEG on Hematology Indices

Indices	Vehicle Control	Dosage (mg/kg/d)[a]				
		10	30	100	300	1000
		Males				
Hematocrit (%)	45.8	45.5	45.0	47.0	46.5	40.0**
Hemoglobin (g/dl)	14.9	14.8	14.7	15.3	15.3	13.3**
Mean cell volume (fl)	50.6	50.6	50.7	50.3	49.9	43.2**
Mean cell hemoglobin (pg)	16.4	16.4	16.4	16.4	16.4	14.4**
Platelets (10^3 µl)	682.0	650.4	663.1	716.9**	781.9**	937.9**
Leukocytes (10^3/µl)	9.14	10.3	9.44	12.0*	9.0	10.1
		Females				
Hematocrit (%)	45.7	45.2	45.3	45.0	44.4	42.9**
Hemoglobin (g/dl)	15.1	14.9	14.9	14.8	14.5*	14.1**

[a] Different from vehicle control *($p \leq 0.05$) or **($p \leq 0.01$).

The effects of MEG on the blood indices are consistent with an anemia. The decreased Hct, Hgb, MCV, and MCH values suggest that a microcytic anemia developed, even though there was no corresponding decrease in RBC counts. Thus, although there was no change in the number of circulating RBCs, the size of the erythrocytes was smaller, which resulted in the lower Hct, Hgb, MCV, and MCH values. Jain (1986a) reported that microcytic anemia is often associated with conditions that cause iron deficiency, disrupted iron metabolism, or altered heme or Hgb production. In addition, the increased platelet counts (thrombocytosis) are consistent with iron deficiency anemia (Jain, 1986b). Rats treated with up to 8000 ppm cupric sulfate for 13 weeks in the feed exhibited an iron deficiency–like anemia that included thrombocytosis (NTP, 1993).

Etiology

The actual mechanism for MEG-induced anemia is unknown. There are at least two possible mechanisms involving direct effects of MEG. One mechanism may involve a defect in iron incorporation into the heme or hemoglobin production, and the other may be associated with iron deficiency due to altered iron absorption. This latter mechanism is certainly possible based on the MEG-induced damage to the stomach glandular cells, which most likely compromised iron absorption and led to the development of an iron-deficient anemia. The effect of MEG on parietal cell gastric acid secretion and the subsequent reduction in iron absorption were described in the section about MEG-related effects on the stomach. Although there were no iron endpoints measured in the Abdo et al. (2001) study using MEG, iron endpoints were measured in a 14-week study with estragole (NTP, 2002), which is closely related to MEG in structure. F344 rats exhibited a decreased serum iron, increased total iron binding capacity (TIBC), and unsaturated iron binding capacity (UIBC) levels for the male and female 300 and 600 mg/kg/d groups. In addition, RBC counts were increased and RBC indices were decreased (Hgb, Hct, MCV, MCH, and MCHC). Thus, MEG may produce microcytic anemia in a similar manner.

Chronic Exposure

No hematology endpoints were measured in the reviewed chronic studies.

Reproductive Organs

Subchronic Exposure

MEG treatment produces reproductive organ toxicity in rodents. Abdo et al. (2001) reported that following 14 weeks of exposure to MEG at doses ranging from 10 to 1000 mg/kg/d, male and female F344 rats exhibited microscopic changes in the testis and uterus, respectively. For the male rats, dilatation of the seminiferous tubules and degeneration of the spermatogenic cells within the seminiferous tubules were observed in all 10 of the 1000 mg/kg/d group animals, whereas no animals in the lower dose groups exhibited this finding. These lesions were scored moderate in severity. However, spermatogonia in the seminiferous tubules appeared morphologically normal. For the female rats, uterine atrophy was observed in 40 and 100% of the 300 and 1000 mg/kg/d group animals, respectively. The atrophy was scored mild to moderate in severity.

Etiology

The reproductive organ toxicity attributed to MEG suggests that it may be the result of primary, rather than secondary, effects. Reproductive organs are often affected when the toxicity of a chemical affects normal growth and development. In such instances, very low BW and BW gain values correspond with organ weights that are also smaller and sometimes immature. In the Abdo et al. (2001) study, the 1000 mg/kg/d male and female group animals exhibited significant decreases relative to the controls in group mean BW — 30 and 16%, respectively — and group mean BW gain, 44 and 30%, respectively. However, the microscopic findings were not indicative merely of undeveloped testes or uterus. Rather, the testicular dilatation and degeneration of spermatogenic cells within the seminiferous tubules and the uterine atrophy were more suggestive of a chemical effect, cell loss, or cell damage. The mechanism for the toxicity on these organs could be direct or indirect. A direct effect of MEG on the Sertoli cells (spermatogenesis) or even on the Leydig cells (steroidogenesis) of the testes could explain the decrease in spermatogenic cells. MEG could also cause these effects indirectly by altering steroidogenesis in the Leydig cell, thereby hindering the development of fully functional Sertoli cells. If MEG has an effect on steroidogenesis, this may also explain the adrenal cortex hypertrophy observed in the 14-week study (Abdo et al., 2001). The steroidogenic pathway in the Leydig cell and the glucocorticoid biosynthetic pathway in the adrenal cortex have many common enzymes and substrates. Any inhibition in either pathway could result in decreased end hormone production, which would stimulate the feedback mechanism in an attempt to return production to normal. The physiological response in such circumstances involves stimulating the affected organ, e.g., cellular hypertrophy. A similar scenario was observed in a 14-week F344 rat study using estragole (NTP, 2002). Estragole-related toxicity was observed in the pituitary gland, testes, and epididymides. In the pituitary gland, the pars distalis (adenohypophysis) showed signs of cytoplasmic alteration and hypertrophy in 10 of 10 male rats in the 300 and 600 mg/kg/d groups. In the testes and epididymides, 10 of 10 rats in the same dose groups exhibited aspermia (average severity level of 3.9 to 4.0). If estragole has a direct effect on Leydig cell–mediated steroidogenesis, then the subsequent decrease in testosterone production could result in aspermia, as well as initiate compensatory pituitary gland cellular responses as a result of feedback effects on the hypothalamus and its enhanced secretion of luteinizing hormone (LH) and follicle stimulating hormone (FSH). Alternatively, if estragole-related induction of CYP enzymes results in increased metabolism of androgens (testosterone) or estrogens (estrone), then low gonadal hormone plasma levels would also lead to stimulation of the hypothalamus and subsequent hypertrophy of the pituitary gland. Also, spermatogenesis could be compromised if testosterone levels remain below the concentrations necessary for normal Sertoli cell function. Thus, regardless of the mechanism, MEG and like molecules appear to alter reproductive function, possibly by disrupting the endocrine system directly or indirectly.

Additional Target Organs

In addition to the target organs listed in the preceding subsections, MEG rodent studies have identified other organs that develop nonneoplastic and neoplastic lesions as a result of exposure (Johnson et al., 2000; Abdo et al., 2001). Table 1.16 summarizes the incidence and severity of these lesions in male F344 rats given MEG for 2 yr at doses from 37 to 150 mg/kg/d and for 1 yr at 300 mg/kg/d, followed by 1 yr of no exposure (stop exposure group), which was included to determine whether discontinuing exposure would reduce the incidence of neoplastic lesions.

In the subchronic study (Abdo et al., 2001), MEG also affected the submandibular salivary gland. Microscopic examination revealed cytoplasmic alteration of the cells in this gland. All the animals receiving 300 or 1000 mg/kg/d had this lesion. The average severity (minimal) did not change with increasing dose.

In the chronic study (Johnson et al., 2000), MEG exposure resulted in significant increases in neoplasms of the kidney, mammary gland, skin, and peritoneum. The incidence of these neoplasms were dose related or were markedly increased at doses greater than the historical control incidences observed for other routes of exposure, thereby indicating that the neoplasms were related to MEG exposure. With respect to the kidney neoplasms, because the tumors were limited to the male rats only and male rat kidney neoplasms are generally associated with α-2_F-globulin (Baetcke et al., 1991), a possible mechanism may involve globulin formation.

TABLE 1.16
Neoplastic Lesions in Male F344 Rats in a 2-Yr Gavage Study of MEG

Finding	Sex	Control	37 mg/kg/d	75 mg/kg/d	150 mg/kg/d	300 mg/kg/d[a]
Renal tubule adenoma or carcinoma	M	4	2	6	6	8*[b]
Malignant mesothelioma	M	1	3	5	12**	5
Mammary gland fibroadenoma	M	5	5	15**	13**	6
Skin fibroma	M	1	9**	8*	5	4
Skin fibroma or fibrosarcoma (combined)	M	1	12**	8**	8**	—

[a] Exposure to MEG was stopped at 52 weeks.
[b] Significantly different at *($p \leq .05$) or **($p \leq .01$).

GENOTOXICITY

MEG has been tested using many of the standard assays developed to evaluate whether a substance has genotoxic activity. The assays employed to date include a variety of prokaryotic assays and eukaryotic nonmammalian and mammalian assays. However, as is often the case when a battery of tests is performed, the results can be equivocal, i.e., some tests are positive, whereas others are negative. An overall evaluation of the assay results suggests that MEG has the potential to be genotoxic. This section reviews and highlights the results of the findings reported in the literature regarding the genotoxicity of MEG.

DNA Mutation

The *Salmonella typhinurium* assay (Ames test) is useful for identifying substances with mutagenic activity (Ames et al., 1975). A number of different investigators, over a period of about 10 yr, used the Ames test as designed or with slight modifications to evaluate whether MEG caused mutations to this type of bacteria (Dorange et al., 1977; Sekizawa and Shibamoto, 1982; Mortelmans et al., 1986; Schiestl et al., 1989). In these experiments, the strains of *S. typhinurium* tested included TA97, TA98, TA100, TA102, TA1535, TA1537, and TA1538 in buffer, with or without S9 metabolic activation. The concentrations of MEG tested in these studies ranged from as low as 0.25 μg/plate to as high as 666 μg/plate. The results of these studies generally showed that MEG does not cause mutations in the absence of metabolic activation, nor was there any evidence of mutations when the S9 liver enzyme fraction was added in order to provide the enzymes necessary for metabolic activation to occur. The only positive induction of mutation by MEG using the Ames test, although the response was extremely weak, was found with strains TA98 and TA102 (Schiestl et al., 1989). Similarly, Green and Muriel (1976) reported that *S. typhinurium* could be replaced with *E. coli* and, if the appropriate amino acid supplement and media were used, then substances with mutagenic activity could also be identified. The *E. coli* assay, with and without metabolic activation, was used by Sekizawa and Shibamoto (1982), and again MEG did not cause any mutation. Regarding these studies, it is important to note that the S9 fraction may not contain the necessary elements for MEG to be mutagenic in this assay. Woo et al. (1997) postulated that the lack of correlation between the mutagenicity test results and the carcinogenic effects of MEG can be attributed to the requirement for sulfation in the metabolic activation of MEG before mutagenicity would be apparent.

DNA Repair Assay

The DNA repair test uses *Bacillus subtilis*, and by comparing the killing zone of a substance on repair-competent and repair-deficient strains of the bacteria, the mutagenic activity of a substance can be assessed (Kada et al., 1980). MEG as well as a number of related analogs and a few MEG-containing essential oils were tested by applying specified amounts of the substance on disks and placing the disks in cultures of one or the other strain (Sekizawa and Shibamoto, 1982). MEG, eugenol, isoeugenol, and safrole, but not estragole, were found to preferentially kill the repair-deficient strain. A similar outcome was obtained using anise oil, clove oil, and pimento oleoresin. These results suggested that MEG had DNA-damaging activity because the repair-competent strain could correct the chemical-induced damage, whereas the repair-deficient strain could not. Thus, this particular prokaryotic cell assay produced a positive result for MEG-induced mutagenesis.

Chromosomal Recombination

A test for genome rearrangement was developed using the yeast *Saccharomyces cerevisiae* (Schiestl et al., 1988). Chemicals that test positive cause a sister-chromatid conversion or intrachromosomal recombination event to occur in the yeast plasmid. In short, homologous recombination between two *HIS3Δ* alleles can cause the deletion of intervening sequences, thereby producing a functional *HIS3* gene. The recombination rate can be determined by measuring the *HIS3+* frequency. When MEG was tested using this assay, it induced intra- and interchromosomal recombination events approximately three and seven times, respectively, greater than the vehicle control (Schiestl et al., 1989). Brennan et al. (1996) reported that MEG produced a 12.5-fold induction in recombination at a concentration of 1.0 mg/ml. Interestingly, the MEG response was not concentration related; rather, a threshold response was observed above 0.3 and 0.6 mg/ml. The authors theorized that the threshold concentration response may reflect what occurs when the defense or repair mechanisms are overwhelmed. In addition, use of S9 did not enhance the MEG-induced recombination. Apparently, the yeast contains a satisfactory enzyme activation system that is sufficient to metabolize MEG to its active form. Furthermore, the allyl moiety is an essential substituent for MEG activity, based on the results obtained using MEG congeners (Brennan et al., 1996). This finding is in good agreement with previously described information regarding the formation of a carbonium ion and the stability conferred on it by the other substituents of MEG.

Chromosomal Aberrations

Chemical-induced alterations to the structural aspects of chromosomes in mammalian cells can be tested using cultured Chinese hamster ovary (CHO) cells. Galloway et al. (1987) incubated MEG with CHO cells, with and without S9, and examined first-division metaphase cells for changes in morphology and completeness of the karyotype. The cells were scored for such aberrations as simple breaks, terminal deletions, rearrangements, translocations, and despiralization of the chromosomes and, with regard to the cell, whether pulverization was observed and the number of cells with greater than ten chromosomal aberrations. The fraction of cells with chromosomal aberrations, when incubated with MEG at concentrations ranging from 50 to 233 μg/ml, ranged from 0 to 1% without the S9 fraction and 1.5 to 4.5% with the S9 fraction. These results were similar to what was obtained with the vehicle (dimethyl sulfoxide, DMSO), thereby indicating that the genotoxic activity of MEG involves a different mechanism of damaging the DNA.

Sister Chromatid Exchange

Sister chromatid exchange (SCE) occurs when, for a given chromosome, there is an exchange at one locus between sister chromatids, which does not result in altering the overall chromosomal morphology (Williams et al., 1983). Galloway et al. (1987) incubated MEG with cultured CHO cells at concentrations ranging from 5 to 167 μg/ml without the S9 fraction and 17 to 250 μg/ml with metabolic activation. MEG in the absence of the S9 fraction did not increase SCEs but, when combined with the S9 fraction, there was an increase in the SCEs per chromosome in treated cells when compared to the SCEs per chromosome in the solvent control cells (relative change of SCEs per chromosome in percent). Table 1.17 illustrates the positive response of MEG on SCEs in CHO cells.

TABLE 1.17
MEG-Induced Sister Chromatid Exchange in Chinese Hamster Ovary Cells

S9 Fraction	Substance Tested	Conc (μg/ml)	SCEs/ Chromosome	Relative Change of SCEs/ Chromosome[a]
Absent	DMSO[b]	—	0.37	—
	Mytomycin-C[c]	0.001/0.004	0.52/0.86	39.8/131
	MEG	5	0.35	–5.03
		17	0.38	3.81
		50	0.42	14.3
Present (Trial 1)	DMSO[b]	—	0.34	—
	Cyclophosphamide[c]	0.125/0.500	0.64/0.98	87.8/186
	MEG	17	0.44	30.5*
		50	0.40	17.5*
		167	0.58	69.6*
Present (Trial 2)	DMSO[b]	—	0.37	—
	Cyclophosphamide[c]	0.125/0.500	0.56/0.99	48.3/163
	MEG	50	0.41	8.35
		167	0.45	20.2*
		250	0.52	38.0*

[a] Solvent control.
[b] Positive control.
[c] SCEs/chromosome in treated cells vs. SCEs/chromosome in solvent control cells.
* Positive response ($p \geq 20\%$ increase over solvent control).

Source: From NTP (1998), Toxicology and Carcinogenesis Studies of Methyleugenol (CAS No. 93-15-12) in F344/N Rats and B6C3F$_1$ Mice (Gavage Studies), Technical Report Series No. 491. NIH Publication No. 98-3950. U.S. DHHS, PHS, NIH, NTP, Research Triangle Park, NC.

Unscheduled DNA Synthesis

Unscheduled DNA synthesis (UDS) is an assay that employs freshly prepared rat hepatocytes in culture and is used to assess the excision repair process that can occur following chemical-induced damage to the DNA (McQueen et al., 1983). MEG was incubated with rat hepatocytes, and the UDS was measured by determining the amount of ^{3}[H]-thymidine incorporated into the hepatocyte nuclear DNA (Howes et al., 1990; Chan and Caldwell, 1992; Burkey, 2000). Earlier studies reported that MEG induced a concentration-related increase in UDS at concentrations ranging from approximately 10^5 to 5×10^3 *M* (concentrations >5×10^3 *M* produced cytotoxicity). More recently, MEG was reported to cause UDS at concentrations ranging from 10 to 500 *μM* using rat hepatocytes and 5 to 500 *μM* using mouse hepatocytes. This assay was also used to evaluate the genotoxicity of MEG's primary metabolite — 1-hydroxyMEG (Chan and Caldwell, 1992; Gardner et al., 1997). A concentration-related increase in UDS was observed at concentrations ranging from 10^5 to 10^4 *M* — concentrations that are about ten times lower than that required by MEG. Thus, MEG was demonstrated to have genotoxic activity based on the UDS assay and, even more revealing concerning the mechanism of MEG-induced genotoxicity, this assay demonstrated the primary metabolite to be more potent than the parent molecule.

Tsai et al. (1994) attributed the stability of the carbonium ion as being one of the key factors for the genotoxicity of MEG. A series of other allylbenzenes and propenylbenzenes were evaluated for their capacity to induce UDS relative to calculated heat of formation (ΔH_R) values. Whether a substance induced UDS was not related to the ΔH_R of the radical species or homolytic cleavage of the C–H bond (propene substituent). However, there was a relationship found between UDS activity and ΔH_R of the carbonium ion. Substances with UDS activity, such as MEG, had ΔH_R values below 231.0 kcal/mol, whereas values above this threshold were not genotoxic. Thus, the genotoxicity of these chemicals appears to be related to the stability of the carbonium ion, which is evident by relevant thermochemical parameters.

Morphological Cell Transformation

SHE cells undergo a morphological transformation and display a progression to neoplasia in the presence of carcinogenic substances (Berwald and Sachs, 1965). Kerckaert et al. (1996) incubated MEG at concentrations ranging from 185 to 250 μg/ml with SHE cells and, after the incubation period, the cells were scored for morphologic transformations. MEG significantly increased ($p \leq .05$) the morphological transformation frequency of the SHE cells three- to fourfold (compared to control) for five of six concentrations tested (the 200 μg/ml concentration had a *p* value of .0615). These results were used to predict that MEG would be eventually identified as a carcinogen in the animal chronic study bioassay, which it was.

In summary, the genotoxic activity of MEG has been reasonably verified based on the results obtained from many of the *in vitro* assays used to evaluate various mechanisms and effects of a substance. Assays that were negative for MEG genotoxic activity included the gene mutation assays using *S. typhinurium* and *E. coli*, with or without metabolic activation, and the chromosomal aberration assay using CHO cells. However, MEG was found to be positive for genotoxic activity when tested with the rec assay for DNA repair using *B. subtilis*, intrachromosomal recombination in yeast (*S. cerevisiae*), SCE assay in CHO cells when incubated with S9, unscheduled DNA synthesis assay using rat and mouse hepatocytes, and the cell morphological transformation assay using SHE cells.

CONCLUSIONS

The use of MEG in foods and beverages is regulated. MEG, a GRAS substance, is permitted for intended use as a synthetic flavoring substance and adjuvant in foods. The primary human use of MEG is as a supplement in dietary and bath products because it imparts a favorable taste or fragrance when used in very low concentrations. Synthetic processes are used to produce MEG, but it is also found naturally in some foods and certain types of spices and essential oils. The average daily intake for humans is estimated to range from 0.3 to 10 μg MEG/kg BW/d. Human serum background levels for MEG averaged 24 pg/g (16 pg/g median).

MEG produces pharmacological effects with potentially therapeutic benefits. These effects include inhibition of histamine release from mast cells, which has an antianaphylactic action; relaxant and antispasmodic action on the smooth muscle of the gastrointestinal tract; analgesic and anesthetic action; and anticonvulsant activity. Effective dose regimens for these effects range from approximately 10 to 250 mg MEG/kg by parenteral administration.

The toxicity spectrum for MEG is varied. Although relatively high doses do not result in severe life-threatening responses (LD_{50} values are measured in grams/kilogram), prolonged exposure to MEG at low doses produces deleterious toxicological and pathological effects when tested in rodents. Target organs following chronic exposure are most evident in the liver and stomach, but also include the blood, reproductive organs, lymph and mammary glands, kidney, skin, and peritoneum. An evaluation of the 2-yr rodent bioassay results led the NTP to indicate that MEG produced clear evidence of carcinogenic activity under the conditions used for testing. The no-observable effect level from the subchronic rodent study was 100 mg/kg/d. In the 2-yr chronic rodent study, nonneoplastic and neoplastic lesions were observed at doses as low as 37 mg/kg/d.

The toxicity of MEG is attributed to the formation of a reactive metabolite. The liver metabolizes MEG to the biologically active metabolite 1-hydroxyMEG. The process involves hydroxylation, followed by sulfation of the allyl substituent. The unstable sulfate ester is hydrolyzed, which forms a carbonium ion. This highly reactive intermediate binds covalently with DNA and proteins to produce genotoxicity and cellular damage.

ACKNOWLEDGMENT

Gratitude and appreciation is extended to Charles D. Alden, NTP/NIEHS, for his editorial comments and review of this manuscript.

LIST OF ABBREVIATIONS

ALT — alanine aminotransferase
BrdU — bromodeoxyuridine
BW — body weight
CAS Number — Chemical Abstract Service Number
CBI — covalent binding index
CDC — Centers for Disease Control
CDCP — Centers for Disease Control and Prevention
C — centigrade
CFR — Code of Federal Register
CHO — Chinese hamster ovary
CHOX — cyclohexane oxide
CNS — central nervous system
CYP — cytochrome P450
D-cell — somatostatin-releasing cell
DMSO — dimethyl sulfoxide
DNA — deoxyribonucleic acid
EC_{50} — effective concentration that produces 50% of the maximal response
ECL — enterochromaffin-like
EEG — electroencephalogram
ELISA — enzyme-linked immunosorbant assay
EPA — Environmental Protection Agency
ETSX — ethosuximide
FDA — U.S. Food and Drug Administration
FSH — follicle stimulating hormone
GABA — gamma amino butyric acid
G-cell — gastrin-releasing cell
GRAS — generally recognized as safe
Hb — hemoglobin
Hct — hematocrit
HDC — L-histidine decarboxylase
HSDB — Hazard Substance Data Base
IARC — International Agency for Research on Cancer
i.p. — intraperitoneal
i.v. — intravenous
KCl — potassium chloride
kDa — kilodalton
K_m — binding constant
LD_{50} — lethal dose for 50 percent
LDH — lactate dehydrogenase
LH — luteinizing hormone
MCH — mean cell hemoglobin
MCHC — mean cell hemoglobin concentration
MCV — mean cell volume
MES — maximum electric shock

mRNA — messenger ribonucleic acid
NADPH — nicotinamide adenine dinucleotide phosphate
NAS — National Academy of Sciences
NIEHS — National Institute of Environmental Health Sciences
NIOSH — National Institute for Occupational Safety and Health
NTP — National Toxicology Program
PCP — pentachlorophenol
PD_{50} — median protective dose
PHYT — phenytoin
PTZ — pentylenetetrazole
RT — room temperature
SCE — sister chromatid exchange
SD — Sprague–Dawley
SDH — sorbitol dehydrogenase
SHE — Syrian hamster embryo
SO — sesame oil
SRI — Stanford Research Institute
SW — Swiss-Webster
TIBC — total iron binding capacity
TP — total protein
UDS — unscheduled DNA synthesis
UIBC — unsaturated iron binding capacity
V_{max} — maximum velocity of a reaction
WBC — white blood cell
WHO — World Health Organization

References

Abdo, K.M., Cunningham, M.L., Snell, M.L., Herbert, R.A., Travlos, G.S., Eldridge, S.R., and Bucher, J.R. (2001), 14-week toxicity and cell proliferation of methyleugenol administered by gavage to F344 rats and B6C3F_1 mice, *Food and Chemical Toxicology*, 39: 303–316.

Ames, B.N., McCann, J., and Yamasaki, E. (1975), Methods for detecting carcinogens and mutagens with *Salmonella*/mammalian microsome mutagenicity test, *Mutation Research*, 31: 347–364.

Baetcke, K.P., Hard, G.C., Rodgers, I.S., McGaughy R.E., and Tahan, L.M. (1991), Alpha$_{2F}$-globulin: Association with Chemically Induced Renal Toxicity and Neoplasia in the Male Rat, Risk Assessment Forum document, U.S. Environmental Protection Agency: Washington, D.C., September, 1991.

Barr, D.B., Barr, J.R., Bailey, S.L., Lapeza, C.R., Jr., Beeson, M.D., Caudill, S.P., Maggio, V.L., Schecter, A., Masten, S.A., Lucier, G.W., Needham, L.L., and Sampson, E.J. (2000), Levels of methyleugenol in a subset of adults in the general U.S. population as determined by high resolution mass spectrometry, *Environmental Health Perspectives*, 108: 1–6.

Beroza, M., Inscoe, M.N., Schwartz, P.H., Jr., Keplinger, M.L., and Mastri, C.W. (1975), Acute toxicity studies with insect attractants, *Toxicology and Applied Pharmacology*, March 31: 421–429.

Berwald, Y. and Sachs, L. (1965), *In vitro* transformation of normal cells to tumor cells by carcinogenic hydrocarbons, *Journal of National Cancer Institute*, 35: 641–661.

Bobin, M.F., Gau, F., Pelltier, J., and Cotte, J. (1991), Etude de L'Arome Basilie, *Rivista Italiana*, EPPOS: 3–13.

Borchert, P., Wislocki, P.G., Miller, J.A., and Miller, E.C. (1973a), The metabolism of the naturally occurring hepatocarcinogen safrole to 1'-hydroxysafrole and the electrophilic reactivity of 1'-acetoxysafrole, *Cancer Research*, March, 33: 575–589.

Borchert, P., Miller, J.A., Miller, E.C., and Shires, T.K. (1973b), 1'-Hydroxysafrole, a proximate carcinogenic metabolite of safrole in the rat and mouse, *Cancer Research*, 33: 590–600.

Burkey, J.L., Sauer, J.M., McQueen, C.A., and Sipes, G.I. (2000), Cytotoxicity and genotoxicity of methyleugenol and related congeners — a mechanism of activation for methyleugenol, *Mutation Research*, 453: 25–33.

Brennan, R.J., Kandikonda, S., Khrimian, A.P., DeMilo, A.B., Liquido, N.J., and Schiestl, R.H. (1996), Saturated and monofluoro analogs of the oriental fruit fly attractant methyl eugenol show reduced genotoxic activities in yeast, *Mutation Research*, 369: 175–181.

Carlini, E.A., Kallmeier, K., and Zelger, J.L. (1981), Methyleugenol as a surgical anesthetic in rodents, *Experientia*, 37: 588–589.

Carlini, E.A., de Oliveira, A.B., and de Oliveira, G.G. (1983), Psychopharmacological effects of the essential oil fraction and the hydrolate obtained from the seeds of *Licoria puchury-major, Journal of Ethnopharmacology*, 8: 225–236.

CDCP/NIEHS, unpublished results. Information referenced in Barr et al., 2000.

CFR (1971), 40 CFR 180 — Part 180 — Tolerances and Exemptions from Tolerances for Pesticide Chemicals in or on Raw Agricultural Commodieties, Promulgated: 36 FR 22540, 11/25/71, U.S. Codes: 21 U.S.C. 346a, 371a.

CFR (1977), 21 CFR 172 — Part 172 — Food Additives Permitted for Direct Addition to Food for Human Consumption, Promulgated: 42 FR 14491 03/15/77. U.S. Codes: 21 U.S.C. 321, 341, 371, 379e.

CFR (1982), 40 CFR 180 — Part 180 Subpart D — Exemptions from Tolerances, Promulgated: 47 CFR 9002, 03/03/82. U.S. Codes: 21 U.S.C. 321(q), 346 (a) and 371.

CFR (1996), 21 CFR 172 — Subpart F — Flavoring Agents and Related Substances, Promulgated: 61 FR 1425, 04/01/96.

Chan, V.S.W. and Caldwell, J. (1992), Comparative induction of unscheduled DNA synthesis in cultured rat hepatocytes by allylbenzenes and their 1-hydroxy metabolites, *Food and Chemical Toxicology*, 30: 831–836.

ChemFinder (2003), Methyleugenol, available at http://www.chemfinder.camsoft.com/, CambridgeSoft (accessed December 2002).

Cheville, N.F. (Ed.) (1994), *Ultrastructural Pathology: An Introduction to Interpretation*, Ames, Iowa: Iowa State University Press, p. 319.

Devereux, T.R., Anna, C.H., Foley, J.F., White, C.M., Sills, R.C., and Barrett, J.C. (1999), Mutation of β-catenin is an early event in chemically induced mouse hepatocellular carcinogenesis, *Oncogene*, 18: 4726–4733.

de Vincenzi, M., Silano, M., Stacchini, P., and Scazzocchio, B. (2000), Constituents of aromatic plants: I. Methyleugenol, *Fitoterapia*, 71: 216–221.

Dockray, G. J. (1999), Topical review: Gastrin and gastric epithelial physiology, *Journal of Physiology*, 518: 315–324.

Dorange, J.L., Delaforge, M., Janiaud, P., and Padau, P. (1977), Mutagenic activity of metabolites of epoxy-diol pathway of safrole and analogues, *C.R. Soc. Biol. (Paris)*, 171: 1041–1048.

Drinkwater, N.R., Miller, E.C., Miller, J.A., and Pitot, H.C. (1976), The hepatocarcinogenicity of estragole (1-allyl-4-methoxybenzene) and 1-hydroxyestragole in the mouse and the mutagenicity of 1-acetoxyestragole in bacteria, *Journal of National Cancer Institute.*, 57: 1323–1331.

Eaton, E.L. and Klaassen, C.D. (1996), Principles of toxicology, in C.K. Klaassen, M.O. Amdur, and J. Doull (Eds.), *Toxicology: The Basic Science of Poisons*, New York: McGraw-Hill, p. 14.

Engelbrecht, J.A., Long, J.P., Nichols, D.E., and Barfknecht, C.F. (1972), Pharmacologic evaluation of 3,4-dimethoxyphenylpropenes and 3,4-dimethoxyphenylpropanediols, *Archives Internationale de Pharmacodynamie*, 199: 226–244.

Farm Chemical Handbook (1992), Meister Publishing, Willoughby, OH, p. C221.

Furia, T.E. and Bellanca, N. (Eds.) (1975), *Fenaroli's Handbook of Flavor Ingredients (1975)*, Cleveland, OH: The Chemical Rubber Co., 2nd ed., Vol. 2, p. 200.

Flavor and Extract Manufacturer Association of the Unites States (FEMA) (1978), Scientific Literature Review of Eugenol and Related Substances in Flavor Usage, Vol. 1. Accession No. PB 283–501. National Technical Information Service, U. S. Department of Commerce, Washington, D.C.

Fukui, H., Kinoshita, T., Maekawa, T., Okada, A., Waki, S., Hassan, M.D.S., Okamao, H., and Chiba, T. (1998), Regenerating gene protein may mediate gastric mucosal proliferation induced by hypergastrinemia in rats, *Gastroenterology*, 115: 1483–1493.

Gallo, M.A. (1996), History and scope of toxicology, in C.D. Klaassen, M.O. Amdur, and J. Doull (Eds.), *Toxicology: The Basic Science of Poisons*, New York: McGraw-Hill, 5th ed., p. 4.

Galloway, S.M., Armstrong, M.J., Reuben, C., Colman, S., Brown, B., Cannon, C., Bloom, A.D., Nakamura, F., Ahmed, M., Duk, S., Rimpo, J., Margolin, B.H., Resnick, M.A., Anderson, B., and Zeiger, E. (1987), Chromosome aberrations and sister chromatid exchanges in Chinese hamster ovary cells: Evaluations of 108 chemicals, *Environmental and Molecular Mutagenesis*, 10 (Suppl. 10): 1–175.

Gardner, I., Bergin, P., Stening, P., Kenna, J.G., and Caldwell, J. (1996), Immunochemical detection of covalently modified protein adducts in livers of rats treated with methyleugenol, CRC, *Critical Reviews in Toxicology*, 9: 713–721.

Gardner, I., Wakazono, H., Bergin, P., de Waziers, I., Beauen, P., Kenna, J.G., and Caldwell, J. (1997), Cytochrome P450 mediated bioactivation of methyleugenol to 1-hydroxymethyleugenol in Fischer 344 rat and human liver microsomes, *Carcinogenesis*, 18: 1775–1783.

Graves, S.W. and Runyon, S. (1995), Determination of methyleugenol in rodent plasma by high-performance liquid chromatography, *Journal of Chromatography B.*, 663: 255–262.

Green, M.H.L. and Muriel, W.J. (1976), Mutagen testing using *trp*+ reversion in *Escherichia coli*, *Mutation Research*, 38: 3–32.

Greenberger, N.J. and Isselbacher, K.J. (1994), Disorders of absorption, in K.J. Isselbacher, E. Braunwald, J.D. Wilson, J.B. Martin, A.S. Fauci, and D.L. Kasper (Eds.), *Harrison's Principles of Internal Medicine*, New York: McGraw-Hill, 13th ed., Vol. 2, p. 1389.

Guenthner, T.M. and Luo, G. (2001), Investigation of the role of the 2, 3-epoxidation pathway in the bioactivation and genotoxicity of dietary allylbenzene analogs, *Toxicology*, 160: 47–58.

Gumbiner, B.M. (1997), Carcinogenesis: A balance between β-catenin and APC, *Current Biology*, 7: R443–R446.

Hall, R.L. and Oser, B.I. (1965), Recent progress in the consideration of flavoring ingredients under the food additives amendment. III. GRAS substances, *Food Toxicology,* 253: 151–197.

Hays, W.J., Jr. and Laws, E.R., Jr. (Eds.) (1991), *Handbook of Pesticide Toxicology,* San Diego: Academic Press, pp. 613–614.

Higham, A., Bishop, A.E., Dimaline, R., Blackmore, C., Dobbins, A.C., Varro, A., Thompson, D.G., and Dockray, G.J. (1999), Mutations of Reg1-α in enterochromaffin-like cell tumors in patients with hypergastrinemia, *Gastroenterology,* 116: 1310–1318.

Hinkle, K.L. and Samuelson, L.C. (1999), Lessons from genetically engineered animal models. III. Lessons learned from gastrin gene deletion in mice, *American Journal of Physiology — Gastrointestinal Liver Physiology,* 277: 500–505.

Howes, A.J., Chan, V.S.W., and Caldwell, J. (1990), Structure-specificity of the genotoxicity of some naturally occurring alkenylbenzenes determined by the unscheduled DNA synthesis assay in rat hepatocytes, *Food and Chemical Toxicology,* 28: 537–542.

Howlett, F.M. (1912), The effect of oil of citronella on two species of *Dacus, Entomol. Soc. Lond. Trans.,* Part II: 412–418.

Howlett, F.M. (1915), Chemical reactions of fruit flies, *Bulletin of Entomological Research,* 6: 297–305.

HSDB (1996), *Methyleugenol,* http://toxnet.nlm. Nih.gov/cgi-bin/sis/htmlgen?HSDB (and type 93-15-2) (accessed December 2002).

IARC (1976), IARC Monographs on the Evaluation of Carcinogenic Risks of Chemicals to Man, Vol. 10. IARC, Lyon, France.

Jain, N.C. (1986a), Clinical and laboratory evaluation of anemias and polycythemias, in N.C. Jain (Ed.), *Schalm's Veterinary Hematology,* Philadelphia: Lea and Febiger, 4th ed., pp. 563–576.

Jain, N.C. (1986b), Qualitative and quantitative disorders of platelets, in N.C. Jain (Ed.), *Schalm's Veterinary Hematology,* Phildelphia: Lea and Febiger, 4th ed., pp. 466–486.

Jenner, P.M., Hagan, E.C., Taylor, J.M., Cook, E.L., and Fitzhugh, O.G. (1964), Food flavourings and compounds of related structure. I. Acute oral toxicity, *Food and Cosmetic Toxicology,* 2: 327.

Johnson, J.D., Ryan, M.J., Toft, J.D., II, Graves, S.W., Hejtmancik, M.R., Cunningham, M.L., Herbert, R. and Abdo, K.M. (2000), Two-year toxicity and carcinogenicity study of methyleugenol in F344/N rats and $B6C3F_1$ mice, *Journal of Agricultural and Food Chemistry,* 48: 3620–3632.

Kada, T., Hirano, K., and Shirasu, Y. (1980), Screening of environmental chemical mutagens by the Rec-assay system with *Bacillus subtilis,* in F.J. DeSerres and A. Hollaender (Eds.), *Chemical Mutagens,* pp. 149–173.

Kanel, G.C. and Korula, J. (Eds.) (2000), *Liver Biopsy Evaluation: Histological Diagnosis and Clinical Correlations,* Philadelphia: W.B. Saunders, p. 4.

Kang, S. and Green, J.P. (1970), Correlation between activity and electronic state of hallucinogenic amphetamines, *Nature,* 226: 645.

Keith, L.H. (1976), Identification of organic compounds in unbleached treated Kraft Paper Mill wastewater, *Current Research,* 10: 555–564.

Keating, J.W. (1972), Report to RIFM, 5 July.

Kerckaert, G.A., Brauninger, R., LeBoeuf, R.A., and Isfort, R.J. (1996), Use of the Syrian Hamster embryo cell transformation assay for carcinogenicity prediction of chemicals currently being tested by the National Toxicology Program in rodent bioassays, *Environmental Health Perspectives,* 104: 1075–1084.

Lawrence, M. and Shu, C.K. (1993), Essential oils as components of mixtures: Their method of analysis and differentiation, in C.T. Ho and C.M. Manley (Eds.), *Flavor Measurement*, New York: Marcel Dekker.

Levy, G.N. and Weber, W.W. (1988), High-performance liquid chromatography analysis of ^{32}P-postlabeled DNA-aromatic carcinogen adducts, *Analytical Biochemistry*, 174: 381–392.

Lewis, R.J., Jr. (Ed.) (2001), *Hawley's Condensed Chemical Dictionary*, New York: John Wiley & Sons, p. 735.

Lide, D.R. (Ed.) (1998), *CRC Handbook of Chemistry and Physics*, Boca Raton: CRC Press, 78th ed., pp. 3–42.

Lima, C.C., Criddle, D.N., Coelho-de-Souze, A.N., Monte, F.J.Q., Jaffar, M., and Cardoso, J.H. (2000), Relaxant and antispasmodic actions of methyleugenol on guinea-pig ileum, *Planta Medica*, 66: 408–411.

Luo, G. and Guenthner, T.M. (1995), Metabolism of allylbenzene 2,3-oxide and estragole 2,3-oxide in the isolated perfused rat liver, *J. Pharmacol. Exp. Ther.*, 272: 588–596.

MacGregor, J.T., Laurence, L.L., and Buttery, R.G. (1974), California bay oil. II. Biological effects of constituents, *Journal of Agricultural and Food Chemistry*, 22: 777–780.

McQueen, C.A., Kreiser, D.M., and Williams, G.M. (1983), The hepatocyte primary culture/DNA repair assay using mouse or hamster hepatocytes, *Environmental Mutagenesis*, 5: 1–8.

Miele, M., Dondero, R., Ciarallo, G., Mazzei., M. (2001a), Methyleugenol in *Ocimum basilicum* L. cv. Genovese Gigante, *Journal of Agricultural and Food Chemistry*, 49: 517–521.

Miele, M., Ledda, B., Falugi, C., and Mazzei, M. (2001b), Methyleugenol and eugenol variation in *Ocimum basilicum* cv. Genovese Gigante grown in greenhouse and *in vitro*, *Boll. Soc. Ital. Biol. Sper.*, 77: 43–50.

Miller, E.C., Swanson, A.B., Phillips, D.H., Fletcher, T.L., Liem, A., and Miller, J.A. (1983), Structure-activity studies of the carcinogenesis in the mouse and rat of some naturally occurring and synthetic alkenylbenzene derivatives related to safrole and estragole, *Cancer Research*, 43: 1124–1134.

Mookherjee, B.D. and Wilson, R.A. (Eds.) (1996), *Kirk-Othmer Encyclopedia of Chemical Technology*, New York: John Wiley & Sons, pp. 1, 82, 94.

Morrison, R.T. and Boyd, R.N. (Eds.) (1974), *Organic Chemistry,* Boston: Allyn and Bacon, 3rd ed., pp. 372–400.

Mortelmans, K., Haworth, S., Lawlor, T., Speck, W., Tainer, B., and Zeiger, E. (1986), *Salmonella* mutagenicity tests: II. Results from the testing of 270 chemicals, *Environmental Mutagenesis*, 8, Suppl. 7: 1–199.

Moslen, M.T. (1996), Toxic responses of the liver, in C.D. Klaassen, M.O. Amdur, and J. Doull, *Toxicology: The Basic Science of Poisons*, New York: McGraw-Hill, pp. 403–416.

National Academy of Sciences (NAS) (1989), Poundage and Technical Effects Update of Substances Added to Food. Committee on Food Additives Survey Data, Food and Nutrition Board, Institute of Medicine, National Academy of Sciences, Washington, D.C.

NIOSH (1990), National Occupational Exposure Survey (1981–1983), unpublished provisional data as of July 1, 1990. NIOSH, Cincinnati, OH.

NTP (1993), NTP Report on the Toxicity Studies of Cupric Sulfate (CAS No. 7758-99-8) in F344/N Rats and B6C3F$_1$ Mice (Drinking Water and Feed Studies), National Toxicology Program. Toxicity Report Series No. 29. NIH Publication No. 93-3352. U.S. Department of Health and Human Services, Public Health Services, National Institutes of Health, Research Triangle Park, NC.

NTP (1998), Toxicology and Carcinogenesis Studies of Methyleugenol (CAS No. 93-15-12) in F344/N Rats and B6C3F$_1$ Mice (Gavage Studies), Technical Report Series No. 491. NIH Publication No. 98-3950. U.S. DHHS, PHS, NIH, NTP, Research Triangle Park, NC.

NTP (2002), The 90-Day Gavage Toxicity Study of Estragole (CAS No. 140-67-0) in Fischer 344 Rats (Gavage Studies), Battelle Final Report (Study Number G004164-R), Columbus, OH.

Needham, L.L., Patterson, D.G., Jr., Burse, V.W., Paschal, D.C., Turner, W.E., and Hill, R.H., Jr. (1996), Reference range data for assessing exposure to selected environmental toxicants, *Toxicology and Industrial Health*, 12: 507–513.

Opdyke, D.L.J. (Ed.) (1979), *Monographs on Fragrance Raw Materials*, New York: Pergamon Press, p. 555.

Osborne, B.E., Plawiuk, M., Graham, C., Bier, C., Losos, G., Broxup, B., and Procter, B.G. (1981), A 91-Day Single Dose Level Dietary Study of Eugenyl Methyl Ether and Isoeugenyl Methyl Ether in the Albino Rat, Bio-Research Laboratories Ltd. Confidential Research Report No. 9203, submitted to FEMA (as described in reference by Smith et al., 2002).

Phillips, D.H., Miller, J.A., Miller, E.C., and Adams, B. (1981), Structures of the DNA adducts formed in mouse liver after administration of the proximate hepatocarcinogen 1-hydroxyestragole, *Cancer Research*, 41: 176.

Pitot, H.C. and Dragon, Y.P. (1996), Chemical Carcinogenesis, in C.D. Klaassen, M.O. Amdur, and J.D. Doull (Eds.), *Toxicology: The Basic Science of Poisons*, New York: McGraw-Hill, p. 201.

Radian (1991), NTP Chemical Repository: Methyleugenol, http://ntp-db.niehs.nih.gov/NTP_Reports/NTP_Chem_H&S/NTP_Chem9/Radian93-15-2.txt, US DHHS, PHS, NIH, National Toxicology Program (accessed December 2002).

Randerath, K., Hagulund, R.E., Phillips, D.H., and Reddy, M.V. (1984), ^{32}P-Post-labeling analysis of DNA adducts formed in the livers of animals treated with safrole, estragole and other naturally-occurring alkenylbenzenes. I. Adult female CD-1 mice, *Carcinogenesis*, 5: 1613–1622.

Riaz, M., Ashraf, C.M., and Chaudhary, M.F. (1989), Studies of the essential oil of the Pakistani *Laurus nobilis* Linn in different seasons, *Pakistan Journal of Scientific and Industrial Research*, 32: 33–35.

Robinson, M. (1999), Review article: current perspectives on hypergastrinemia and enterochromaffin-like-cell hyperplasia, *Aliment Pharmacology and Therapy*, 13: 5–10.

Rompelberg, C.J., Ploemen, J.H., Jespersen, S., van der Greef, J., Verhagen, H., and van Bladeren, P.J. (1996), Inhibition of rat, mouse, and human glutathione *S*-transferase by eugenol and its oxidation products, *Chemical and Biological Interactions*, January 5: 99: 85–97.

Rozengurt, E. and Walsh, J.H. (2001), Gastrin, CCK, signaling, and cancer, *Annual Review of Physiology*, 63: 49–76.

Sayyah, M., Valizadeh, J., and Kamalinejad, M. (2002), Anticonvulsant activity of the leaf essential oil of *Laurus nobilis* against pentylenetetrazole- and maximal electroshock-induced seizures, *Phytomedicine*, 9: 212–216.

Schiestl, R.H., Igarashi, S., and Hastings, P.J. (1988), Analysis of the mechanism for reversion of a disrupted gene, *Genetics*, 119: 237–247.

Schiestl, R.H., Chan, W.S., Gietz, R.D., Mehta, R.D., and Hastings, P.J. (1989), Safrole, eugenol and methyleugenol induce intrachromosomal recombination in yeast, *Mutation Research*, 224: 427–436

Sekizawa, J. and Shibamoto, T. (1982), Genotoxicity of safrole-related chemicals in microbial test systems, *Mutation Research*, 101: 127–140.

Sell, A.B. and Carlini, E.A. (1976), Anesthetic action of methyleugenol and other eugenol derivatives, *Pharmacology*, 14: 367–377.

Shaver, T.N. and Bull, D.L. (1980), Environmental fate of methyl eugenol, *Bulletin of Environmental Contamination and Toxicology*, 24: 619–626.

Shin, B.K., Lee, E.H., and Kim, H.M. (1997), Suppression of L-histidine decarboxylase mRNA expression by methyleugenol, *Biochemical and Biophysical Research Communications*, 232: 188–191.

Smith, R.L., Adams, T.B., Doull, J., Feron, V.J., Goodman, J.I., Marnett, L.J., Portoghese, P.S., Waddell, W.J., Wagner, B.M., Rogers, A.E., Caldwell, J., and Sipes, I.G. (2002), Safety assessment of allylalkoxybenzene derivatives used as flavouring substances — methyl eugenol and estragole, *Food Chemical Toxicology*, 40: 851–870.

Solheim, E. and Scheline, R. R. (1973), Metabolism of alkenebenzene derivatives in the rat. I. *p*-Methoxyallylbenzene (estragole) and *p*-methoxypropenylbenzene (anethole), *Xenobiotics*, 3: 493–510.

Solheim, E. and Scheline, R.R. (1976), Metabolism of alkenebenzene derivatives in the rat. II. Eugenol and isoeugenol methyl ethers, *Xenobiotics*, 6: 137–150.

SRI International (1990), *Directory of Chemical Producers, United States of America*, SRI, Menlo Park, CA., pp. 654, 8, 291, 292, 391.

Steiner, L.F. (1952), Methyl eugenol as an attractant for oriental fruit fly, *Journal of Economic Entomology*, 45: 241–248.

Steiner, L.F., Mitchell, W.C., Harris, E.J., Kozuma, T.T., and Fujimoto, M.S. (1965), Oriental fruit fly by male annihilation, *Journal of Economic Entomology*, 58: 961–964.

Steiner, L.F., Hart, W.G., Harris, E.J., Cunningham, R.T., Ohinata, K., and Kamakahi, D.C. (1970), Eradication of the oriental fruit fly from the Mariana Islands by the methods of male annihilation and sterile insect release, *Journal of Economic Entomology*, 63: 131–135.

Stillwell, W.G., Carman, J., Bell, L., and Horning, M.G. (1974), The metabolism of safrole and 2′,3′-epoxysafrole in the rat and guinea pig, *Drug Metabolism and Disposal*, 2: 489–98.

Stofberg, J. and Grundschober, F. (1987), Consumption ratio and food predominance of flavoring materials, *Perfumer and Flavorist,* 12: 27.

Szabadics, J. and Erdelyi, L. (2000), Pre- and post-synaptic effects of eugenol and related compounds on *Helix pomatia L. neurons*, *Acta Bioligica Hungarica*, 51: 265–273.

Thacke, D.C., Iatrapoulos, M.J., Hard, G.C., Hotz, K.J., Wang, C.X., Williams, G.M., and Wilson, A.G.E. (1995), A study of the mechanism of butachlor-associated gastric neoplasms in Sprague-Dawley rats, *Experimental and Toxicological Pathology.*, 47: 107–116.

Thompson, D.C., Teodosiu, D., Egestad, B., Mickos, H., and Moldeus, P. (1990), Formation of glutathione conjugates during oxidation of eugenol by microsomal fractions of rat liver and lung, *Biochemical Pharmacology*, 39: 1587–1595.

Thompson, D.C., Constantin-Teodosiu, D., and Moldeus, P. (1991), Metabolism and cytotoxicity of eugenol in isolated rat hepatocytes, *Chemico-Biological Interactions*, 77: 137–147.

Thompson, D.C., Perera, K., Krol, E.S., and Bolton, J.L. (1998), O-Methoxy-4-alkylphenols that form quinone methides of intermediate reactivity are the most toxic in rat liver slices, *Chemical Research in Toxicology*, 8: 323–327.

Tsai, S.J. and Sheen L.Y. (1987), Essential oil of *Ocimum basilicum* L. cultivated in Taiwan, in L.W. Sze and F.C. Woo (Eds.), *Trends in Food Science, Proceedings of the 7th World Congress of Food Science and Technology,* Singapore: Institute of Food Science and Technology, pp. 66–70.

Tsai, R.S., Carrupt, P.A., Testa, B., and Caldwell, J. (1994), Structure-genotoxicity relationships of allylbenzenes and propenylbenzenes: a quantum chemical study, *Chemical Research in Toxicology*, 7: 73–76.

Wang, T.C., Koh, T.J., Varro, A., Cahill, R.J., Dangler, C.A., Fox, J.G., and Dockray, G.J. (1996), Processing and proliferative effects of human progastrin in transgenic mice, *Journal of Clinical Investigation*, 98: 1918–1929.

WHO (1981), Evaluation of Certain Food Additives and Contaminants, *Twenty-sixth Report of the Joint FAO/WHO Expert Committee on Food Additives*, Technical Report Series 669, WHO, Geneva, pp. 92–94.

Wie, M.B., Won, M.H., Lee, K.H., Shin, J.H., Lee, J.C., Suh, H.W., Song, D.K., and Kim, Y.H. (1997), Eugenol protects neuronal cells from excitotoxic and oxidative injury in primary cortical cultures, *Neuroscience Letters*, 225:93–96.

Williams, G.M., Dunke, V.C., and Ray, V.A. (1983), Cellular systems for toxicity testing, *Annals of the NewYork Academy of Sciences*, 407: 1–484.

Woo, Y., Lai, D., Arcos, J., Argus, M., Cimino, M., DeVito, S., and Keifer, L. (1997), Mechanism-based structure-activity relationship (SAR) analysis of carcinogenic potential of 30 NTP test chemicals, *Environment Carcinogenesis and Ecotoxicology Reviews*, C15: 139–160.

2 Methyl Mercury Toxicity: Pharmacokinetics and Toxicodynamic Aspects

Janusz Z. Byczkowski

CONTENTS

ABSTRACT

The current information on pharmacokinetic and toxicodynamic aspects of methyl mercury toxicity, which may be helpful in understanding the risk of exposure and potential adverse health outcomes of its ingestion with food, is reviewed. The following topics are discussed: (1) methyl mercury in the environment, (2) accumulation of methyl mercury in the food chain, (3) residues of methyl mercury in freshwater fishery and seafood, (4) exposure to methyl mercury, (5) sensitive subpopulation, (6) intake assessment, (7) pharmacokinetics in sensitive subpopulation,

0-8493-3516-7/05/$0.00+$1.50

(8) pharmacokinetics in experimental animals, (9) mathematical models of transplacental and lactational transfer, (10) potential health effects, (11) mode of action, (12) toxicodynamics, and (13) risk-based guidance values. It is estimated that for about 5% of the U.S. general population of women there may be some risk of exposing their unborn and nursing infants to methyl mercury levels that potentially may cause some adverse effects in the central nervous system (CNS). This estimate is based on the most conservative reference dose of 0.1 μg MeHg/kg/d, which bears a substantial uncertainty and is skewed mostly in the health-protective direction. It is concluded that any potential adverse effects of exposure to methyl mercury in seafood must be carefully balanced against the obvious nutritional and health benefits of seafood and breast feeding.

INTRODUCTION

Methyl mercury (MeHg) is a ubiquitous chemical pollutant which is redistributed in the environment through both natural processes and human activities. Naturally, MeHg is produced by biomethylation of inorganic mercury (Hg), and it can be accumulated up the food chain, especially in aquatic systems. An extensive review of environmental and health aspects of Hg and its compounds, including MeHg, has been published by the U.S. Environmental Protection Agency (U.S. EPA, 1997). The eight volumes of the U.S. EPA Mercury Study Report to Congress, available on line through the Internet, provided inventory of the quantity of Hg emissions from a number of anthropogenic sources, evaluation of the health and environmental effects of these emissions, and described possible methods of mitigating the Hg emissions from these sources (for the executive summary, see U.S. EPA, 2002). Another extensive review of the current scientific literature on toxic effects of MeHg has been compiled and published by the U.S. National Academy of Sciences (NAS-NRC, 2000). Many other rich resources of information about toxicity, epidemiology, and environmental health aspects of MeHg, with excellent compilations of critical literature are also available on line, e.g., Toxicological Profile for Mercury (ATSDR, 1999), Environmental Health Criteria for Methylmercury (IPCS-INCHEM, 1990) or WHO Food Additives Series: 44; Methylmercury (WHO-JECFA, 2000). As these compilations, prepared and reviewed by the best internationally recognized experts, are freely accessible in the public domain, the present review is not thought to be either exhaustive or replicative of those previous reports. Rather, the author of this review intends to focus on pharmacokinetic and toxicodynamic aspects of MeHg, which may be helpful in understanding the risk of exposure and potential adverse health outcomes of MeHg ingestion with food.

METHYL MERCURY IN THE ENVIRONMENT

Biomethylation is a final step in the global cycling of mercury in the environment. At this step, microorganisms in sediments of both fresh and seawater metabolize inorganic compounds of Hg, which are produced in the atmosphere by oxidation of vapors, emitted naturally from land and sea (including emissions from volcanoes), and also emitted anthropogenically by burning of fossil fuel (especially coal) and municipal waste. In addition, mining and chemical industry release both metallic Hg and its compounds into the environment. For example, large amounts of Hg compounds were discharged into the Minamata Bay, Japan, over time in the 1950s. Metallic Hg, which was used as a catalyst in the acetaldehyde plant near the bay, was converted to MeHg and accumulated in the seafood, causing massive poisoning among local fishermen and their families. A similar outbreak of "Minamata disease" was seen later in Nigata prefecture, Japan, from 1965 to 1974 (Tsubaki and Irukayama, 1977).

Around the world, substantial amounts of Hg compounds were used as slimicides by paper industry and were discharged into natural bodies of water. Even though this practice has now been mostly stopped, still hundreds of tons of Hg are being released into the environment every year, especially in goldmining operations (Clearly, 1990). Whereas the natural global emission of Hg to the atmosphere, due to volcanic eruptions, degassing, etc., is estimated to be around 112 t/yr (Nriagu and Becker, 2003), mining and refining processes may contribute an additional 10 to 30 t/yr in direct flux of Hg to the atmosphere (Hylander and Mili, 2003).

Mercury is a common pollutant found in hazardous waste sites around the world. In the U.S., the most hazardous sites were included in the National Priorities List (NPL) and have been targeted for long-term federal clean-up activities. Mercury has been found in about 50% of the 1467 current or former NPL sites. An estimated 3 to 4 million children, a subpopulation most vulnerable to adverse effects of Hg, live within 1 mi of at least one of the 1300 active NPL sites in the U.S. (Aschner, 2002).

Historically, man-made alkylated Hg compounds, including MeHg, were used in large quantities in agriculture and were responsible for serious poisoning outbreaks — in rural Iraq in 1956, 1960, and in the more recent massive epidemics from 1971 to 1972 (Bakir et al., 1973); in Guatemala from 1963 to 1965; in Pakistan from 1961 to 1969; and in Ghana in 1967 (Klaassen, 1996). Typically, in those cases, seed grain treated with antifungal agent containing alkylated Hg was misused by peasant farmers to prepare homemade bread (Clarkson Clarkson, 2002).

ACCUMULATION IN THE FOOD CHAIN

Whereas environmental damage from alkylated Hg-treated grain was first noticed in predatory birds that prey on small mammals which consumed freshly planted treated grain in the fields, the nutritional exposure of humans to MeHg occurs primarily via the aquatic food chain due to accumulation of MeHg in fish and other seafood (Ahmed, 1991). Methyl mercury released from microorganisms into the water and sediments is subsequently sequestrated by fish, shellfish (mainly lobster and crab), whales, and other potential seafood species (U.S. FDA-CFSAN, 2001).

As fish absorb most of MeHg from their food, the predatory fish (tilefish, swordfish, king mackerel, shark, walleye, or northern pike) are generally more contaminated than, for example, bluegills or crappies. In general, piscivorous fish at the top of a food chain in any aquatic system, lake, river, or ocean, accumulate the highest concentration of MeHg, mostly in their fat. Typically, the content of

MeHg in fish parallels their food preference, with piscivorous and omnivorous species at the top and herbivorous species at the bottom of both aquatic food chain and MeHg inventory list (U.S. EPA, 1997).

RESIDUES IN FRESHWATER FISHERY AND SEAFOOD

As the aquatic species absorb MeHg from their food, water, and sediment throughout their life, the old fish are more likely to be heavily contaminated than the young. Obviously, the level of contamination with MeHg depends also on the region where the fish live and feed. For example, by the mid-1990s, 46 states in the U.S. had fish consumption advisories (U.S. EPA, 2002a) covering 40% of the nation's rivers, lakes, and streams, and 60% of those advisories were due to MeHg contamination (Aschner, 2002). On the other hand, in some regions of the U.S., declines in MeHg concentration in freshwater fisheries have been reported recently, in parallel to decreases in atmospheric deposition of H^+, $SO_4{}^2$, and deacidification of water bodies (Hrabik and Watras, 2002).

According to the U.S. FDA-CFSAN (2001) inventory, the Hg concentrations ranged between 0.05 to 4.54 mg/kg in fish with highest MeHg levels (tilefish, swordfish, king mackerel, and shark) and between nondetectable concentration to 1.35 mg/kg in fish and shellfish with low MeHg levels. For the seafood species most consumed in the U.S., the mean MeHg concentrations (expressed in mg Hg/kg) were: swordfish, 1.0; shark, 0.96; lobster (Northern American), 0.31; halibut, 0.23; sablefish, 0.22; Pollock, 0.2; tuna (canned), 0.17; crab Dungeness, 0.18; crab tanner, 0.15; crab king, 0.09; scallop, 0.05; catfish, 0.07; and salmon, oysters, and shrimps, nondetectable concentrations. Additionally, in small samples (with $n \leq 11$) of other most consumed species, the reported mean concentrations (also expressed in mg Hg/kg) were: red snapper, 0.6; cod (Atlantic), 0.19; ocean perch, 0.18; and spiny lobster, 0.13; whereas clams had nondetectable concentrations (U.S. FDA-CFSAN, 2001).

Substantial concentrations of MeHg were detected in American alligators. Recent reports from the Florida Everglades show total Hg concentrations between 0.1 to 1.8 mg Hg/kg in tail muscle and between 0.6 to as much as 17 mg Hg/kg in alligator liver (Rumbold et al., 2002). Even higher concentrations were detected in marine mammals (toothed whales and dolphins), still traditionally marketed for human consumption in Japan. In boiled liver, kidney, and lungs, on an average, as much as 370, 40.5 and 42.8 mg of total Hg/kg was found, respectively (Endo et al., 2002).

EXPOSURE TO METHYL MERCURY

Although the most severe, massive MeHg poisonings of humans resulted from the consumption of homemade bread prepared from seedwheat treated with MeHg-containing fungicides, the most common source of silent exposure of humans to MeHg is contaminated seafood. For many subpopulations, seafood, especially fish, is a staple diet. For example, as identified in the major studies on exposure of humans to MeHg, in the Seychelles, deep-sea fish and reef fish are staples in diet of local residents. In the Faroe Islands, pilot whale meat and fish (mainly cod) are the traditionally consumed food. In the Amazon, fish is the most consumed source of MeHg (especially along the Tapajos River). In New Zealand, 935 out of nearly 11,000 women surveyed reported eating fish more than three times per week during pregnancy (NAS-NRC, 2000). However, any potential adverse effects of exposure to MeHg in seafood must be carefully balanced against the obvious nutritional and health benefits of seafood (U.S. EPA and TERA, 1999; Kris-Etherton, 2002) and breast feeding.

SENSITIVE SUBPOPULATION

Developing fetuses and infant children are the human groups that are most susceptible to MeHg toxicity by virtue of their developing CNS. Damage done to the forming CNS during the early windows of susceptibility is irreversible. However, attempts to measure quantitatively the subtle effects of neurodevelopmental toxicity in this sensitive subpopulation exposed to low doses of MeHg turned out to be extremely difficult.

The difficulties that arose in interpretation of the epidemiological studies of neuropsychological consequences of exposures of developing infants and children to marine fish and mammals may be illustrated by comparing the results reported for two cohorts of children, one on the Faroe Islands (Grandjean et al., 1995, 1997) and the other on the Seychelles Islands (Davidson et al., 1998; Myers et al., 1997). An association between fish consumption and developmental deficits was found in the study of the Faroe Islands but not of the Seychelles Islands, although the body burdens of MeHg were roughly equivalent in these two cohorts. The major confounding factor was the cocontamination of marine fish and mammals in the Faroe Islands with polychlorinated biphenyls (PCBs), which, similarly to MeHg, also have a potential of causing developmental neurotoxicity (Newland, 2002). It was suggested recently that some of the neurodevelopmental deficits observed in children from the Faroe Islands might be due to interaction between PCBs and MeHg, rather than due to exposure to MeHg alone (Grandjean et al., 2001; Seegal and Bowers, 2002).

Interpretation of the results from studying developmental milestones in Faroe Islands infants exposed to MeHg in breast milk (Grandjean et al., 1995) appear even more difficult. Thus, the breast-fed infants who, at 12 months of age, had significantly higher Hg concentrations in their hair reached developmental milestone criteria earlier than those with lower concentrations. This unexpected association, contrary to what would be expected from potentially neurotoxic effects of MeHg, can be explained by the beneficial effects of breast milk — even contaminated milk. The early milestone development was clearly associated with breast feeding, which was, in turn, related to increased hair Hg levels (Grandjean et al., 1995).

Although there is no reason to question that developing fetuses and infant children indeed represent sensitive subpopulations, this assumption is based on extrapolation of the results from animal studies (including nonhuman primates) and our current understanding of developmental neurotoxicity and the biological effects of MeHg, as well as a few highly exposed cases. Because it is unlikely that a direct epidemiological evidence from low-dose exposures to MeHg can ever be demonstrated due to so many confounding factors and lack of a specific biomarker, it seems that the risk assessment in children can be aided by a quantitative pharmacokinetic and pharmacodynamic modeling of MeHg neurodevelopmental toxicity (Lewandowski et al., 2001; Faustman, 2001).

INTAKE ASSESSMENT

An example of a stochastic quantitative modeling of MeHg intake was recently presented by Carrington and Bolger (2002), who developed and partly validated an exposure model that related seafood consumption (estimated from the U.S. Department of Agriculture Continuing Survey of Food Intake by Individuals, 1998) to concentrations of MeHg in the blood and hair of the U.S. population, including two subpopulations of concern: children 2 to 5 years old and women 18 to 45 years old. The model was built using data from cases of exposures of humans to MeHg and the information from the U.S. FDA-CFSAN (2001) inventory. The modeling results compared favorably to the results of the National Health and Nutritional Examination Survey (NHANES) conducted by the Center for Disease Control (CDC) and Prevention, Atlanta, GA. (CDC-MMWR, 2001).

The CDC's NHANES is a continuous survey of the health and nutritional status of the U.S. population. Until recently, estimates of the MeHg intake in the general population were based on Hg concentrations in fish tissue and dietary recall survey data. However, the 1999 survey included the blood and hair Hg concentration data, the first nationally representative human tissue measures of the U.S. population's exposure to Hg. In the 1999 representative population sample, the geometric mean of total blood Hg concentration for all women aged 16 to 49 years was 1.2 ppb and for children aged 1 to 5 years was 0.2 ppb. The 90th percentile of blood Hg for women and children was 6.2 ppb and 1.4 ppb, respectively. Almost all inorganic Hg levels were undetectable, suggesting that these measures indicate blood MeHg levels. The 90th percentile of hair Hg for women and children was 1.4 ppm and 0.4 ppm, respectively (geometric mean values [median] were not calculated for hair Hg concentrations; CDC-MMWR, 2001). Even though the sample size of NHANES 1999 was small and the survey was conducted in only 12 locations, the results were similar to the outputs from the stochastic population model of Carrington and Bolger (2002), who estimated median blood Hg for women aged 18 to 44 years as 0.7 ppb and for children aged 2 to 5 years as 0.3 ppb. Also, upper ends of distributions in NHANES (1999) survey and Carrington and Bolger (2002) simulations were similar. For example, the simulated median 90th percentile of total blood Hg concentration for women aged 18 to 44 years was 3.7 ppb and for children aged 2 to 5 years was 2.1 ppb. The simulated median 90th percentile of hair Hg for women and children was 1.2 ppm and 0.8 ppm, respectively.

These similarities between the results of simulation and survey attest to the credibility of the estimates by Carrington and Bolger (2002) for distributions of the

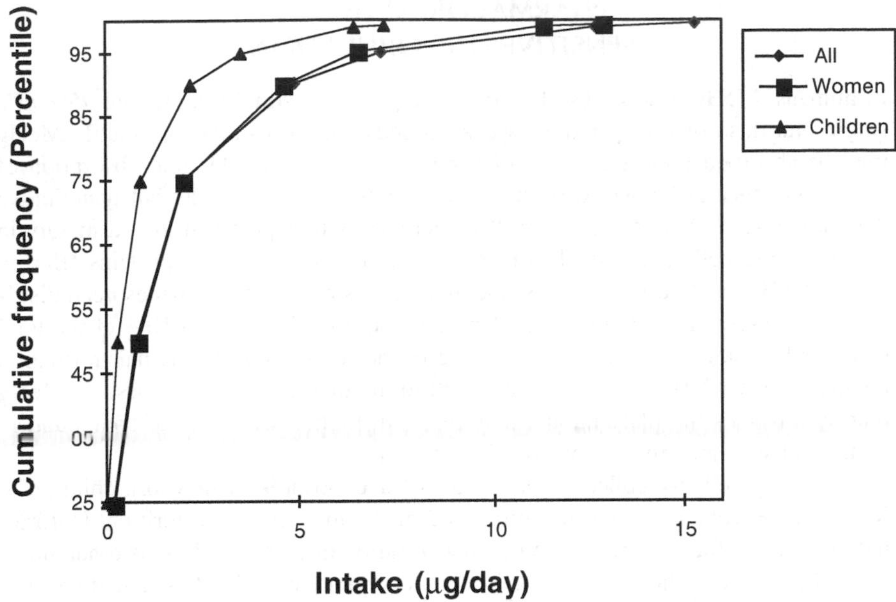

FIGURE 2.1 Estimated distributions of the *per capita* daily MeHg intake from seafood for the three simulated U.S. subpopulations: all persons, women aged 18 to 44 years, and children aged 2 to 5 years (according to simulations by Carrington and Bolger, 2002). The results are presented as median intakes selected from cumulative frequencies (percentile) between 25th and 99.5th percentile. Each point is a median value for the entire uncertainty distribution at each selected frequency.

per capita daily MeHg intake from seafood, as presented in Figure 2.1. Distribution median (50th percentile) per person MeHg intakes from seafood for the three simulated U.S. subpopulation were: all persons, 0.8 μg MeHg/d (uncertainty confidence interval, 0.6, 1.0); women aged 18 to 44 years, 0.8 μg MeHg/d (uncertainty confidence interval, 0.7, 0.8); and children aged 2 to 5 years, 0.3 μg MeHg/d (uncertainty confidence interval, 0.2, 0.3). Obviously, the stochastic population study did not address the seasonal variations in seafood consumption or bolus intakes on specific occasions.

Unfortunately, both studies, the survey and the simulation, did not address the subpopulation of particular concern — newborns and nursing infants to 1 year of age. Even though some data exist for the newborn children from sampling the cord blood for Hg concentration, the reliable estimate of MeHg intake for this subpopulation of particular concern should consider both continuous and bolus exposures and changes in physiology and pharmacokinetics during the entire period of embryo and fetal development and breast feeding, not just a point value at the moment of parturition. These exposures could be quantified only by applying physiologically based pharmacokinetic modeling of combined gestational and lactational transfers of MeHg from mother to infant.

PHARMACOKINETICS IN SENSITIVE SUBPOPULATION

In humans, MeHg is well absorbed from the gastrointestinal tract (about 95% of a single oral dose of MeHg nitrate was absorbed) and up to 80% of volatile MeHg may be absorbed upon inhalation of vapors (Berlin, 1983). Dermal absorption of methyl mercury is known to occur in both humans and animals, but quantitative data are lacking (Young, 1992). Methyl mercury is transported in blood mainly in the red blood cells but a small fraction is also bound to plasma proteins (Berlin, 1983). MeHg easily diffuses across the membranes resulting in a widespread distribution in tissues. It accumulates preferentially in the CNS (up to 10% of the total dose) and in hair. Accumulation of MeHg in the follicle during the hair formation results in up to 250 times greater concentrations than that in other tissues. MeHg readily traverses the placenta and may result in higher levels of the compound in fetal relative to maternal blood (ATSDR, 1999).

The only reliably collected survey data for exposures of newborn children to MeHg came from sampling the cord blood at the moment of parturition. Unfortunately, it was difficult to relate the Hg concentration in cord blood to maternal intake of MeHg because the ratio of Hg in cord and maternal blood is uncertain. The National Academy of Sciences-National Research Council (NAS-NRC) (2000) committee summarized some studies that suggested that cord blood values may be 20 to 30% higher than corresponding maternal blood levels. However, other studies suggested that the ratio may be anything between 1:1 and 2:1 (Peixoto-Boischio and Henshel, 2000). Based on a Monte Carlo meta-analysis of 10 studies, Stern and Smith (2003) estimated cord to maternal blood Hg ratio of 1.7:1 (central tendency with coefficient of variation of 0.56 and a 95th percentile of 3.4:1). Also, it is still unknown how the concentration of Hg in cord blood correlates with the concentration in a newborn's brain due to continuous accumulation during the pregnancy. Moreover, in breast-fed infants, the exposure continues after parturition due to lactational transfer of MeHg from maternal intakes. Thus, the cord blood data do not seem to be very reliable MeHg dosimetrics in children.

PHARMACOKINETICS IN EXPERIMENTAL ANIMALS

Unlike the limited and uncertain human data, a wealth of information on the pharmacokinetics of Hg and MeHg comes from experiments in animals (extensively reviewed by U.S. EPA, 1997). Inorganic Hg compounds and MeHg behave quite differently during transplacental and lactational transfer *in vivo*. For example, Sundberg et al. (1998) demonstrated that the plasma clearance values for MeHg were significantly higher in lactating vs. nonlactating mice and that the transfer of total Hg to nursing pups was significantly higher when the dams were dosed with inorganic Hg than MeHg (the peak concentration of total Hg in milk was fivefold higher in dams exposed to inorganic Hg compounds vs. MeHg). When investigated over 9 d following an I.V. dose, the terminal elimination half-life ($t_{1/2}$) for Hg in milk from dams exposed to inorganic Hg was 107 h, whereas the concentration in milk from dams exposed to MeHg remained almost constant over the examination period

(Sundberg et al., 1998). These results indicated that lactation may represent a significant elimination pathway for dams exposed to Hg compounds and demonstrated a preferential lactational transfer of inorganic Hg *vs.* MeHg.

Several studies of lactational transfer of MeHg have been conducted in rats. For example, Magos et al. (1980) evaluated the elimination and toxicity of MeHg administered to lactating, nonlactating, and virgin female rats. Their results also demonstrated that the body burden of Hg decreased much more rapidly in lactating animals than in nonlactating ones. Sundberg et al. (1991a) dosed Sprague–Dawley rats on day 9 postparturition via oral gavage with 0.5 to 9.4 mg Hg as MeHg. Mercury content of milk, blood (plasma and erythrocytes), and solid tissues (liver, brain, and kidney) were assessed in dams and pups for 72 h. Results indicated a linear dose–response relationship in the transfer of Hg from plasma to milk over the dose range evaluated, with milk concentrations approximately 10% of the level in plasma. These authors demonstrated a significant correlation between the concentrations of Hg in milk and neonatal brain, liver, and kidney tissues, as well as between concentrations of Hg in maternal and pup erythrocytes. Together, these data suggested a linear transfer of MeHg from maternal blood to milk and hence to pups, over the range of doses evaluated. The authors suggested also that lactational transfer prefers an inorganic Hg, and they indicated that 25% of the Hg observed in plasma on days 2 to 10 following a dose of MeHg was in the inorganic ionized form of Hg^{2+}.

Sundberg et al. (1991b), in a study designed to evaluate the impact of dietary selenium on the lactational transfer of Hg, administered single oral doses of radiolabeled mercuric acetate at 0.1 to 5.8 mg Hg/kg to lactating rats on days 8 to 11 postparturition. Radioactivity was assessed in milk, erythrocytes, and plasma for 72 h postdose. These authors reported that, in blood, a greater portion of the dose was attained in erythrocytes than in plasma and Hg concentrations declined much more rapidly in plasma than in erythrocytes. A highly significant correlation was observed (at 24 h) between the concentration of Hg in milk and maternal plasma (Sundberg et al., 1991b). The Hg level in milk at 24 h was 10% of that in whole blood and at 72 h was approximately 8% that of whole blood concentration. The uptake of Hg into pup tissues was linearly related to the milk concentrations. These authors demonstrated that the long-term exposure to selenite (through diet) increased the plasma and milk levels of Hg. However, these differences could not be discerned in pups (the only alteration reaching the level of significance was a selenium-dependent decrease in pup liver Hg).

In a unique, cross-fostering study, Sundberg and Oskarsson (1992), examined lactational transfer of Hg from the dietary exposure of dams to MeHg. The study design featured a diet enriched with MeHg, fed to female Sprague–Dawley rats beginning 11 weeks prior to mating. Rat pups were separated into four groups so that groups were exposed to Hg through gestation and lactation, through gestation only, through lactation only, and not exposed. Pups were observed until postnatal day 15. When exposure was through gestation only, the resulting Hg levels in blood were twice those reported for pups exposed through lactation only, and brain concentrations were four times as high as corresponding levels in pups exposed through lactation only (Sundberg and Oskarsson, 1992). The authors noted that the MeHg concentrations in blood of pups exposed both pre- and postnatally approximated to the sum of the concentrations of MeHg observed in the blood of pups exposed either through gestation only and through lactation only. The authors noted that pups exposed only through lactation demonstrated a higher blood-to-brain ratio (4.6 ± 0.8, mean ± S.D.) when compared to pups exposed through gestation only (blood to brain ratio 2.8 ± 0.7) and pups exposed through both gestation and lactation (blood to brain ratio 3.5 ± 0.9).

These results indicate that Hg from maternal exposures during gestation is much more readily absorbed by the developing fetal brain than the Hg ingested by the pup at the time of nursing. Given the age difference during prenatal and postnatal exposures, the continued functional development of the blood–brain barrier may reduce the rate of Hg transfer from pup blood to brain when compared to those in fetus. However, the concentration of Hg in the brain of pups exposed both pre- and postnatally could be predicted by combining the brain Hg concentration of pups exposed only through gestation with the brain Hg concentrations observed in pups exposed only through lactation. After 15 d of nursing, brain tissue from pups exposed only through lactation demonstrated roughly one fourth the level of Hg in brain tissue from pups exposed only through gestation. Likewise, the blood of pups exposed only through lactation contained roughly one third the MeHg and total Hg when compared to that of pups exposed only through gestation (Sundberg and Oskarsson, 1992).

More recently, Yoshida et al. (1994) examined the lactational transfer of inorganic and MeHg in guinea pigs. Their experimental design included the intraperitoneal administration of 1 mg Hg/kg as either $HgCl_2$ or MeHg to dams 12 h postparturition and the examination of pups (for total Hg only) on days 3, 5, and 10 postparturition. Their data (in MeHg-exposed dams) indicated that the proportion of MeHg to total Hg in milk remained fairly constant at approximately 50%. Mercury concentrations in whole blood of pups declined gradually from approximately 18 ng Hg/g blood after 3 d to approximately 14 ng/g after 10 d. Some of this decrease may have been the result of dilution, given the increase of pup body weight during this time. However, the Hg content in brain tissue of pups whose mothers were exposed to MeHg actually increased from day 3 (approximately 12 ng Hg/g) to day 5 (approximately 20 ng Hg/g) and remained at that level until the day 10. The MeHg profile in pup liver basically paralleled the profile in pup brain, except when it reached a level almost threefold higher. Together, these data suggested a fairly constant exposure of the nursing guinea pig pups to MeHg from a single dose to the dam, even though MeHg concentrations in milk fell from 30 ng Hg/g on day 3 to 10 ng Hg/g on day 10.

Together, the data from experimental animals indicated that the extent of lactational transfer of Hg during the first 15 d of lactation in rodents may be equivalent roughly to one third of the transfer of Hg during gestation. However, this 1:3 transfer ratio for lactation *vs.* gestation may not be relevant to primates. Given that plasma Hg from a dose of MeHg has much longer half-life than a from a dose of inorganic Hg, and that the concentration of total Hg in milk remains fairly constant at roughly 10% of the concentration in blood, and that the breast-feeding (with associated lactational transfer of MeHg) may continue in primates for periods longer than gestation, it appears possible that in humans the extent of lactational transfer of

MeHg may equal or even exceed the dosing encountered by the developing fetus due to gestational transfer of MeHg.

MATHEMATICAL MODELS OF TRANSPLACENTAL AND LACTATIONAL TRANSFER

Utilizing the extensive information on disposition of Hg and MeHg in rodents, Farris et al. (1993) and then Gray (1995) developed physiologically based pharmacokinetic (PBPK) models for rats. An elaborated PBPK model of a pregnant rat, developed by Gray (1995), included both maternal and fetal compartments, separated by the placental barrier (Figure 2.2). In the fetal unit, five individual organs were modeled, including the fetal brain. A flow scheme of MeHg disposition in the Gray (1995) model is shown in Figure 2.2. This PBPK model described adequately the diffusion-limited distribution of MeHg between plasma, red blood cells, and the main tissue compartments, as well as its transplacental transfer. Although this model characterized gestational transfer of MeHg to the growing fetus, it did not address the lactational transfer. Later, Sundberg et al. (1998) described the lactational transfer of both MeHg and inorganic Hg in mice, using a simplified three-compartment linear pharmacokinetic model. However, due to the differences in disposition of MeHg in rodents and primates, the applicability of these rodent models to simulate pharmacokinetics in humans has been limited (Byczkowski and Lipscomb, 2001).

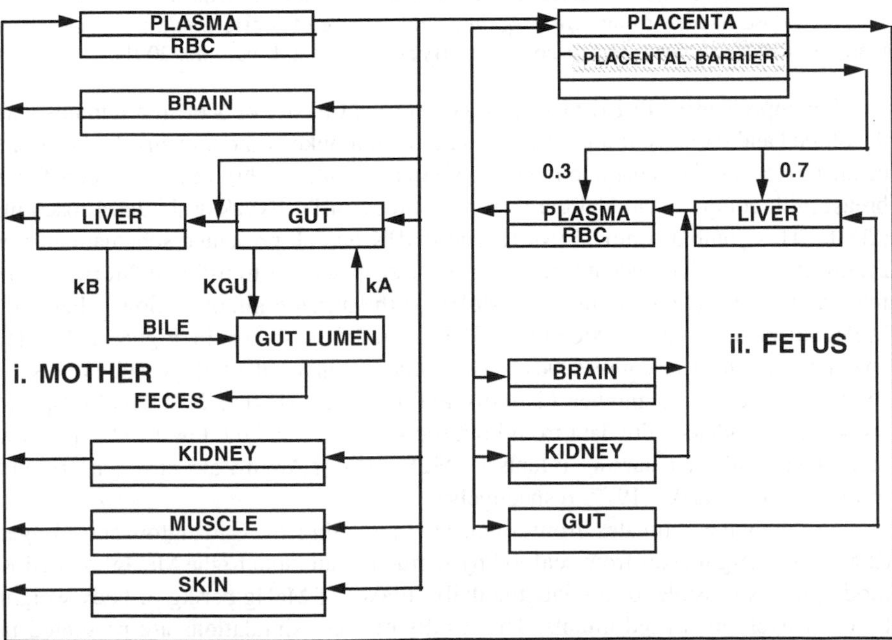

FIGURE 2.2 Flow scheme of the physiologically based pharmacokinetic model for MeHg disposition in the pregnant rat and fetus (Gray, 1995). The symbols kB, kA and KGU represent rate constants for biliary secretion, gut absorption, and gut cells shedding, respectively.

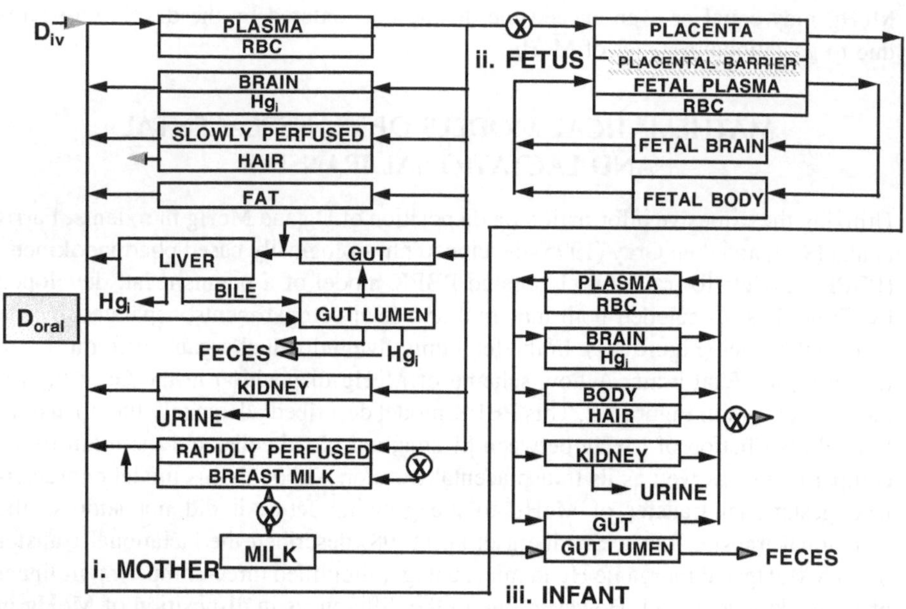

FIGURE 2.3 A diagram of the conceptual framework for the physiologically based pharmacokinetic (PBPK) model of methylmercury (MeHg) lactational transfer from the exposed mother to her nursing infant (MeHgLac PBPK model). X, time-dependent on–off switches; D_{oral}, oral dose of MeHg (mg/kg/d); D_{iv}, intravenous dose of MeHg (mg/h); Hg_i, inorganic mercury; RBC, red blood cells (according to Byczkowski and Lipscomb, 2001).

A comprehensive PBPK model for gestational transfer of MeHg in humans was developed and partially validated (with data from monkeys and humans) by Gearhart et al. (1995, 1996), reparametrized by Clewell et al. (1999), and expanded for a breast milk compartment and the nursing infant unit by Byczkowski and Lipscomb (2001). This combined model (MeHgLac PBPK model, presented schematically in Figure 2.3), described adequately both the transplacental transfer of MeHg to the fetus and the lactational transfer of MeHg to the nursing infant. A flow scheme of MeHg disposition in the MeHgLac PBPK model is shown in Figure 2.3, after Byczkowski and Lipscomb (2001). This model consisted of three submodels or PBPK modules for (1) mother, (2) fetus, and (3) infant (Figure 2.3). The MeHgLac model was validated with data from both the high-dose and the low-level exposures of mothers and their nursing infants to MeHg (from Amin-Zaki et al., 1976, and Fujita and Takabatake, 1977, respectively).

Applying the estimates from stochastic modeling by Carrington and Bolger (2002) of MeHg intakes from seafood by women as an input to the MeHgLac PBPK model, it was possible to simulate the daily intakes of MeHg per kg of body weight by exclusively breast-fed infants. The results of these simulations are presented in Figure 2.4, for children younger than 2 yr of age, at 50th, 75th, 90th, and 95th percentile of maternal MeHg intakes. Initially, after parturition, the daily MeHg intake by infants, expressed per kg body weight, exceeded the maternal intake. It was equal to the maternal intake at the age of about 7.5 to 8 months, but within 20 months

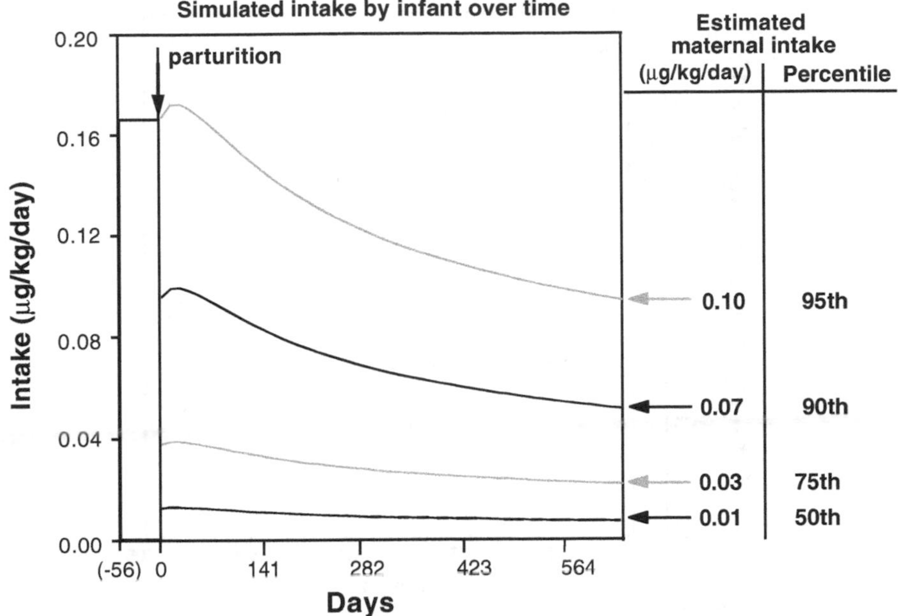

FIGURE 2.4 Results of the MeHgLac PBPK model simulations of the daily intake of MeHg (in µg/kg/d) by typical nursing female infants, whose mothers were exposed to 50th, 75th, 90th, and 95th percentiles of the estimated U.S. MeHg intake from seafood (also in µg/kg/d), according to Carrington and Bolger (2002). The PBPK simulations were performed using the MeHgLac model of Byczkowski and Lipscomb (2001), which included growth functions for both physiological infant body growth and increased intake of breast milk over the 20 months from parturition.

from parturition, it decreased to about 70% of the maternal MeHg daily intake, expressed per kg of body weight. This effect was mostly due to the dilution of MeHg by the rapidly growing infant's body mass (Byczkowski and Lipscomb, 2001).

Whereas at the end of a nursing period (assumed in this simulation to last for as long as 20 months), MeHg intake per kg body weight in the exclusively breast-fed infant was about 30% lower than that in the mother, at the same time, the Hg concentration in the infant's blood was about 44% higher and, consequently, the total Hg concentration in the infant's brain was about 24% higher than that in the mother's brain (Figure 2.5). These pharmacokinetic differences between the mother and her nursing infant were mainly due to the fact that maternal bile secretion, in concert with gastrointestinal clearance and excretion with milk, kept disposing the absorbed MeHg at the rate of about 12% of the intake, whereas, in the exclusively breast-fed infant, a very slow bile flow and inability to metabolize MeHg by the immature intestinal microflora limited the efficiency of Hg clearance (Byczkowski and Lipscomb, 2001; for review, see U.S. EPA, 1997).

Although quantification of the timing and the dose of MeHg accumulated in the target tissue is crucial for the assessment of risk of exposure, dosimetry alone is not sufficient without the understanding of what a particular dose can do to a developing organism at a specific time.

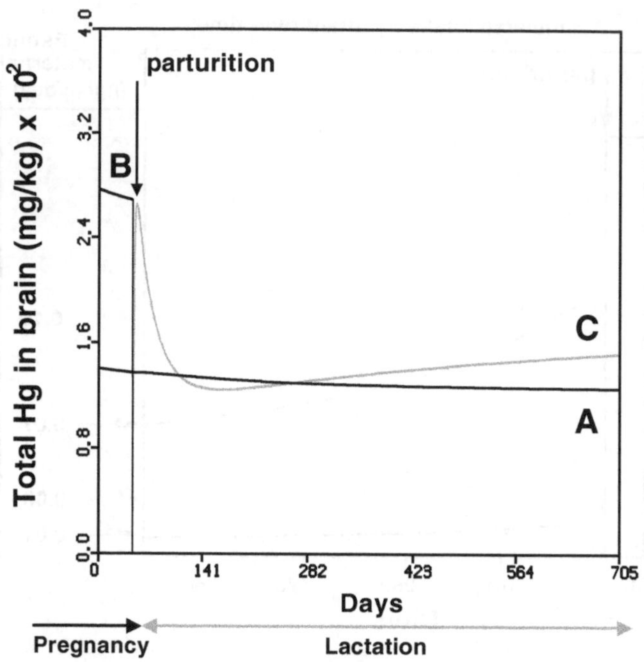

FIGURE 2.5 Combined results of MeHgLac PBPK model simulations of total Hg concentrations in the brain of (A) mother, (B) her gestationally exposed fetus at the last months of pregnancy, and (C) lactationally exposed exclusively breast-fed infant, under scenario of continuous near steady-state maternal intake of MeHg with seafood at 95th percentile estimated for the U.S. (according to Carrington and Bolger, 2002, estimate — 0.1 μg/kg/d), starting 500 d before pregnancy and continuing at a constant intake throughout the pregnancy (9 months) and over the lactation period (assumed here to last for 20 months). The PBPK simulations were performed using the MeHgLac model of Byczkowski and Lipscomb (2001).

POTENTIAL HEALTH EFFECTS

The most important, yet, still unanswered question in risk assessment of MeHg is whether the MeHg intake from a diet high in seafood can cause the aberrant CNS function. Despite a gradual progress in defining the dose–response relationships for MeHg (Lewandowski et al., 2001), a mechanism-based toxicodynamic model for this chemical still does not exist.

The most spectacular clinical effects of MeHg reported in humans were observed following the high-dose poisoning episodes such as those in Japan and Iraq (Tsubaki and Irukayama, 1977; Bakir et al., 1973). At a very high dose (*per os*, a range between 10 and 60 mg/kg body weight; by inhalation, about 5 mg/kg body weight), the acute exposure to MeHg may be lethal. However, the onset of MeHg poisoning may be delayed for weeks or even months, depending on the total body accumulation of the compound (Young, 1992). The LD_{50} values for various rodent species range from 21 to 57.6 mg/kg (RTECS, 2001). At lower doses, subchronic and chronic oral exposures to MeHg produced noninflammatory changes in the CNS (Eto, 2000), nephritis and a tubular degeneration in the kidney, nonischemic cardiovascular abnormalities, irritation of the gastrointestinal tract, and fatty changes in the liver (for review, see ATSDR, 1999). The clinical effects included mental retardation, cerebral palsy, deafness, blindness, and dysarthria in individuals who were exposed *in utero*, and ataxia and sensory and motor impairment in the exposed adults (NAS-NRC, 2000). Neurological and behavioral disorders (including hand tremors, emotional lability, and performance deficits in tests of cognitive and motor function) were also noted in humans, following ingestion of seafood contaminated with MeHg (ATSDR, 1999). The CNS has been identified as the primary site of Hg toxicity in humans and animals, following exposures to both high- and low-doses of inorganic Hg and MeHg (Klaassen, 1996). Effects on neurological development occurred at lower doses of MeHg than those producing adverse effects in organs other than CNS: the kidney, cardiovascular system, the immune system, gastrointestinal tract, and reproductive organs.

Experimental studies in animals exposed to inorganic Hg or MeHg have demonstrated multiple changes in neurological function and behavior, accompanied by disturbed morphology of the nervous system and neurochemistry of the brain. Animal studies confirmed also that neurological development can be adversely affected by *in utero* exposure to MeHg. The other adverse effects of MeHg were observed in the kidney, including degeneration or necrosis of the proximal convoluted tubules and signs of mercury-induced autoimmunity in genetically predisposed animals. Depending on the genetic characteristic and the dose, an autoimmune stimulation or a suppression of the immune system was observed. In orally exposed animals, decreased sperm motility, decreased spermatogenesis, and degeneration of seminiferous tubules were noted, as well as reduced litter size, decreased fetal survival, fetal resorptions, and fetal malformations (for review, see ATSDR, 1999; and U.S. EPA-IRIS, 2001).

MODE OF ACTION

The most sensitive target organ for MeHg toxicity is the CNS, especially the brain, and apparently some adverse effects in humans may occur at a dose level as low as 3 μg/kg. Manifestation of toxic effects (neurobehavioral alterations and degenerative changes in the central and peripheral nervous system) is probably a function of accumulation of critical levels of Hg^{2+} in the neuronal tissue (Young 1992). However, the exact mechanism by which MeHg causes neurotoxic effects is not known (NAS-NRC, 2000).

To cause neurotoxicity, MeHg must be transported from the blood into the brain across the capillary endothelial cells that constitute the blood–brain barrier (BBB). MeHg may cross this barrier either as an organic complex or as a partly ionized species that possesses affinity for specific carrier-mediated transport systems within the endothelial cell plasma membrane. The transport across the BBB is a crucial step in the accumulation of MeHg in the brain; thus, the rate and extent of MeHg transport across the BBB are ultimately related to its toxicity. *In vitro* studies suggest that MeHg is transported into the CNS, at least in part as a conjugate with cysteine (Aschner, 2002).

The high-affinity binding activity of the accumulated inorganic Hg^{2+} ion to thiol compounds or sulfhydryl groups of proteins is thought to be a central molecular mode of action involved in the various toxic effects of both inorganic Hg and MeHg. It was postulated that the Hg-induced damage to neuronal or renal tissues may involve oxidative stress resulting from depletion of reduced glutathione by Hg^{2+}, osmotic swelling and deenergization of mitochondrial membranes (which may lead to formation of reactive oxygen species), and depletion of the reduced pyridine nucleotide pool. It has been further postulated that neurons are particularly sensitive to Hg because of their low endogenous glutathione content and inefficient glutathione reduction activity (for review, see ATSDR, 1999).

The effects of mercurials on mitochondrial structure and function were extensively studied both *in vivo* and *in vitro* (for review, see Byczkowski and Sorenson, 1984). It was demonstrated *in vitro* that, depending on their polarity, both inorganic Hg^{2+} and organic mercurials such as MeHg may increase the passive permeability of mitochondrial membranes to cations and anions, increase the energy-linked accumulation of potassium cations (K^+), affect mitochondrial respiration, and at high concentrations, may inhibit phosphate and substrate transport across the mitochondrial membranes. It was demonstrated in human and rat liver mitochondria that MeHg at concentrations between 10 and 50 nmol/mg of mitochondrial protein inhibited the state 3 respiration, oxidative phosphorylation, and the adenosine diphosphate:inorganic phosphate (ADP/P_i^-) exchange reaction. Swelling of mitochondria, stimulation of "latent" adenosine triphosphatase (ATPase) activity, and energy-dependent H^+ ejection induced by MeHg were accompanied by mitochondrial K^+ uptake and dissipation of the transmembrane potential (these effects are shown schematically in Figure 2.6B). At higher MeHg concentrations the mitochondrial respiratory chain was also inhibited, apparently at the site located between the flavin and the Cytochrome b (Byczkowski and Sorenson, 1984). It was suggested that mercurials may affect mitochondrial membranes causing permeability transition and releasing Cytochrome c (Araragi et al., 2003). All those biochemical effects were noted at Hg concentrations several orders of magnitude higher than those attained in human tissues under environmentally-relevant exposures.

In fetal hepatocytes of rats treated with MeHg *in utero*, a dose-related decrease in the volume density of mitochondria was observed (Fowler and Woods, 1977). Moreover, the treatment of mothers with drinking water containing 5 or 10 μg of MeHg per ml, before and during pregnancy, resulted in pups with decreased synthesis of mitochondrial proteins, loss of respiratory control, suppressed state 3 respiration, and decreased activities of outer and inner mitochondrial membrane marker enzymes (as shown schematically in Figure 2.6A) (without change in malate dehydrogenase activity, a mitochondrial matrix marker enzyme). These effects were attributed to depression by MeHg of mitochondrial membrane synthesis rather than a direct enzyme inhibition. In contrast, in adult rats, treated with 5 or 10 μg of MeHg per ml of drinking water, a swelling of the renal proximal tubular cell mitochondria was observed, associated with decreased respiratory control ratios and dose-dependent decrease of monoamine oxidase and cytochrome oxidase activities in cortical mitochondria. Unlike fetal rat liver mitochondria, in adult rat kidney the delta-aminolevulinic acid synthetase activity was increased, although malate dehydrogenase activity was also unaffected (reviewed by Byczkowski and Sorenson, 1984). The effect of MeHg on mitochondrial membrane synthesis in the fetal rats may parallel the

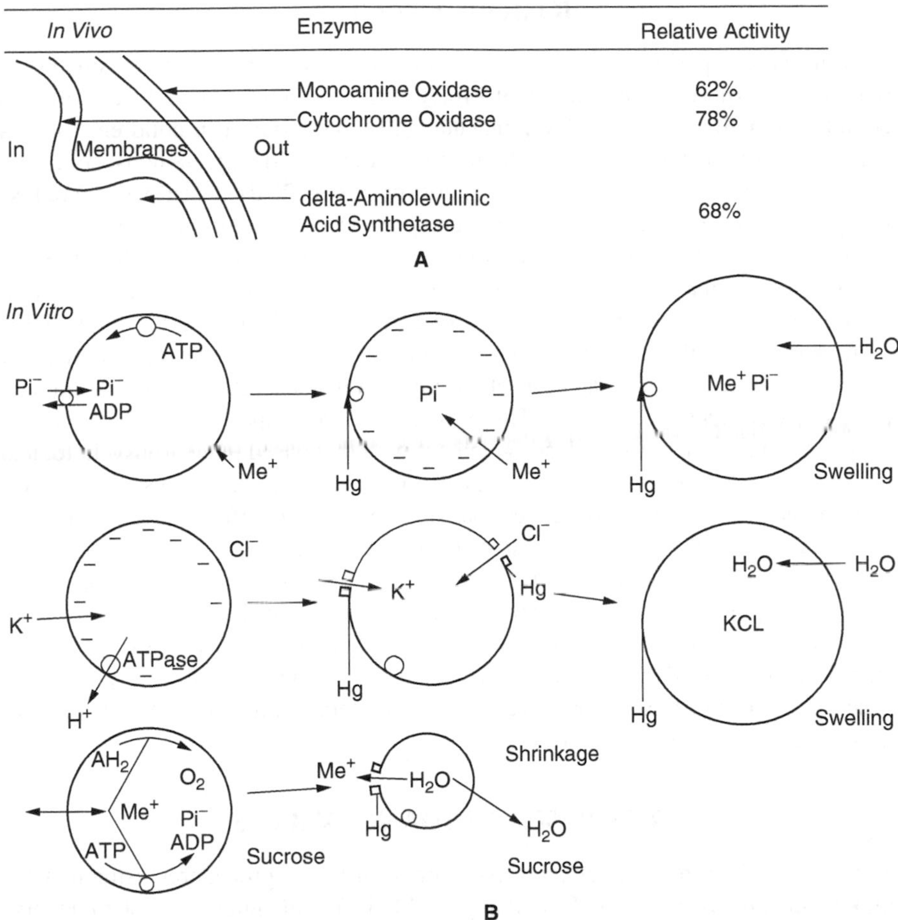

FIGURE 2.6 Action of methylmercury (MeHg) on mitochondria (according to Byczkowski and Sorenson, 1984). A. Effects of MeHg on rat liver mitochondrial marker enzymes following *in utero* exposure and location of the affected marker enzymes. Dams were treated with drinking water containing 10 μg of MeHg/ml for 4 weeks prior to mating and through day 19 of pregnancy. Enzymatic activities were measured in liver mitochondria isolated from pups (Fowler and Woods, 1977). In, inside mitochondria; Membranes, mitochondrial inner and outer membranes; Out, outside mitochondria. B. Osmotic effects of MeHg on isolated mitochondria treated *in vitro*. P_i^-, inorganic phosphate; ADP, adenosine diphosphate; ATP, adenosine triphosphate; Me^+, osmotically active monovalent cation; Hg, organic mercury; K^+, potassium cation; ATPase, vectorial adenosine triphosphatase; H^+, proton; Cl^-, chloride anion; AH_2, substrate for mitochondrial respiratory chain; O_2, molecular oxygen.

mechanism of action on the developing nervous system, which apparently include inhibitory effects of MeHg on mitosis through impairment of microtubule assembly. Moreover, MeHg and inorganic Hg inhibit enzymes such as protein kinase C and decrease transport mechanisms in developing brain cells (ATSDR, 1999).

TOXICODYNAMICS

How all those experimental findings translate into toxicodynamics in sensitive subpopulations exposed to low doses of MeHg is still quite uncertain. The scores on cognitive function tests and developmental milestone criteria in children, used as the end point to quantify neurodevelopmental toxicity of MeHg, are nonspecific and notoriously biased by confounders (e.g., coexposure to PCBs in seafood). Paradoxically, some extensive clinical studies found enhanced performance associated with both prenatal and postnatal exposures to fish contaminated with MeHg (Myers et al., 2000), thus, emphasizing that the beneficial nutritional effects of fish consumption (as well as breast feeding) may actually overcome some adverse effects of low doses of MeHg (U.S. EPA and TERA, 1999). On the other hand, the measurement of delays in evoked potential latencies at specific frequencies (20 and 40 Hz), claimed to be objective and specific to MeHg, provided a dose–response relationship in the exposed children that was in close agreement with that based on neuropsychological test performance (Murata et al., 2002). A 14-yr follow-up of the Faroese birth cohort by Grandjean et al. (2004) suggested that some neurotoxic effects from intrauterine exposure to MeHg may be irreversible (Murata et al., 2004).

It seems that even though, at this state of knowledge, the toxicodynamics of neurodevelopmental effects of low-dose exposures to MeHg cannot be reliably established, it is possible, at least, to estimate the daily dose of MeHg that is without significant risk of adverse effects in sensitive subpopulations. However, this risk-based estimate must bear a relatively high uncertainty, reflecting the gaps in our knowledge (Dourson et al., 2001).

RISK-BASED GUIDANCE VALUES

Three U.S. governmental agencies using three different procedures estimated the maximum daily exposure to MeHg that probably is free of significant risk of adverse effects over the course of a person's life (Table 2.1). Each of these estimated health guidance values was a subject of heated discussions and criticism. Given that these values were derived using different data sets and understanding the extent of uncertainty in derivation of those risk-based guidance values, it is remarkable how close the reference dose (RfD: 0.1 μg/kg/d), the minimal risk level (MRL: 0.3 μg/kg/d), and the allowable daily intake (ADI: 0.4 μg/kg/d) are to each other. The provisional tolerable weekly intake (PTWI: 1.6 μg/kg/week) most recently derived by FAO of the UN-WHO (JECFA, 2003) fits in the middle of this range. The most conservative reference dose of 0.1 μg MeHg/kg/d bears a substantial uncertainty (Rice et al., 2003), spanning at least an order of magnitude, and is skewed mostly in a health-protective direction (Dourson et al., 2001).

CONCLUSIONS

Seafood is the most significant source of nutritional exposure to MeHg. Developing fetuses and infant children have been identified as a subpopulation that is the most

TABLE 2.1
The Estimates by the Three Different U.S. Governmental Agencies of Maximum Daily Intake of MeHg that Probably Is Free of Significant Risk for Adverse Effects over the Course of a Person's Life

U.S. Governmental Agency	Health Guidance	Reference Value (μg/kg/d)
Food and Drug Administration Agency[a]	ADI	0.4
Agency for Toxic Substances and Disease Registry[b]	MRL	0.3
Environmental Protection Agency[c]	RfD	0.1

[a] From U.S. FDA-CFSAN (2001), Mercury Levels in Seafood Species (Center for Food Safety and Applied Nutrition, Office of Seafood, U.S. Food and Drug Administration, May 2001). Online. Available HTTP: <http://www.cfsan.fda.gov/~frf/sea-mehg.html> (accessed September 1, 2002).

[b] From ATSDR (1999), Toxicological Profile for Mercury (Agency for Toxic Substances and Disease Registry, U.S. Department of Health and Human Services, Public Health Service, Atlanta, GA.) Online. Available HTTP: <http://www.atsdr.cdc.gov/toxprofiles/tp46.html> (accessed July 23, 2002).

[c] From U.S. EPA-IRIS (2001), Methylmercury (MeHg) (CASRN 22967-92-6), U.S. Environmental Protection Agency, Integrated Risk Information System. Online. Available HTTP: <http://www.epa.gov/iris/subst/0073.htm> (accessed November 24, 2002).

susceptible to MeHg toxicity. Transplacental and lactational transfer of MeHg are responsible for exposures of the unborn and the nursing infants to MeHg ingested by mothers. The developing CNS represents target organ for the MeHg toxicity. The mechanism of MeHg toxicity and its dose-response are not fully understood. The risk-based guidance values, developed by governmental agencies, represent the maximum daily intake of MeHg that probably is free of significant risk of adverse effects over the course of a person's life. Even though the estimated 95% of the U.S. general population of women have daily MeHg intakes from seafood lower than the most health-protective reference dose of 0.1 μg/kg/d, for the remaining 5% there may be some risk of exposing their unborn and nursing infants to MeHg levels that potentially may cause some adverse effects in the CNS. Any potential adverse effects of exposure to MeHg in seafood must be carefully balanced against the obvious nutritional and health benefits of seafood and breast feeding.

ACKNOWLEDGMENTS

The information contained in this paper was presented in poster form at the 2003 Society for Risk Analysis Annual Meeting, Baltimore, MD (Final Program M21.8, p. 54, 2003) and the 2004 Toxicology and Risk Assessment Conference, Cincinnati, OH (Abstract Book, p. 15, 2004).

Glossary

ADI — allowable daily intake.
Ataxia — inability to coordinate voluntary muscular movements; in common usage, unsteady gait. Broadly speaking, the word ataxia means unsteadiness and clumsiness at the earliest symptoms. As the disorder progresses with damage to both nerve and muscle cells, people with ataxia usually lose the ability to walk and can become totally disabled. Vision (and in some cases hearing) and speech may also be affected.
CDC — Centers for Disease Control and Prevention, Atlanta, GA.
CNS — central nervous system.
Dysarthria — speech disorder resulting from a weakness or incoordination of the speech muscles. Speech is slow, weak, imprecise, or uncoordinated.
MRL — minimal risk level.
NHANES — National Health and Nutrition Examination Survey.
ppb — parts per billion (μg/kg).
ppm — parts per million (mg/kg).
RfD — reference dose.

REFERENCES

Ahmed, F.E. (Ed.) (1991), Seafood Safety (U.S. National Academy of Sciences, Committee on Evaluation of the Safety of Fishery Products, Food and Nutrition Board, Institute of Medicine). National Academy Press, Washington, D.C. Online. Available HTTP: <http://books.nap.edu/books/0309043875/html/> (accessed July 23, 2002).

Amin-Zaki, L., Elhassani, S., Majeed, M.A., Clarkson, T.W., Doherty, R.A., Greenwood, M.R., and Giovanoli-Jakubczak, T. (1976), Perinatal methylmercury poisoning in Iraq, *American Journal of Diseases of Children.* 130:1070–1076.

Araragi, S., Kondoh, M., Kawase, M., Saito, S., Higashimoto, M., and Sato, M. (2003), Mercuric chloride induces apoptosis via a mitochondrial-dependent pathway in human leukemia cells, *Toxicology* 184:1–9.

Aschner, M. (2002), Neurotoxic mechanisms of fish-borne methylmercury, *Environmental Toxicology and Pharmacology* 12:101–104.

ATSDR (1999), Toxicological Profile for Mercury (Agency for Toxic Substances and Disease Registry, U.S. Department of Health and Human Services, Public Health Service, Atlanta, GA.) Online. Available HTTP: <http://www.atsdr.cdc.gov/toxprofiles/tp46.html> (accessed July 23, 2002).

ATSDR (2002), Draft Interaction Profile for: Persistent Chemicals Found in Fish (Chlorinated Dibenzo-p-Dioxins, Hexachlorobenzene, p,p'-DDE, Methylmercury, and Polychlorinated Biphenyls), U.S. Department of Health and Human Services, Public Health Service Agency for Toxic Substances and Disease Registry, September 2, 2002. Online. Available HTTP: <http://www.atsdr.cdc.gov/interactionprofiles/IP-fish/ip01.pdf> (accessed September 2, 2002).

Bakir, F., Damluji, S.F., Amin-Zaki, L., Mutada, M., Khalidi, A., al-Rawi, N.Y., Tikriti, S., Dhahir, H.I., Clarkson, T.W., Smith, J.C., et al. (1973), Methylmercury poisoning in Iraq, *Science* 181:230–241.

Berlin, I. (1983), Organic compounds of mercury, in *Encyclopedia of Occupational Health and Safety,* L. Parmeggiani (Ed.), 3rd ed., Geneva: International Labour Organization.

Byczkowski, J.Z. and Sorenson, J.R.J. (1984), Effects of metal compounds on mitochondrial function: a review, *The Science of the Total Environment* 37:133–162.

Byczkowski, J.Z. and Lipscomb, J.C. (2001), Physiologically based pharmacokinetic modeling of the lactational transfer of methylmercury, *Risk Analysis* 21:869–882.

Carrington, C.D. and Bolger, M.P. (2002), An exposure assessment for methylmercury from seafood for consumers in the United States, *Risk Analysis* 22:689–699.

CDC-MMWR (2001), Blood and Hair Mercury Levels in Young Children and Women of Childbearing Age — United States, 1999, Center for Disease Control and Prevention, *Morbidity and Mortality Weekly Report*, March 02, 2001, 50(08); 140–143. Online. Available HTTP: <http://www.cdc.gov/mmwr/preview/mmwrhtml/mm5008a2.htm> (accessed September 8, 2002).

Clarkson, T.W. (2002), The three modern faces of mercury, *Environmental Health Perspectives* 110(Suppl. 1):11–23.

Clearly, D. (1990), *Anatomy of the Amazon Gold Rush*, p. 245, MacMillan: Oxford.

Clewell, H.J., Gearhart, J.M., Gentry, P.R., Covington, T.R., VanLandingham, C.B., Crump, K.S., and Shipp, A.M. (1999), Evaluation of the uncertainty in an oral reference dose for methylmercury due to interindividual variability in pharmacokinetics, *Risk Analysis* 19:547–558.

Davidson, P.W., Myers, G.J., Cox, C., Axtell, C., Shamlaye, C., Sloane-Reeves, J., Cernichiari, E., Needham, L., Choi, A., Wang, Y., Berlin, M., and Clarkson, T.W. (1998), Effects of prenatal and postnatal methylmercury exposure from fish consumption on neurodevelopment: outcomes at 66 months of age in the Seychelles child development study, *Journal of American Medical Association* 280:701–707.

Dourson, M.L., Wullenweber, A.E., and Poirier, K.A. (2001), Uncertainties in the reference dose for methylmercury, *Neurotoxicology* 22:677–689.

Endo, T., Haraguchi, H., and Sakata, M. (2002), Mercury and selenium concentrations in the internal organs of toothed whales and dolphins marketed for human consumption in Japan, *The Science of the Total Environment* 300:15–22.

Eto, K. (2000), Minamata disease, *Neuropathology* 20(S1):14–19.

Farris, F.F., Dedrick, R.L., Allen, P.V., and Smith, J.C. (1993), Physiological model for the pharmacokinetics of methyl mercury in the growing rat, *Toxicology and Applied Pharmacology* 119:74–90.

Faustman, E.M. (2001), Toxicodynamic considerations in understanding children's health risks from environmental agents, *Abstract of Meeting Paper, Society for Risk Analysis 2001 Annual Meeting*. Online. Available HTTP: <http://www.riskworld.com/Abstract/AB5ME001.HTM> (accessed October 21, 2002).

Fowler, B.A. and Woods, J.S. (1977), The transplacental toxicity of methyl mercury to fetal rat liver mitochondria: morphometric and biochemical studies, *Laboratory Investigations* 36:122–130.

Fujita, M. and Takabatake, E. (1977), Mercury levels in human maternal and neonatal blood, hair and milk, *Bulletin of Environmental Contamination and Toxicology* 18:205–209.

Gearhart, J.M., Clewell, H.J., III, Crump, K.S., Shipp, A.M., and Silvers, A. (1995), Pharmacokinetic dose estimates of mercury in children and dose-response curves of performance tests in a large epidemiological study, *Water, Air, and Soil Pollution* 80: 49–58.

Gearhart, J., Covington, T., and Clewell, H., III (1996), Application of a physiologically based pharmacokinetic model for MeHg in a dose reconstruction of the Iraqi accidental exposures, Presented at *the Fourth International Conference on Mercury as a Global Pollutant*, August 4–8, Congress Centre, Hamburg, Germany.

Grandjean, P., Murata, K., Budtz-Jorgensen, E., and Weihe, P. (2004), Cardiac autonomic activity in methylmercury neurotoxicity: 14-year follow-up of a Faroese birth cohort, *The Journal of Pediatrics* 144:169–176.

Grandjean, P., Weihe, P., and White, R.F. (1995), Milestone development in infants exposed to methylmercury from human milk, *Neurotoxicology* 16:27–33.

Grandjean, P., Weihe, P., White, R.F., Debes, F., Araki, S., Yokoyama, K., Murata, K., Sorensen, N., Dahl, R., and Jorgensen, P.J. (1997), Cognitive deficit in 7-year-old children with prenatal exposure to methylmercury, *Neurotoxicology Teratology* 17: 417–428.

Grandjean, P., Weihe, P., Burse, V.W., Needham, L.L., Storr-Hansen, E., Heinzow, B., Debes, F., Murata, K., Simonsen, H., Ellefsen, P., Budtz-Jorgensen, E., Keiding, N., and White, R.F. (2001), Neurobehavioral deficits associated with PCB in 7-year-old children prenatally exposed to seafood neurotoxicants, *Neurotoxicology Teratology* 23:305–17.

Gray, D.G. (1995), A physiologically based pharmacokinetic model for methyl mercury in the pregnant rat and fetus, *Toxicology and Applied Pharmacology* 132:91–102.

Hrabik, T.R. and Watras, C.J. (2002), Recent declines in mercury concentration in a freshwater fishery: isolating the effects of de-acidification and decreased atmospheric mercury deposition in Little Rock Lake, *The Science of the Total Environment* 297:229–237.

Hylander, L.D. and Mili, M. (2003), 500 years of mercury production: global annual inventory by region until 2000 and associated emissions, *The Science of the Total Environment* 304:13–27.

IPCS-INCHEM (1990), Environmental Health Criteria 101, Methylmercury (International Programme on Chemical Safety, United Nations Environment Programme, the International Labour Organisation, and the World Health Organization), Geneva. Online. Available HTTP: <http://www.inchem.org/documents/ehc/ehc/ehc101.htm> (accessed July 24, 2002).

JECFA (2003), Summary and conclusions of the sixty-first meeting of the Joint FAO/WHO Expert Committee of Food Additives, Rome, June 10–19, 2003. Online. Available HTTP: <http://www.who.int/pcs/jecfa/Summary61.pdf> (accessed February 20, 2004).

Klaassen, C.D. (Ed.) (1996), *Casarett and Doull's Toxicology: The Basic Science of Poisons*, 5th ed., New York: McGraw-Hill.

Kris-Etherton, P.M. (2002), Fish consumption, fish oil, omega-3 fatty acids, and cardiovascular disease, *Circulation* 106:2747–2757.

Lewandowski, T.A., Hoeft, J.M., Bartell, S.M., Wong, E.Y., Griffith, W.C., and Faustman, E.M. (2001), Modeling developmental neurotoxicity and mechanism based dose response models, *Abstract of Meeting Paper, Society for Risk Analysis 2001 Annual Meeting.* Online. Available HTTP: <http://www.riskworld.com/Abstract/AB5ME001.HTM> (accessed October 21, 2002).

Magos, L., Peristianis, G.C., Clarkson, T.W., and Snowden, R.T. (1980), The effect of lactation on methylmercury intoxication, *Archives of Toxicology* 45:143–148.

Murata, K., Budtz-Jorgensen, E., and Grandjean, P. (2002), Benchmark dose calculations for methylmercury-associated delays on evoked potential latencies in two cohorts of children, *Risk Analysis* 22:465–474.

Murata, K., Weihe, P., Budtz-Jorgensen, E., Jorgensen, P.J., and Grandjean, P. (2004), Delayed brainstem auditory evoked potential latancies in 14-year-old children exposed to methylmercury, *The Journal of Pediatrics* 144:177–183.

Myers, G.J., Davidson, P.W., Shamlaye, C.F., Axtell, C.D., Cernichiari, E., Choisy, O., Choi, A., Cox, C., and Clarkson, T.W. (1997), Effects of prenatal methylmercury exposure from a high fish diet on developmental milestones in the Seychelles child development study, *Neurotoxicology* 18:819–829.

Myers, G.J., Davidson, P.W., Cox, C., Shamlaye, C., Cernichiari, E., and Clarkson, T.W. (2000), Twenty-seven years studying the human neurotoxicity of methylmercury exposure, *Environmental Research Section A* 83:275–285.

NAS-NRC (2000), Toxicological Effects of Methylmercury (U.S. National Academy of Sciences, Committee on the Toxicological Effects of Methylmercury, Board on Environmental Studies and Toxicology, Commission on Life Sciences, National Research Council). National Academy Press, Washington, D.C. Online. Available HTTP: <http://www.nap.edu/books/0309071402/html/> (accessed July 23, 2002).

NHANES (1999), Third National Health and Nutrition Examination Survey, NHANES III. Online. Available www.cdc.gov/ncns/nhanes.htm (accessed November 12, 2004).

Newland, M.C. (2002), Neurobehavioral toxicity of methylmercury and PCBs. Effects-profiles and sensitive populations, *Environmental Toxicology Pharmacology* 12:119–128.

Nriagu, J. and Becker, C. (2003), Volcanic emissions of mercury to the atmosphere: global and regional inventories. *The Science of the Total Environment* 304:3–12.

Peixoto-Boischio, A.A. and Henshel, D.S. (2000), Linear regression models of methyl mercury exposure during prenatal and early postnatal life among Riverside People along the Upper Madeira River, Amazon, *Environmental Research Section A* 83:150–161.

Rice, D.C., Schoney, R., and Mahaffey, K. (2003), Methods and rationale for derivation of a reference dose for methylmercury by the U.S. EPA, *Risk Analysis*. 23:107–115.

RTECS (2001), Registry of Toxic Effects of Chemical Substances, U.S. National Institute of Occupational Safety and Health (NIOSH), Cincinnati, OH. Available HTTP: <http://ccinfoweb.ccohs.ca/rtecs/search.html> (accessed December 4, 2002).

Rumbold, D.G., Fink, L.E., Laine, K.A., Niemczyk, S.L., Chandrasekhar, T., Wankel, S.D., and Kendall, C. (2002), Levels of mercury in alligators (Aligator mississippiensis) collected along a transect through the Florida Everglades, *The Science of the Total Environment* 297:239–252.

Seegal, R.F. and Bowers, W.J. (2002), Consequences and mechanisms of action of fish-borne toxicants: what we do not know and why, *Environmental Toxicology and Pharmacology* 12:63–68.

Stern, A.H. and Smith, A.E. (2003), An assessment of the cord blood:maternal blood methylmercury ratio: implications for risk assessment, *Environmental Health Perspectives* 111:1465–1470.

Sundberg, J., Oskarsson, A., and Albanus, L. (1991a), Methylmercury exposure during lactation: milk concentration and tissue uptake of mercury in the neonatal rat, *Bulletin of Environmental Contamination and Toxicology.* 46:255–262.

Sundberg, J., Oskarssson, A., and Bergman, K. (1991b), Milk transfer of inorganic mercury to suckling rats — interaction with selenite, *Biological Trace Element Research* 28(Suppl. 1):27–38.

Sundberg, J. and Oskarsson, A. (1992), Placental and lactational transfer of mercury from rats exposed to methylmercury in their diet: speciation of mercury in the offspring, *Journal of Trace Elements and Experimental Medicine.* 5:47–56.

Sundberg, J., Jonsson, S., Karlsson, M.O., Hallen, I.P., and Oskarsson, A. (1998), Kinetics of methylmercury and inorganic mercury in lactating and nonlactating mice, *Toxicology and Applied Pharmacology* 151:319–329

Tsubaki, T. and Irukayama, K. (Eds.) (1977), *Minamata Disease*, Kodansha, Tokyo.

U.S. Department of Agriculture Continuing Survey of Food Intake by Individuals (1998),1998 CSFII Dataset. Online. Available www.barc.usda.gov/bhnrc/foodsurvey/ (accessed November 12, 2004).

U.S. EPA (1997), Mercury Study Report to Congress (U.S. Environmental Protection Agency, Office of Air Quality Planning and Standards, and Office of Research and Development, EPA-452/R-97-007, 1997). Online. Available HTTP: <http://www.epa.gov/oar/mercury.html> (accessed July 21, 2002).

U.S. EPA (2002), EPA Mercury Study Report to Congress, white paper, U.S. Environmental Protection Agency. Online. Available HTTP: <http://www.epa.gov/oar/merwhite.html> (accessed July 21, 2002).

U.S. EPA (2002a), Fish Advisories, U.S. Environmental Protection Agency. Online. Available HTTP: <http://www.epa.gov/ost/fish/> (accessed September 8, 2002).

U.S. EPA-IRIS (2001), Methylmercury (MeHg) (CASRN 22967-92-6), U.S. Environmental Protection Agency, Integrated Risk Information System. Online. Available HTTP: <http://www.epa.gov/iris/subst/0073.htm> (accessed November 24, 2002).

U.S. EPA and TERA (1999), Comparative Dietary Risks: Balancing the Risk and Benefits of Fish Consumption, *Results of a Cooperative Agreement between The U.S. Environmental Protection Agency and Toxicology Excellence for Risk Assessment (TERA),* Final August 1999. Online. Available HTTP: <http://www.tera.org/pubs/cdrpage.htm> (accessed August 26, 2002).

U.S. FDA-CFSAN (2001), Mercury Levels in Seafood Species (Center for Food Safety and Applied Nutrition, Office of Seafood, U.S. Food and Drug Administration, May 2001). Online. Available HTTP: <http://www.cfsan.fda.gov/~frf/sea-mehg.html> (accessed September 1, 2002).

WHO-JECFA (2000), Safety Evaluation Of Certain Food Additives And Contaminants — WHO Food Additives Series: 44, *Methylmercury* (Prepared by the Fifty-third meeting of the Joint FAO/WHO Expert Committee on Food Additives-JECFA), World Health Organization and International Programme on Chemical Safety (IPCS), Geneva. Online. Available HTTP: <http://www.inchem.org/documents/jecfa/jecmono/v44jec13.htm> (accessed July 24, 2002).

Yoshida, M., Watanabe, C., Satoh, H., Kishimoto, T., and Yamamura, Y. (1994), Milk transfer and tissue uptake of mercury in suckling offspring after exposure of lactating maternal guinea pigs to inorganic or methylmercury, *Archives of Toxicology* 68:174–178.

Young, R.A. (1992), Toxicity summary for Methyl Mercury, *Oak Ridge Reservation Environmental Restoration Program*, February, 1992. Online. Available HTTP: <http://risk.lsd.ornl.gov/tox/profiles/methyl_mercury_f_V1.shtml> (accessed November 26, 2002).

3 Uranium in Food and Water: Actual and Potential Effects

Edmond J. Baratta

CONTENTS

ABSTRACT

Uranium is a naturally radioactive element that exists in the form of two naturally occurring isotopes, uranium-238 and uranium-235. A third isotope, uranium-234, is the decay product of uranium-238. These three radioactive isotopes are normally found in a certain percentage, and there are some areas in the world where uranium is more abundant than other elements. It had only limited use prior to 1939; however, after the development of the atomic bomb, it was mined extensively. Enriched uranium-235 was used to make atomic weapons and as a power source (in nuclear reactors). Uranium-235 is also used to produce radiopharmaceuticals for medical use. The depleted uranium-238 (product of the enrichment of uranium-235) has been used in the production of armor-piercing projectiles. It is also used as a radiation shielding material. The exposure of the body to uranium via ingestion and inhalation was initially of concern to the health of the workers handling this chemical. Later, it became of concern to the health of the general population because it could find its way into the body through ingestion via food and water. This chapter discusses the exposure limits set by various agencies and their possible effects.

INTRODUCTION

The three isotopes of uranium are part of the actinide series. Uranium-234 is normally found in nature in equilibrium with its precursor, uranium-238. Uranium-238

0-8493-3516-7/05/$0.00+$1.50

TABLE 3.1
Radiation Properties of Uranium Isotopes Found in Nature

Isotope	Half-Life (Years)	Abundance (%)	Alpha Energies
Uranium-238	4.5E+09	99.276	4.2 (75%), 4.15 (25%)
Uranium-234	2.6E+05	0.0057	4.77 (72%), 4.72 (28%)
Uranium-235	8.8E+08	0.7196	4.58 (8%), 4.4 (57%), 4.37 (18%)

and uranium-234 are part of the uranium–radium decay series. Uranium-235 is in the uranium–actinium series. Table 3.1 [from the *Radiological Health Handbook* (1970)] shows the half-lives, major abundances, and alpha energies of these isotopes. Uranium-238 decays by emitting alpha particles to become thorium-234, which has a 24.1-d half-life; this in turn decays to form two short-lived radionuclides, protoactinium-234m and protoactinium-234. These three radionuclides decay by emitting beta particles and, in the process, emit gamma rays. Protoactinium-234 decays to uranium-234. Uranium-234 then decays to long-lived thorium-230, which is an alpha particle emitter with a half-life of 8.0E+04 yr. Uranium-238 emits alpha particles and its decay products emit beta particles and gamma rays. Uranium-235 decays to short-lived thorium-231 (a beta emitter with low-abundance gamma rays) and decays to protoactinium-231. Protoactinium-231 is long-lived, with a half-life of 3.5E+04 yr.

Uranium is highly toxic as a chemical and presents a radiation hazard if ingested or taken into the lungs. The maximum allowable concentration (MAC) for soluble compounds as promulgated by the American Conference of Government Industrial Hygienists (ACGIH) (September 1978) is 0.5 mg/m^3 of air. For insoluble compounds it is 0.25 mg/m^3 of air. There are many references regarding uranium toxicity which refer to its toxicity as a chemical. These are listed in the Material Safety Data Sheets and U.S. government publications, and are not discussed here. This chapter deals only with the effects of uranium in water and food products.

The annals of the International Commission on Radiation Protection (ICRP) Publication 30 (1979) on the "Annual Limits on Intake of Radionuclides by Workers Based on the 1979 Recommendations" lists the limits for inhalation and ingestion for the three radionuclides as shown in Table 3.2. The report also gives the metabolic data for uranium in Reference Man (ICRP, 1975). It reports that the uranium content of the body is 90 µg; for the skeleton, 59 µg; and for the kidneys, 7 µg. It estimates the daily intake of uranium from food and fluids to be 1.9 µg.

These limits were revised in 1990 based on the additional data that was available to the Commission (Sowby, 1990). It should be noted that the values given are for soluble compounds of uranium, which would more likely be found in food and water. The inhalation ALIs are shown for comparison in Table 3.3.

URANIUM IN DRINKING WATER

In 1976, the U.S. Environmental Protection Agency (EPA) did not regulate uranium in drinking water as a specific radionuclide but only as part of the gross alpha content.

TABLE 3.2
International Commission on Radiation Protection — Publication 30, Annual Limits of Intake for Workers (Supplement to Part 1)

Isotope	Half-Life (Years)	Limits, Bq	
		Ingestion	Inhalation
Uranium-238	4.5E+09	5.0E+05	5.0E+04
Uranium-234	2.6E+05	4.0E+05	5.0E+04
Uranium-235	8.8E+08	5.0E+04	5.0E+04

TABLE 3.3
International Commission on Radiation Protection — Publication 61, Annual Limits of Intake for Workers

Isotope	Half-Life (Years)	Limits, Bq	
		Ingestion	Inhalation
Uranium-238	4.5E+09	8.0E+05	9.0E+04
Uranium-234	2.6E+05	7.0E+05	8.0E+04
Uranium-235	8.8E+08	7.0E+05	8.0E+04

The EPA recently published in the Federal Register, the "National Primary Drinking Water Regulations: Radionuclides: Final Rule" (December 7, 2000). 40 Code of Federal Regulations (CFR) parts 9, 141, and 142. Uranium was one of the radionuclides addressed in these Regulations. The 40 CFR part 141, subpart Q, "Standard Health Effects Language for Public Notification" lists the maximum contamination level goal (MCLG) for uranium to be zero. The maximum contamination level (MCL) was set at 30 μg/l. This is equivalent to approximately 3.7E-02 Bq/μg or 1.11 Bq/l. This is based on the consumption of 2 l of water per day. The annual limit of intake (ALI) for the MCL would then be 8.1E+02 Bq or 2.19E+04 μg. The reasoning is that some people who drink water containing uranium in excess of the MCL over many years may have an increased risk of getting cancer and kidney toxicity. The calculated dose for this exposure is 0.5 mSv/yr.

Uranium being a carcinogenic agent, the calculations for Uranium limits in drinking water are based on the health effects of kidney toxicity. In 1991 the EPA had proposed a limit of 20 μg/l, which was approximately equal to 30 pCi/l or 1.11 Bq/l. This was based on the assumed mass ratio of 1.5 pCi/l (4.11E-02 Bq/μg). Later data showed the mass ratio to be near 1:1 (in terms of pCi/l). The 1991 limit was based on the drinking water equivalent level (DWEL) exposure to the kidneys for a lifetime. The 1991 limit of 1.11 Bq/l was based on the estimated lifetime cancer risk of approximately 1 in 10,000.

Uranium has been identified as a nephrotoxic (kidney toxic) metal. It is, however, less toxic than some other heavy metals such as cadmium, lead, and mercury. The discontinued exposure to uranium in animals has shown that the results may be reversible. Recent studies have shown the DWEL to be 20 μg/l. This is based on a daily ingestion to the population, which includes various subgroups at risk. The value of the lowest observed adverse effects levels of 60 μg/d has been supported by animal studies (rats). The EPA believes that 30 μg/l is protective against kidney toxicity, although it states that 20 μg/l is its best estimate of the DWEL.

Hess et al. (1985) published in *Health Physics* "The Occurrence of Radioactivity in Public Water Supplies in the United States," focusing on the distribution of uranium in ground and surface waters. They collected data for surface waters from all the states, excluding Hawaii, South Carolina, and Florida. The highest concentration of uranium in μg/l was found in Alaska, which had an average of 7.5, but ground water in Alaska was not included in this study. In the lower continental states, the values ranged from 0.01 to 0.75 μg/l. The values for ground water were from 0.01 to 3.5 μg/l. The domestic supplies found the range to be from 0.01 to 3.5 μg/l.

The U.S. Nuclear Regulatory Commission (NRC) limits the ingestion and inhalation of uranium for radiation workers through 10 CFR part 20 (2002). The ALI for uranium (in various chemical forms) for ingestion by workers is set at 10 μCi or 3.7E+05 Bq. Converted to micrograms, the amount is 1.0E+07 μg. The daily ingestion would be 3.85E+04 μg (based on 8-h exposure for 260 d). The amount of any radionuclide (Hess et al., 1988) allowed for the general population, based on what is allowed for radiation workers, is generally divided by 100. Therefore the amount allowed for the general population would be 126.7 μg/d. This is not in line with the EPA value of 60 μg. The NAS (NAS83, 1983), however, had recommended a chronic suggested no-adverse-response level (SNARL) for uranium, based on chemical toxicity. They assumed a minimum observed-effect level of 1 mg/kg/d, with an uncertainty factor of 100, for a 70-kg adult who consumes 2 l/d of water, which provides 10% of the daily uranium intake. At equilibrium this would be equivalent to 4.29E-01 Bq of uranium per day. The EPA values for drinking water do not agree with this.

The International Commission on Radiation Protection (Sowby, 1990) too has values for workers (Table 3.3). The ICRP-61 lists the ALI for ingestion for workers as 8.0E+05 Bq. This is slightly more than the NRC values for the U.S. worker.

URANIUM IN FOOD

The U.S. Food and Drug Administration (FDA) is responsible for the wholesomeness of the food supply. This includes regulating radionuclides in bottled water, but the same limits are used as the EPA has set for radionuclides in drinking water. The FDA has not set a limit to the amount of uranium allowed in the food supply. The Federal Radiation Council's (FRC) Background Material for the Development of Radiation Protection Standards Report No. 2 (1961) did not address this radionuclide. However, the FRC does state that "for radionuclides not considered in this report, Federal agencies should use concentration values in air, water, or items of food which are consistent with recommended Radiation Protection Guides and the general guidance on intake." The FRC Report No. 2 at that time did recommend 5 mSv/yr for the general population. Since then, this has been lowered to 2 mSv. The FDA's latest document (1998) does not address uranium. This document, Accidental Radioactive Contamination of Human Food and Animal Feeds: Recommendations for State and Local Agencies, addresses accidents or other events involving releases of radioactivity. The levels the FDA recommended would be based on EPA guidelines for drinking water as compared with the levels for radium-226, addressed in FRC Report No. 2.

The Tri-City study, one of the most extensive studies of uranium in the human diet, was done by Welford and Baird (1967) as reported in *Health Physics*. The cities in the study were New York, Chicago, and San Francisco. The data, reported in μg of uranium/yr, showed that the daily intake for the three cities was constant at 1.3 μg/d for New York and San Francisco, and 1.4 μg/d for Chicago. In Becquerels, this is 3.367E-02/d and over 12.29 Bq/yr. The measurements were taken using two methods, neutron activation and fluorimetric. The ICRP at that time had accepted the results of 2 μg/d, but this and other data did not reflect the general population's daily intake of uranium. Welford and Baird also made measurements of random samples of air, bone, and lung from New York and urine samples from the Chicago area. The annual intakes showed that the products with the highest amounts of uranium (>50 μg/yr) were bakery products, meat, (excluding poultry), fresh fruits (all types), and potatoes. They also analyzed tap (drinking) water in the New York area and the mean was 0.032 μg/l ± 0.2%, with a mean range of 0.024 to 0.041 μg/l. They analyzed the deionized water in their facility and found the results to be 0.002 μg/l ± 2% with a mean range of 0.0018 to 0.0034 μg/l.

Subsequently, Baratta and Mackill (2001) reported in the *Journal of Radioanalytical and Nuclear Chemistry* on analyzed selected samples for isotopic uranium. Their limit of detection was 1.0E-04 Bq (2.7E-03 μg), which was approximately the limit of the method in Standard Methods for the Determination of Water and Waste Water (Clesceri et al., 1998). Welford and Baird's limit was 3.7E-04 Bq (1.0E-02 μg). It must be remembered that the latter were not analyzing for isotopic uranium, but "total" Uranium. Baratta and Mackill's study found that fodder used for dairy cows showed the concentrations of isotopic uranium to be in naturally occurring ratios. An analysis of milk showed that the isotopic uranium was below detectable levels. The same was the case of well water that was filtered to be used in food production. Unfiltered well water contained isotopic uranium. The content of uranium in filtered water was again below detectable levels.

These studies were in keeping with the FRC's mandate of Report No. 2. The latter study was charged with investigating concerns that enriched uranium in one case and depleted uranium in the other could perhaps be contaminating the food supply.

CONCLUSIONS: ACTUAL AND POTENTIAL EFFECTS

Uranium has been shown to be toxic and carcinogenic in the workplace. It has been reported to concentrate in the kidneys and bones (ICRP-30). The EPA levels and the recommendations by the FRC are designed to limit the intake of this radionuclide as much as possible.

The combined concentrations of uranium in water and food in the U.S. have been shown to be well below the MCL for drinking water (30 μg/l or 60 μg/d) as set by the EPA or even its DWEL of 20 μg/l or 40 μg/d. According to a report by the World Health Organization (WHO) (1998), the ingestion of uranium in food accounts for the main source of this radionuclide. The drinking water guidelines are very conservative as they contribute only a minor portion of the ingestion uranium. The WHO estimates that 90% of the daily intake is from food. The Welford and Baird study showed that for three major cities the amount consumed in food was 1 to 1.5 μg/d. The Hess et al. study showed that the concentration of uranium in drinking water varied in the U.S. and other countries. From their data, it is difficult to calculate the ratio or percentage of uranium in food to drinking water. The uptakes of both, in most cases, are below EPA's DWEL of 20 μg/l.

Limson et al. (2002) did a study between two cities in Canada, measuring the uptake of uranium from the ingestion of food and water in the gastrointestinal tract in animals. The uptake in one city was well below EPA limits, whereas the other contained levels ranging from well below the limits to very high concentrations. The study also included testing of urine and feces. In the latter city, the amount of uranium in food that contributed to uranium intake was in the area of 20%. Under the EPA limits, these communities would have to lower the uranium concentration in drinking water to the 30 μg/l level or find another water source.

The ingestion of uranium via food and drinking water is well below the levels set by the U.S. Environmental Protection Agency and Health and Human Services (Food and Drug Administration) and the value report by WHO (1998). At these levels, any effect should be minimal. Effects from accidental ingestion or inhalation by workers would be the greatest potential hazard. Although uranium exposure is of concern, data in most cases have shown that there does not appear to be any actual or potential hazard to the general population through the ingestion of food and water.

References

Baratta, E.J. and Mackill, P. (2001), Determination of isotopic uranium in food and water, *Journal of Radioanalytical and Nuclear Chemistry*, 248, 2:473–475.

Clesceri, L., Greenberg, A., and Eaton, A. (Eds.) (1998), Standard Methods for the Determination of Water and Waste Water, American Public Health Association, 20th ed., 7–48 to 7–51.

Code of Federal Regulations — Energy (2002), Annual Limits on Intake (ALIs) and Derived Air Concentrations (DACs) of Radionuclides for Occupational Exposure; Effluent Concentrations for Release to Sewerage, 10 CFR 20 Appendix B, Energy: 391.

EPA (1998), Environmental Protection Agency, National Primary Drinking Water Regulations: Radionuclides; Final Rule, 40 Code of Federal Regulations Part 9, 141 Federal Register, pp. 76708–76753.

Federal Radiation Council (1961), Background Material for the Development of Radiation Protection Standards, Report No.2, *Department of Health, Education, and Welfare.*

Hess, C.T., Michel, J., Horton, T.R., Prichard, H.M., and Coniglio, W.A. (1985), The occurrence of radioactivity in public water supplies in the United States, *Health Physics* 48, 5:553–586.

ICRP (1975), Drinking Water and Health Report of the Task Group on Reference Man, Annals of the ICRP, ICRP Publication 23.

Limson, M., Zielinski, J.M., Meyerhof, D.P., and Tracy, B.L. (2002), Gastrointestinal absorption of uranium in humans, *Health Physics*, 83, 1: 35–45.

National Academy of Sciences (1983), Drinking Water and Health, NAS83, National Academy of Sciences, 5: 147–151.

Sowby, F.D. (Ed.) (1979), Limits for Intakes of Radionuclides by Workers, *Annals of the ICRP, ICRP Publication 30*, 2, Part 1: 101–102.

Sowby, F.D. (Ed.) (1979), Limits for Intakes of Radionuclides by Workers, Annals of the ICRP, ICRP Publication 30, 3, Supplement to Part 1: 375–378.

Sowby, F.D. (Ed.) (1990), Annual Limits of Intakes of Radionuclides by Workers, Based on the 1990 Recommendations, Annals of the ICRP, ICRP Publication 61, 2:37.

U.S. Department of Health and Human Services, Food and Drug Administration (1998), Guidance on Accidental Radioactive Contamination of Human Food Feeds: Recommendations for State and Local Agencies, Federal Register: 43402–43403.

USDHHS/PHS (1978), Occupational Health Guide for Uranium and Insoluble Compounds (as Uranium), *U.S. Department of Health and Human Services and Department of Labor.*

Welford, G.A. and Baird (1967), Uranium levels in human diet and biological materials, *Health Physics,* 13: 1321–1324.

World Health Organization (WHO) (1998), *Uranium: Guidelines for Drinking Water Quality and Health Criteria and Other Supporting Information,* World Health Organization, 2nd ed.

4 Polychlorinated Biphenyls, Oxidative Stress, and Diet

Bernhard Hennig, Michal Toborek, Pachaikani Ramadass, Gabriele Ludewig, and Larry W. Robertson

CONTENTS

0-8493-3516-7/05/$0.00+$1.50

ABSTRACT

Polychlorinated biphenyls (PCBs) are polyhalogenated aromatic hydrocarbons that are persistent and widely dispersed in the environment. Diet is the major route of exposure to PCBs, and PCBs distribute themselves into tissues, especially adipose, where they are in dynamic equilibrium with blood. Excretion of PCBs from the body is very slow, resulting in an increased PCB body burden with age. PCBs have broad adverse effects, including cancer induction, immune system suppression, nervous system effects, disruption of normal hormone function, and developmental and reproductive abnormalities, as well as increased cardiovascular and inflammatory diseases. PCBs stimulate the production of reactive oxygen species through metabolic events and via cellular signaling pathways. PCBs may also lower the levels of enzymatic and nutrient antioxidants. A resulting imbalance in the cellular oxidative stress or antioxidant status may be a common denominator in PCB-induced disease processes. There is evidence that diet and selected nutrients can modify the cytotoxicity of PCBs. For example, certain dietary fats can upregulate, and nutrients with antioxidant properties can downregulate, cellular signaling pathways involved in PCB-mediated cell dysfunction. This chapter provides evidence that nutrition can modify PCB toxicity, which may have critical implications in numerous age-related diseases such as cancer, cardiovascular disease, or neurological disorders. Nutrition may affect risk assessment of environmental contaminants beyond PCBs. Thus, proper nutrition counseling should be considered by health officials and the medical community to reduce the overall risk for environmental hazards such as PCB toxicity and disease development. More research is needed to understand observed interactions of PCB toxicity with nutritional interventions.

INTRODUCTION

POLYCHLORINATED BIPHENYLS

Polychlorinated Biphenyls (PCBs) are industrial chemicals that were produced and sold in the U.S. for approximately 50 years (Hansen, 1999). Chemically they are biphenyl molecules with 1 to 10 chlorines attached to the carbons (see Figure 4.1; the numbering of the carbon atoms is shown for clarification). Considering the number and position of the chlorines, 209 individual PCB isomers and congeners are possible. Chlorines may be present in the 2, 2′, 6, and 6′-positions, *ortho* to the biphenyl bridge. Such congeners are called ortho-substituted PCBs. Chlorines in 3, 3′, 5, and 5′-positions are *meta* to the biphenyl bridge, whereas chlorination at the

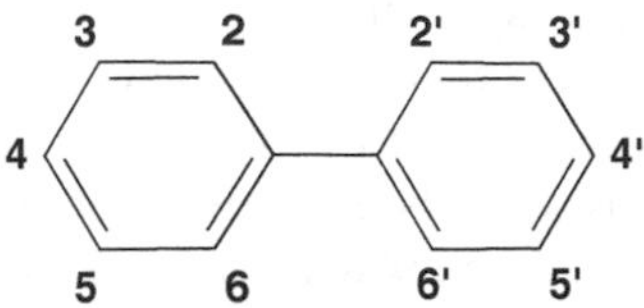

FIGURE 4.1 Structure of biphenyl showing sites for chlorine atom attachment.

4 and 4′-carbon atoms is *para* to the biphenyl bridge. Ortho chlorines have a direct impact on the three-dimensional structure of the molecule by diminishing the coplanarity of the phenyl rings, which has significant influence on the biological effects of the congeners. To illustrate this, PCB congeners with no *ortho*-substituents such as 3, 3′, 4, 4′-tetrachlorobiphenyl are also called coplanar PCBs, whereas those with *ortho*-chloro-substituents are called noncoplanar PCBs. Although often individual PCBs are named by indicating the site of chlorine substitution, a simpler, one- to three-digit, nomenclature has come into common use whereby the individual PCBs are named sequentially from 1 to 209 (see pages 445–447 of Robertson and Hansen, 2001). For example 3, 3, 4, 4-tetrachlorobiphenyl is then referred to as PCB 77.

Although the U.S. sales and distribution of PCBs ended in the late 1970s, a significant portion of PCBs produced by industry are still in use, mostly within capacitors and transformers. PCBs were not sold as individual compounds or congeners, but rather as complex mixtures containing scores of individual compounds. Typically, these PCB mixtures were sold under various trade names (such as Aroclor in the U.S., Kanechlor in Japan, or Clophen in Germany) with numerical designations indicating percent chlorine by weight (e.g., Aroclor 1254 is 54% and Clophen A60 is 60%). Many isomers and congeners may be found in multiple PCB formulations, even those formulations differing considerably in chlorine content (Silberhorn et al., 1990). The PCB isomers and congeners present, either individually or in combination, determine the biologic and toxic effects of PCB mixtures and commercial formulations. However, the composition of PCB mixtures is not static but may be influenced by the biologic or chemical breakdown of individual components or by selective transport or uptake.

In 1976 Congress passed the Toxic Substances Control Act that banned the manufacture of PCBs and PCB-containing products and established strict regulations regarding their future use and sale. Presently, the amount of PCB that is distributed in the global environment is estimated at about 400,000 t. The majority of this is found in the open ocean water sediment, but contamination of rivers, creeks, lakes, and Superfund sites is also frequent. Overall, PCBs are now a ubiquitous contaminant that is found in places as far away as the lakes in the Himalaya mountains or the ice of the arctic (Robertson and Hansen, 2001, Chapter 1). Unfortunately, they have become normal constituents in human and animal tissues.

PCBs have been classified as probable human carcinogens. They also have a variety of other adverse effects including immune system suppression, nervous system effects, disruption of normal hormone function, and reproduction abnormalities such as reduced birth weight and reduced fertility, as well as recent evidence of increased cardiovascular and inflammatory diseases. Much evidence now exists that many types of PCBs, and especially coplanar PCBs, exhibit their toxicity and disease potential via oxidative stress-sensitive mechanisms. This is of particular interest to health officials because many diseases, including cancer and cardiovascular and neurological diseases, are believed to be initiated through imbalances of the body's oxidative stress and antioxidant status (Butterfield et al., 2002; Chen et al., 2003). Furthermore, the cellular oxidative stress and antioxidant balance can be manipulated through the environment, which includes toxins and nutrients.

In the following text, we will attempt to summarize the mechanisms by which PCBs may cause oxidative stress and demonstrate the involvement of PCB-induced oxidative stress in disease processes. Citations and references are limited to those that focus on more recent studies.

PCBs and Nutrition

The diet is the primary source of PCBs for animals and humans and is, therefore, the predominant route of exposure. To protect the public and to provide a level of safety, various federal agencies, such as the Environmental Protection Agency (EPA), the Occupational Safety and Health Administration (OSHA), and the Food and Drug Administration (FDA), are charged with developing regulations for toxic substances, including PCBs. "The EPA standard for PCBs in drinking water is 0.5 parts of PCBs per billion parts (ppb) of water. For the protection of human health from the possible effects of drinking the water or eating the fish or shellfish from lakes and streams that are contaminated with PCBs, the EPA regulates that the level of PCBs in these waters be no greater than 0.17 parts of PCBs per trillion parts (ppt) of water .... The FDA has set residue limits for PCBs in various foods to protect from harmful health effects. FDA-required limits include 0.2 parts of PCBs per million part (ppm) in infant and junior foods, 0.3 ppm in eggs, 1.5 ppm in milk and other dairy products (fat basis), 2 ppm in fish and shellfish (edible portions), and 3 ppm in poultry and red meat (fat basis)" (ATSDR 2000).

Although the higher halogenated PCBs are more resistant to metabolic attack and are sequestered in fatty tissues, their levels can be influenced by a number of factors. In mammals, PCBs are secreted into milk (particularly in early lactation) and hair oil, excreted into urine and feces, and residues may pass from pregnant females to their offspring across the placenta. Two of these routes of "excretion" are the direct transfer of PCBs (via milk or placenta) to other individuals in the population. Also, within an individual, levels of PCBs are not static but are influenced by numerous physiologic factors. Lactation in particular accelerates the elimination of PCBs, whereas dieting may mobilize PCBs from fat storage with a concomitant increase in blood levels.

PCBs, Fatty Acids, and Vitamins and Minerals

Nutrition, especially the consumption of fatty foods, is the major source of PCBs for humans. On the other hand, if nutrition could modify the toxicity of PCBs, then this would provide a means for populations at risk — i.e., sports fishermen or populations residing near Superfund sites or areas of contamination — to reduce their overall risk for PCB toxicity and disease development by dietary means. Very little is known about the effects of different diets on the adverse health effects by environmental contaminants such as PCBs, and such interactions can be very complex. For example, Greenlanders consume a high amount of marine food or fish that is high in protective omega-3 fatty acids. However, PCBs can concentrate in the marine food chain, resulting in much higher PCB concentrations in tissues of Greenlanders compared to other populations (Mulvad et al., 1996). Whether a marine-based diet can ameliorate the disease-causing toxicity of PCBs is not known, but our data (see the section titled "PCBs, Diet, and Implications in Cardiovascular Disease") clearly show that nutrition can modulate PCB toxicity and that specific fatty acids rich in plant oils, such as linoleic acid (the parent omega-6 fatty acid), can amplify PCB toxicity in vascular endothelial cells.

Our data also suggest (see the section titled "Diet and Protection Against PCB-Induced Cytotoxicity") that antioxidants, and especially vitamin E, can protect against PCB toxicity. The body's primary lipophilic antioxidant is alpha-tocopherol (vitamin E). Its main function involves the termination of free radical-induced lipid peroxidation cascades of peroxyl radicals. The resulting alpha-tocopherol radical can be reduced back to alpha-tocopherol by ascorbic acid, the body's prime hydrophilic antioxidant. If the radical remains unreduced and interacts with another peroxyl radical, then alpha-tocopheryl quinone will be formed. This alpha-tocopheryl quinone can be subsequently reduced to its corresponding alpha-tocopherol hydroquinone (Figure 4.2). The alpha-tocopherol hydroquinone may also possess antioxidant activity. Therefore, alpha-tocopherol may act as a reservoir of this antioxidant; the relative concentrations of alpha-tocopherol, alpha-tocopheryl quinone, and alpha-tocopheryl hydroquinone may provide additional pertinent information about the possible mechanisms of toxicity of compounds such as PCBs by pointing towards oxidative stress as intermediate. Changes in the levels of vitamin E and its metabolites may also indicate the importance of antioxidant defense mechanisms in the protection against PCB toxicity. Our studies clearly demonstrate that vitamin E has potent protective properties against PCB-induced vascular endothelial cell dysfunction, a critical issue in understanding the protective mechanisms of vitamin E against PCB-mediated cardiovascular diseases.

Among the minerals, selenium is so far the only compound that has consistently been shown to have chemoprotective activity against cancer development. Selenium is an essential trace element, and deficiency in selenium is now proposed to be a possible risk factor for cancer predisposition and development. Selenium also seems to have major beneficial effects for the cardiovascular system, mainly a reduced incidence of atherosclerosis (Mulvad et al., 1996). Therefore, it is not surprising that selenium is gaining more and more attention as a beneficial nutrient and is also considered as one possible option for chemoprotection against health damage by food contaminants, including PCBs. However, selenium is a two-edged sword because high selenium concentrations are toxic. Thus, several studies focused on selenium as a contaminant and potential health hazard. Results seem to indicate that often a higher PCB body burden is accompanied by elevated selenium levels, a correlation that was reported, for example, for cord blood measurements in a European study (Janousek et al., 1994). This may in part be caused by the tendency of

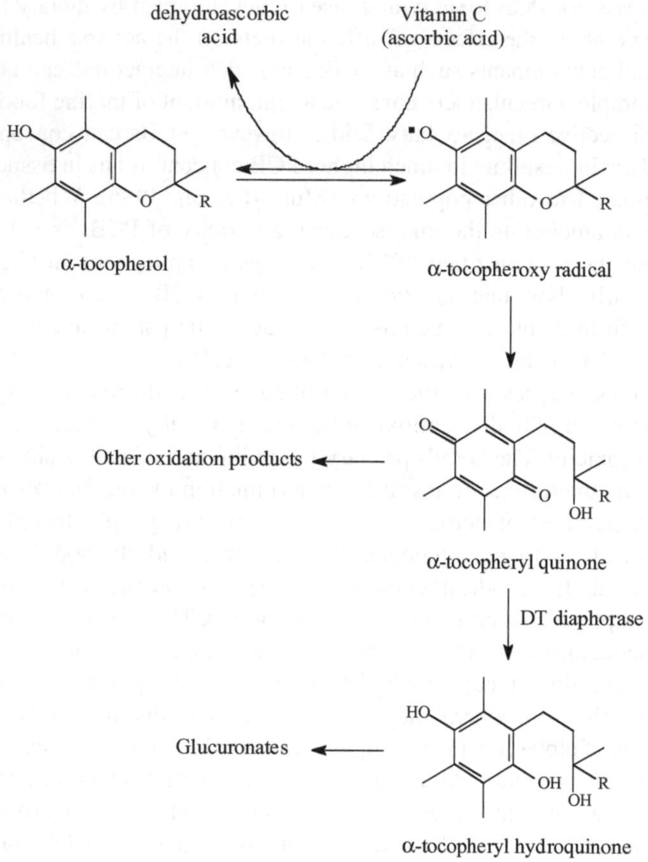

FIGURE 4.2 Metabolism of alpha-tocopherol (vitamin E).

metals to accumulate in the same food sources as PCBs, mainly fish. At least one study, however, did not find a correlation between selenium and seafood consumption, although this study found a clear correlation between PCB levels and frequency of whale dinners (Grandjean et al., 1995). Another reason why PCB and selenium levels seem to correlate could be because the body burden of both PCBs and selenium increased with age, as one study reported (Bjerregaard and Hansen, 2000). None of these studies detected dangerously high selenium levels in humans and thus this coexposure may be beneficial. Our recent work suggests that selenium may provide protection against PCB toxicity through regulation of critical selenium-dependent enzymes such as selenium-dependent glutathione peroxidase (Se-GPx). Others have reported that PCB 77 induces a selenium-binding 54-kDa protein in the cytosol of rats (Ishii et al., 1996). The function of this protein and the significance or consequences, and the role this protein may play in the overall reduction of liver selenium and enzymes such as selenium-dependent glutathione peroxidase (Se-GPx) needs to be determined.

Overall, because exposure to PCBs occurs to a large part through fish, which is also high in *n*-3-polyunsaturated fatty acids, monounsaturated fatty acids, selenium, and other beneficial nutrients, all believed to provide protection against cardiovascular, neurological, and cancer-causing damage, it may be difficult if not impossible to decipher these complicated interactions and networks in human epidemiological studies. Therefore, systematic *in vivo* and *in vitro* experiments may be required to understand the mechanisms of effects and to develop recommendations for chemoprotection through dietary modulation. Our groups have performed systematic experiments with individual PCB congeners to gain further insight into protective phenomena of selected nutrients, especially fatty acids, vitamin E, and selenium, the results of which are described in more detail in the following text.

OXIDATIVE STRESS AND PCBS

It is unquestionable that PCBs induce health damage, but the mechanisms of this damage induction are not clear. One possibility could be through the generation of oxidative stress, defined as an increased production of reactive oxygen species (ROS) or free radicals, or reduced antioxidant capacity. Increased oxidative stress during PCB exposure or metabolism has been detected in cell-free systems (McLean et al., 1996b; Oakley et al., 1996a; Schlezinger et al., 2000; Schlezinger et al., 1999), in cells in culture (Ludewig et al., 1998; Slim et al., 1999; Slim et al., 2000), and in laboratory animals (Dogra et al., 1988; Kamohara et al., 1984; Pelissier et al., 1990; Saito, 1990). Reported targets of PCB-induced oxidative stress are lipids (Dogra et al., 1988; Kamohara et al., 1984; Pelissier et al., 1990), proteins (Shimada and Sawabe, 1983), and DNA (McLean et al., 1996b; Oakley et al., 1996b). Possible mechanisms and consequences of PCB-induced oxidative stress for the organism are discussed in the following text.

OXIDATIVE STRESS THROUGH RECEPTOR BINDING AND GENE REGULATION

Individual PCB congeners are ligands for a number of cellular and nuclear receptors. The earliest description was of PCBs as ligands for the aryl hydrocarbon receptor (AhR) (Bandiera et al., 1982). This binding preceded the efficacious induction of a broad spectrum of xenobiotic-metabolizing enzymes, most noticeably certain cytochrome P-450-dependent monoxygenases (CYPs), especially of the CYP1A subfamily, and microsomal epoxide hydrolase (Parkinson et al., 1983), glutathione transferases (Schramm et al., 1985), and UDP-glucuronosyltransferases (Püttmann et al., 1990). Individual PCB congeners that were the best ligands for the AhR were identified as those isomers and congeners in which halogens are present in the meta and para positions of biphenyl but not in the ortho positions. These PCBs were often referred to as coplanar congeners, typical examples of which are PCBs 77, 126, and 169. Other PCBs, characterized by halogen substitution in the ortho and para positions of biphenyl (e.g., 2,2′,4,4′,5,5′-hexachlorobiphenyl, PCB 153), interact with the constitutively active receptor (CAR). PCBs in this group induce CYP 2B1/2 and as such resemble the drug phenobarbital in their mode of CYP induction (Denomme et al., 1983). Other acute adverse biochemical and toxic effects of PCBs may be mediated by the interactions of various PCBs with other sites and receptors within cells. Several such interactions have been recently described and, interestingly, involve PCBs with multiple ortho-chlorine atoms. Biochemical and toxic effects of these ortho-substituted PCBs, which are believed to be mediated through their interactions with receptors, include estrogenicity and anti-estrogenicity through binding to the estrogen receptor (Gierthy et al., 1995), effects on calcium channels via activation of the ryanodine receptor (Wong et al., 1997), their ability to cause insulin release from cells in culture (Fischer et al., 1998), their potency in lowering cellular dopamine levels (Seegal, 1995), and their ability to activate neutrophils to produce superoxide (Fischer et al., 1998).

An imbalance in metabolizing enzymes may lead to increased oxidative stress through at least 3 mechanisms: (1) It has been demonstrated that the persistent induction of CYPs, in the absence of substrate, may lead to the production of ROS (Schlezinger et al., 1999; Schlezinger et al., 2000). (2) An increase or induction of certain metabolizing enzymes, especially CYPs and epoxide hydrolase (EH), may steer the metabolism of endogenous and exogenous compounds, estradiole, PCBs, etc., toward more redox-reactive intermediates, and increase redox cycling (CYP reductase, DT-diaphorase). (3) A reduction of antioxidant enzymes, such as SeGPx, may cause an increase in oxidative stress through reduction in antioxidant defenses (see the following text).

Metabolic Activation of Lower Halogenated Biphenyls

Recently, we have focused on the metabolic activation of lower halogenated biphenyls to electrophiles that react with cellular substituents such as proteins and DNA, the production of oxygen-centered radicals during PCBs metabolism, and the biologic and toxic consequences of these reactions. It is apparent that at least two possible electrophiles may be produced as a consequence of CYP-catalyzed hydroxylation of PCBs (Figure 4.3). The first is an intermediate of the hydroxylation reaction, namely an arene oxide. Intermediacy of arene oxides gives rise to sulfur-containing metabolites of PCBs (attack by glutathione catalyzed by glutathione transferase, and subsequent reactions) and dihydrodiol metabolites (as a result of the hydrolysis of the arene oxide by epoxide hydrolase). A second electrophile is a quinone, the oxidation product of catechols and hydroquinones.

Studies of the metabolism of 4-chlorobiphenyl have been described in detail (McLean et al., 1996a). Collectively, these and other data from our laboratories clearly show that hepatic microsomes are competent to catalyze the metabolic conversion of lower halogenated PCBs, i.e., mono-, di-, and tri-chlorobiphenyls to catechols (the hydroxyl groups are ortho or next to each other) and hydroquinones (the hydroxyl groups are para or opposite to each other; see Figure 4.4). Dihydroxy

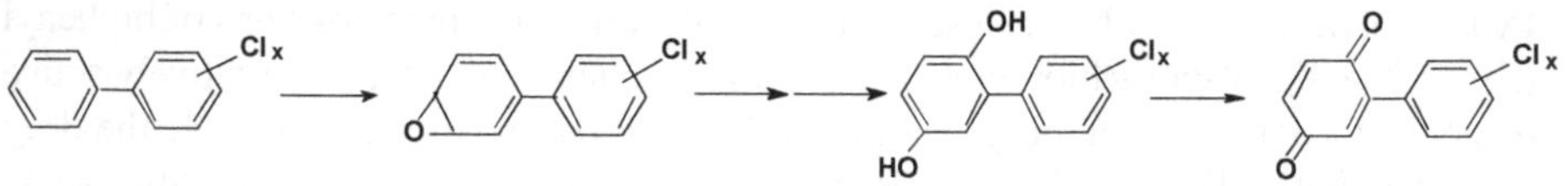

FIGURE 4.3 Metabolic activation of polychlorinated biphenyls (PCBs).

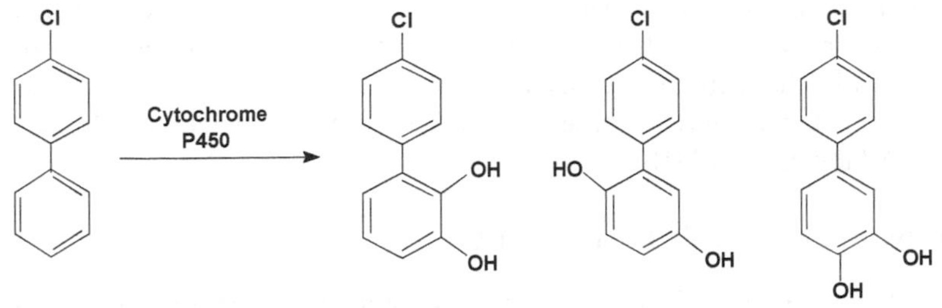

FIGURE 4.4 Metabolic conversion of 4-chlorobiphenyl to catechols and hydroquinone, precursors of ortho and para quinones.

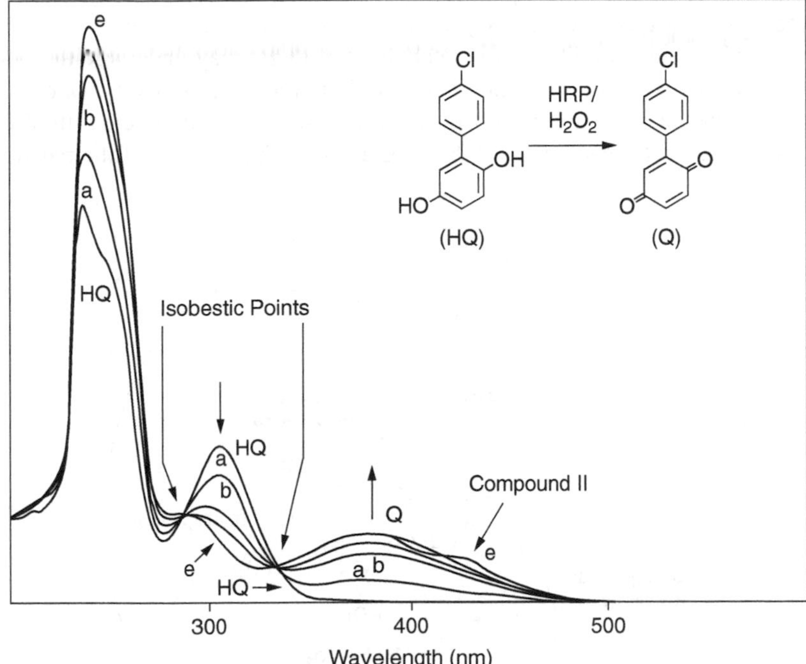

FIGURE 4.5 Absorption spectrum depicting the oxidation of 2-(4-chlorophenyl)-1,4-hydroquinone (HQ) to 2-(4-chlorophenyl)-1,4-benzoquinone (Q) by horseradish peroxidase (HRP) and hydrogen peroxide.

metabolites of 4-chlorobiphenyl (and other lower halogenated biphenyls) may be oxidized *in vitro* by peroxidases and H_2O_2 to products that have characteristic visible absorption maxima. The formation of these compounds can be followed spectrometrically. The stepwise conversion of 2,5-dihydroxy-4-chlorobiphenyl to the corresponding quinone, as limiting increments of H_2O_2 were sequentially added, is shown in Figure 4.5. The primary hydroquinone absorbance at 300 nm diminishes in a stepwise fashion as the quinone absorbance at 380 nm increases. After all of the hydroquinone is oxidized, an additional peak at approximately 415 nm appears, which corresponds to Compound II of horseradish peroxidase (HRP). Oxidation of the hydroquinone is, therefore, dependent on both the enzyme and the oxidant. Besides HRP, other peroxidases, such as myeloperoxidase and the peroxidase component of prostaglandin H synthase, can also catalyze these reactions (McLean et al., 1996b; Oakley et al., 1996b).

ROS Production by PCB Metabolites

During this oxidation of dihydroxylated PCB metabolites, semiquinones and quinones are formed concomitant with the production of superoxide (see Figure 4.6). One method to quantify superoxide production *in vitro* is with the help of the dye nitroblue tetrazolium, which forms a blue formazan upon reaction with superoxide and can be measured spectrophotometrically. An analysis of 6 synthesized PCB-hydroquinones showed that all of them produced superoxides (Srinivasan et al., 2001).

The ortho- and para-dihydroxy metabolites of PCBs may undergo either a one- or two-electron oxidation to semiquinones and quinones, respectively. One or both oxidation products may react with nucleophiles such as amino acids, thiol compounds, and DNA (Amaro et al., 1996) (Figure 4.6). The product of the reaction of

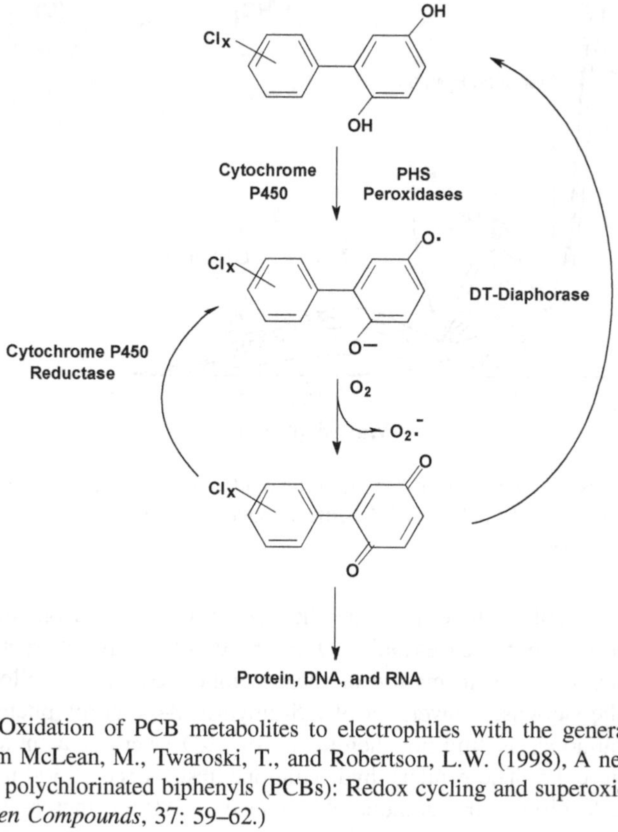

Cytochrome P450

PHS Peroxidases

DT-Diaphorase

Cytochrome P450 Reductase

O_2

$O_2^{\cdot -}$

Protein, DNA, and RNA

FIGURE 4.6 Oxidation of PCB metabolites to electrophiles with the generation of ROS. (Modified from McLean, M., Twaroski, T., and Robertson, L.W. (1998), A new mechanism of toxicity for polychlorinated biphenyls (PCBs): Redox cycling and superoxide generation, *Organo-halogen Compounds*, 37: 59–62.)

a PCB quinone with thiol compounds such as glutathione (GSH) is a reduced dihydroxy-PCB glutathionyl. This GSH adduct may be oxidized again to the corresponding quinone, with superoxide production, and then subjected to another round of GSH adduction and oxidation. This repeated oxidation and GSH binding would not only produce significantly more superoxide in the presence of GSH, as we have shown *in vitro* (Srinivasan et al., 2001), but also decrease the intracellular levels of this important antioxidant (Srinivasan et al., 2002), which may leave the cell vulnerable to additional attacks from electrophiles or ROS.

Another possible reaction of PCB-quinones is to undergo repeated reduction-oxidation steps with the formation of superoxide, so called redox cycling. To test this hypothesis we incubated seven different PCB quinones with and without NADPH-CYP reductase (rat liver microsomes) in the presence of oxidized acetylated cytochrome c and an NADPH-regenerating system. Radical production was monitored by the reduction of acetylated cytochrome c at 550 nm and the apparent kinetic constants K_m, V_{max}, k_{cat}, and k_{cat}/K_m were determined according to the method of Azzi et al., 1975. The incubations were carried out both with and without superoxide dismutase (SOD). Major components and their interactions are pictured in Figure 4.7. We found a reduction of acetylated cytochrome c in the presence of a quinone (1) with NADPH alone, which was greatly enhanced, and (2) with microsomes (reductase) in the presence of NADPH. These data demonstrate that the PCB-like quinones will stimulate electron flow from NADPH either by 2-electron reduction to the dihydroxy-PCB or by an enzymatic 1-electron reduction to the semiquinone. Presumably, within an intact cellular environment, this may contribute to a pro-oxidant environment via depletion of cellular reducing equivalents (McLean et al., 1998; McLean et al., 2000).

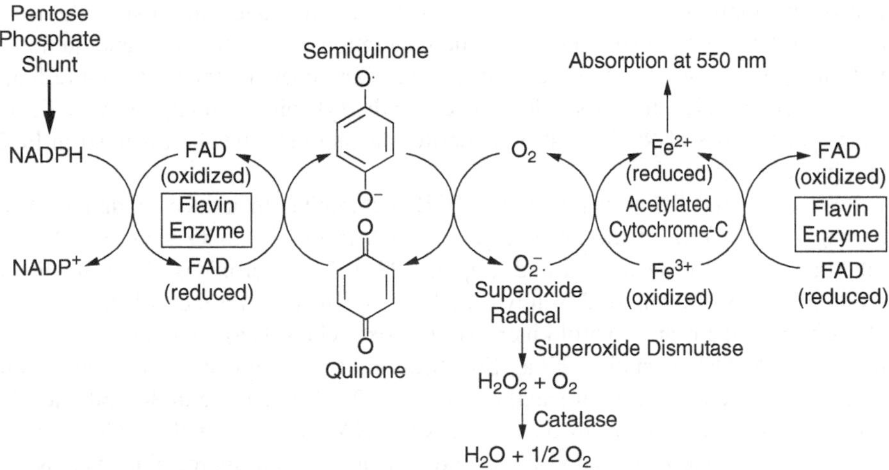

FIGURE 4.7 Redox cycling of polychlorinated biphenyl (PCB)-derived metabolites. (Modified from McLean, M., Twaroski, T., and Robertson, L.W. (1998), A new mechanism of toxicity for polychlorinated biphenyls (PCBs): Redox cycling and superoxide generation, *Organo-halogen Compounds*, 37: 59–62.)

Whatever the mechanism, these enzymatic and nonenzymatic oxidation and reduction reactions of the PCB metabolites produce ROS *in vitro*. Our studies with synthesized metabolites and cells in culture show that exposure to these PCB metabolites also results in an increase in ROS in cells and in cytotoxicity (Srinivasan et al., 2001).

GENERATION OF OXIDATIVE STRESS: IMPLICATION IN CARCINOGENESIS

The process of chemical-induced carcinogenesis is believed to occur over several stages, such as initiation, promotion, and progression. Initiation is believed to be caused by a genotoxic event, whereas promotion is characterized by an increase in the number of initiated cells through long-term epigenetic mechanisms. ROS seem to be involved in both, damage to the DNA and thus initiation of carcinogenesis and changes in the proliferation of a subset of cells, which are the hallmarks of promotion. PCBs are carcinogens in rodents and probable carcinogens in humans (Silberhorn et al., 1990). There is ample evidence that the carcinogenic activity of PCBs may be at least in part due to their ability to produce oxidative stress, as will be discussed in the following text.

PCB-Induced Oxidative DNA Damage *In Vitro*

Chronic exposure to PCBs produces chromosome aberrations, including chromosome breaks and sister chromatid exchanges (reviewed in Silberhorn et al., 1990; Ludewig, 2001). One class of agents that causes these types of genotoxic lesions is oxygen radicals. We have shown in the preceding text that lower halogenated PCBs can be metabolized to mono- and dihydroxylated compounds and that further oxidation of these compounds can lead to the generation of semiquinones and quinones with the production of superoxide. Superoxide can generate other, even more reactive, oxygen species such as singlet oxygen, hydroxyl anions, and hydroxyl radicals. These latter types of ROS are strong genotoxins and may thus be involved in PCB carcinogenesis.

We examined the ability of these PCB metabolites to induce oxidative DNA damage in the form of the promutagenic DNA lesion 8-oxodG using a ^{32}P-postlabeling assay (Devanaboyina and Gupta, 1996). Lactoperoxidase/H_2O_2-mediated oxidation of 3′,4′-dichloro-2,5-dihydroxybiphenyl in the presence of calf thymus DNA produced a nearly twofold increase in 8-oxodG levels as compared to vehicle-treated DNA (Oakley et al., 1996a). Substitution of the enzymatic oxidation system with $CuCl_2$ resulted in a significantly greater (>22-fold) increase in 8-oxodG levels; however, $FeCl_3$ was found to be ineffective. DNA incubated with $CuCl_2$ or 3′,4′-dichloro-2,5-dihydroxybiphenyl alone did not produce an appreciable increase in 8-oxodG levels as compared to vehicle treatment. 2,5-Dihydroxy derivatives of several other PCBs (2-, 3-, 4-, 2,5- and 3,4,5-) were also found to be as good or better substrates in the $CuCl_2$-mediated induction of 8-oxodG (Oakley et al., 1996a). When lactoperoxidase/H_2O_2 oxidation of 3′,4′-dichloro-2,5-dihydroxybiphenyl was performed in the presence of the free-radical inhibitors such as catalase, sodium azide, or SOD, only catalase and sodium azide resulted in significant (>65%) inhibition of 8-oxodG, indicating the involvement of H_2O_2 or H_2O_2-derived ROS species, but not superoxide, in the oxidative DNA damage induced by the oxidation of PCB dihydroxy metabolite.

We also tested the ability of these PCB metabolites to produce DNA strand breaks *in vitro* using supercoiled plasmid DNA. Unbroken plasmid DNA has a supercoiled structure. Induction of DNA strand breaks results in circular and linear DNA, which can be visualized by gel electrophoresis. All 6 dihydroxy-PCB metabolites tested induced DNA strand breaks *in vitro*, but only in the presence of Cu(II) (Srinivasan et al., 2001). PCB-quinones were only active when GSH was present as well. This is in agreement with the hypothesis that ROS are the genotoxic agents in PCB-induced genotoxicity and with previous reports that superoxide does not induce strand breaks but that other copper-derived ROS species do. The hydroxyl radical scavenger DMSO and thiourea, and the singlet oxygen scavengers sodium azide and Tris, protected the DNA from strand break induction by the PCB metabolites, indicating that these two ROS species are the ultimate genotoxin (Srinivasan et al., 2001). With respect to structure–activity relationships, all of the compounds tested produced strand breaks with similar strength; however, the ortho and meta chloro-substituted biphenyls seemed to be slightly more active than the para chloro-substituted. This is in agreement with our findings concerning 8-oxodG formation by various para-dihydroxybiphenyls, in which the order of activity was 2 > 2,5 > 3 > 4. The position of the dihydroxy groups also influenced the activity in the strand break assay, with 2,5-dihydroxy being more active than the 3,4- or 2,3- metabolites. This is also in agreement with our findings in the 8-oxodG assay.

ROS Production, GSH Depletion, and Cytotoxicity in Cells in Culture

These effects of PCB metabolites *in vitro* were also observed in various types of cells in culture. Cytotoxicity experiments with yeast and bacteria revealed that the number and position of chlorines as well as the position of the two hydroxyl groups also have an influence on cytotoxicity. The para-dihydroxy metabolites (2,5-) were less toxic than either of the ortho-dihydroxy metabolites (2,3-; 3,4-). Chlorines in para- and meta-position increased the toxicity compared to chlorines in the ortho position. When the compounds are ordered according to increasing cytotoxicity, an inverse relationship was seen compared to strand break induction and the formation of 8-oxodG by these compounds (see the preceding text). This suggests that cytotoxicity and ROS production, and ROS-mediated genotoxicity are not related and are probably due to opposite physicochemical properties of the compounds tested.

Exposure of human leukemia cells (HL-60) to either PCB quinones or PCB-hydroquinones caused a strong increase in intracellular oxidative stress in HL-60 cells (Srinivasan et al., 2001). In contrast to this, PCB-quinones strongly reduced cell viability (Srinivasan et al., 2002), whereas PCB-hydroquinones had no significant effect. PCB quinones, but not hydroquinones, also strongly reduced intracellular

GSH levels. Pretreatment of cells with *N*-ethylmaleimide, a GSH-depleting agent, significantly increased the cytotoxicity of PCB quinones but had no effect on hydroquinone-treated cells. This indicates that it is the GSH depletion, not ROS production, which is cytotoxic to mammalian cells in culture.

To summarize the results, we found that lower chlorinated dihydroxybiphenyls (1) oxidize with the formation of superoxide, (2) undergo redox reactions *in vitro* with more ROS formation and consumption of cofactors such as NADPH and GSH, (3) result in strand break induction in the presence of Cu(II) *in vitro*, which is mediated by the formation of hydroxyl radicals and singlet oxygen, (4) cause a decrease in intracellular GSH levels, (5) produce an increase in intracellular oxidative stress in cells in culture, and (6) can be cytotoxic. We have further found that (7) structure–activity relationships exist, with ortho-chloro metabolites being more genotoxic, but less cytotoxic, than their corresponding para chlorinated metabolites. These results support the involvement of ROS in the genotoxicity of PCBs but also indicate that a possible reduction in protective cofactors and reducing equivalents may leave exposed cells more vulnerable to attacks by cytotoxins and genotoxins. These results were obtained with synthesized PCB metabolites, but as described in the following text, findings based on *in vivo* experiment with PCBs support the hypothesis that oxidative stress may be involved in PCB carcinogenesis.

PCBs Produce a Decrease in the Levels of Vitamin E and GSH

If PCBs induce oxidative stress *in vivo*, then they may also cause a depletion of important intracellular radical scavengers such as GSH and antioxidants such as vitamin E in target organs. In one rodent experiment with repeated PCB injections over three weeks, only the combination of PCB 77 and PCB 153 lowered liver total GSH in male rats, and only after the first week (female rats were more sensitive in this experiment). In a time-course study with PCB 77 in which a single injection of the PCB (300 μmole/kg) was given in the livers of male rats and the rats were killed at various times thereafter, reduced glutathione, determined by HPLC, decreased from 12 hr through 3 weeks after the injection (unpublished data).

All PCB treatments reduced alpha-tocopherol levels compared to control (Twaroski et al., 2001a and 2001b). Furthermore, a significant increase in oxidized vitamin E, alpha-tocopheryl quinone, was observed with PCB 38 and PCB 153. Also in the time-course with PCB 77, significant increases in the hepatic levels of alpha-tocopheryl quinone were observed in male rats (at all time points from 4 d through 3 weeks, unpublished data), and in female rats (at 2 weeks and at 3 weeks). These data suggest (1) that the cells experience increased oxidative stress, and (2) that the ability of the liver to recycle or excrete oxidized vitamin E was somehow impaired or insufficient. PCB 77 increases DT-diaphorase levels (also called quinone oxido-reductase). DT-diaphorase reduces quinones in a 2-electron reduction to the corresponding dihydroxy-metabolite with consumption of NADPH; hence the increased alpha-tocopheryl quinone levels in the presence of increased activity of DT-diaphorase were a surprising finding.

We do not know the mechanisms of this increase in alpha-tocopheryl quinone or the biological consequences for the cell. Moreover, the effect of some PCBs on vitamin E may be stronger in other organs, and other organs may be more sensitive to interference in vitamin E status. These effects certainly warrant a closer look, especially because dietary status may not reflect the condition in these organs as exposed populations may have silent vitamin E (and selenium) deficiencies due to the effects of PCBs and other environmental contaminants.

PCB Exposure Reduces Selenium Levels by Yet Unknown Mechanisms

Our studies have shown that PCB exposure significantly reduces selenium in the liver of rats (Twaroski et al., 2001b). Animals were injected with 100 μmol/kg/injection or vehicle alone (controls) twice per week and sacrificed after receiving 2, 4, or 6 injections within 1, 2, or 3 weeks, respectively. The selenium concentration in the liver was reduced to 80% or less after the first week of treatment. We are currently investigating the mechanisms by which this reduction occurs, and whether this occurs in other organs also. Possible mechanisms are increased storage, binding to selenoproteins in extrahepatic organs, or increased metabolism and excretion. The selenium-depleting effect of PCBs could possibly be more pronounced or reversed in other organs. Selenium, however, is considered to be a very important antioxidant and the most promising cancer chemopreventive agent in clinical trial today. If selenium levels are also reduced in other organs, it could result in unpredictable, dangerously low tissue selenium levels despite normal dietary uptake of selenium. Low selenium levels were found to be a prediagnostic indicator for increased risk of prostate cancer (Yoshizawa et al., 1998). Supplementation of the normal diet with selenium was found to reduce prostate cancer risk by 63% (Clark et al., 1998), indicating that even a normal diet provides only suboptimal levels of selenium. Taken together, these data indicate that some PCBs alter selenium and vitamin E homeostasis in the livers of rats. Studies are under way to further investigate these phenomena.

PCBs Repress the Selenium-dependent GSH Peroxidase in the Liver

Significant decreases in antioxidant defenses, i.e., antioxidant enzyme levels, have been reported after exposure to PCBs (Hori et al., 1997; Schramm et al., 1985). Our data indicate that some PCB congeners may have a significant effect on glutathione-related enzymes, i.e., glutathione transferase (GST), glutathione reductase (GR), and glutathione peroxidase (GPx) (Twaroski et al., 2001a and 2001b). These enzymes play an important role in the protection of cells against electrophiles, including ROS. Of special interest is the reduction in Se-GPx activity, which could leave the cells more prone to oxidative damage.

In an experiment in which animals received PCB injections twice per week and were sacrificed at different time points, PCB77 caused an increase in CYP1A activity and GST (Twaroski et al., 2001a and 2001b). PCB77 and PCB153 also increased GR at the earliest time point tested. Whether increased GST or GR have an effect or are only indicators of exposure to intracellular oxidative stress is not known. More important and surprising, however, was the effect on GPx. PCB77 resulted in a significantly reduced activity, especially evident in the Se-dependent form of GPx.

This decrease was mirrored in reduced mRNA levels for the enzyme and in reduced total selenium.

PCBS AND DIET — IMPLICATIONS IN CARDIOVASCULAR DISEASE

From epidemiological studies, there is substantial evidence that cardiovascular diseases are linked to environmental pollution. Epidemiological studies found that exposure to polycyclic aromatic hydrocarbons can lead to human cardiovascular toxicity. For example, there was a significant increase in mortality from cardiovascular diseases among Swedish capacitor manufacturing workers exposed to PCBs for at least 5 yr (Gustavsson and Hogstedt, 1997), and most of the deaths were due to cardiovascular disease in power workers exposed to phenoxy herbicides and PCBs in waste transformer oil (Hay and Tarrel, 1997). Furthermore, studies on the Seveso population (as an outcome of the industrial accident that occurred in the town of Seveso, Italy, in 1976) detected an increase in cardiovascular disease (Bertazzi et al., 1998). Finally, a report by Tokunaga et al., (1999) confirmed many other studies on chronic Yusho patients (accidental ingestion of rice-bran oil contaminated with PCBs), which showed that in this population elevated serum levels of triglycerides and total cholesterol were significantly associated with the blood PCB levels.

Dysfunction of endothelial cells is a critical underlying cause of the initiation of many vascular diseases, including cardiovascular diseases such as atherosclerosis. In addition, endothelial activation and dysfunction is an important factor in the overall regulation of vascular lesion pathology. The lining of blood vessels is protected by the endothelium, and endothelial cells play an active role in physiological processes such as regulation of vessel tone, blood coagulation, and vascular permeability. In addition, vascular endothelial cells protect underlying tissues against various harmful blood-borne agents such as plaque-forming materials, lipids, or cellular debris. In addition to endothelial barrier dysfunction, another functional change in atherosclerosis is the activation of the endothelium that is manifested as an increase in the expression of specific cytokines and adhesion molecules. These cytokines and adhesion molecules are proposed to mediate the inflammatory aspects of the disease by regulating the vascular entry of leukocytes, namely macrophages and lymphocytes. Thus, severe endothelial cell activation and insult or injury can lead to necrotic cytotoxicity, as well as to apoptotic or programmed cell death and ultimately to disruption of endothelial integrity. Maintenance of endothelial integrity is critical not only for protection against adverse metabolic activities that may be damaging to blood vessels but also for performance of the normal barrier function of endothelial cells, which limit the entry of plasma components such as cholesterol-rich lipoproteins into the vessel wall.

Even though PCBs are linked to numerous age-related diseases, little is known about the underlying molecular events. Complex mixtures of PCB isomers and congeners, such as Aroclor 1254, appear to modulate expression of genes involved in cellular differentiation and stress (Borlak and Thum, 2002). Oxidative stress is a critical underlying event in mechanisms of endothelial cell dysfunction mediated by certain environmental contaminants, especially AhR ligands such as coplanar PCBs and 2,3,7,8-tetrachlorodibenzo-*p*-dioxin (TCDD).

Endothelial Toxicity of Coplanar PCBs

The mechanisms by which environmental chemicals alter endothelial cell metabolism are not fully understood, and little is known about how PCB-mediated cell dysfunction can be prevented or blocked. Several studies indicate that a critical underlying mechanism of PCB-mediated endothelial cell activation and dysfunction is an increase in cellular oxidative stress (reviewed in Hennig et al., 2002a). Evidence suggests that the oxidative stress induced by contaminants such as PCB 77, TCDD, and acyclic substances such as arachidonic metabolites is due to the interaction of these compounds with the AhR and activation of the CYP1A subfamily (Alsharif et al., 1994; Safe and Krishnan, 1995; Schaldach et al., 1999; Hennig et al., 1999). Schlezinger et al. (1999) demonstrated that PCB 77 can uncouple the catalytic cycle of CYP1A1, allowing heme iron within the active site of this enzyme complex to undergo cycles of oxidation and reduction and act as a Fenton catalyst, generating hydroxyl radicals from hydrogen peroxide. Other coplanar PCBs such as PCB 126 and PCB 169 markedly increased cellular oxidative stress in vascular endothelial cells, and brain and hepatic tissues (Hennig et al., 2002b; Hassoun et al., 2002), with superoxide anions being a major reactive oxygen species. Thus, induction of CYP1A1 or CYP1A2 may lead to generation of excess levels of reactive oxygen species resulting in cell injury.

Coplanar PCBs and Proinflammatory Events

Transcriptional regulation of metabolic events leading to endothelial cell dysfunction, and an inflammatory response induced by AhR agonists such as coplanar PCBs, are not well understood but appears to involve an imbalance in the cellular oxidative stress or antioxidant status.

In vascular endothelial cells, we found that oxidative stress was induced only by the coplanar PCB 77 and not by the diortho-substituted PCB 153 (Toborek et al., 1995; Hennig et al., 1999). Presumably, this is due to the interaction of PCB 77 with the AhR and activation of the CYP1A subfamily. In a recent study we demonstrated that, in addition to PCB 77, other coplanar PCBs, i.e., PCB 126 and PCB 169, also can induce oxidative stress in vascular endothelial cells (Hennig et al., 2002b). Our data also suggest that all three coplanar PCBs tested can induce cellular oxidative stress in a concentration-dependent manner, and that the concentration needed for the apparent maximal induction of oxidative stress differs among these coplanar PCBs tested. Interestingly, PCB 126 exhibited maximal oxidative stress already at 0.5 μM, whereas the other two PCBs required 2.5 μM for maximal induction of cellular oxidative stress. These results may be explained by the fact that PCB 126 has comparatively the highest binding affinity for the AhR. Much of the oxidative stress appears to be generated as a result of uncoupling of CYP1A1, and this excessive oxidative stress in fact downregulates CYP1A1 (Schlezinger et al., 1999). This suggests that chronic exposure to low concentrations of coplanar PCBs may be even more detrimental in promoting inflammatory diseases than previously thought.

Evidence indicates that the oxidative stress–induced transcription factors nuclear factor-kappaB (NF-κB) and activator protein-1 (AP-1), which regulate inflammatory cytokine and adhesion molecule production, play critical roles in the induction of inflammatory responses. Transcriptional regulation of metabolic events leading to endothelial cell dysfunction, and an inflammatory response induced by AhR agonists such as PCBs, are not well understood. Recent evidence suggests that the AhR agonist TCDD can activate NF-κB and AP-1. Puga et al., (2000) proposed that CYP1A1-dependent and AhR complex–dependent oxidative signals are partially responsible for the observed activation of these transcription factors. Furthermore, a dose-dependent activation of NF-κB was observed in fish exposed to coplanar PCBs (Schlezinger et al., 2000). Pretreating a human leukemic mast cell line with the NF-κB pathway inhibitor pyrrolidine dithiocarbomate suppressed PCB-induced NF-κB activation and reduced cyclooxygenase-2 and interleukin-6 (IL-6) mRNA expression (Kwon et al., 2002), suggesting that the PCB-mediated induction of cyclooxygense-2 and proinflammatory cytokine expression is dependent on a functional NF-κB signaling pathway. Our data strongly support the hypothesis that PCBs, especially coplanar PCBs, which function as AhR agonists, may be proinflammatory and atherogenic by activating NF-κB in vascular endothelial cells (Hennig et al., 1999). Interestingly, PCBs 126 and 169 were as efficacious as PCB 77 in disrupting endothelial barrier function, in causing oxidative stress and activation of NF-κB, in production of IL-6, and in expression of the vascular cell adhesion molecule-1 (VCAM-1) gene in cultured cells (Hennig et al., 2002b), suggesting that proinflammatory properties of these coplanar PCBs are not related to their chlorination level.

Using an AhR-deficient mouse model, our recent work provides *in vivo* evidence that a functional AhR is critical for the proinflammatory events mediated by coplanar PCBs and possible other AhR agonists in the vascular endothelium (Hennig et al., 2002b). Other *in vivo* studies with AhR knockout mice indicate that AhR activation is involved in the inflammatory response to TCDD hepatotoxicity (Thurmond et al., 1999). Furthermore, a recent study demonstrated that inductions of xanthine oxidase or xanthine dehydrogenase activity by TCDD and enhanced lipid peroxidation in livers of mice treated with TCDD were not found in AhR-deficient mice (Sugihara et al., 2001).

Critical molecular events involved in PCB-mediated endothelial cytotoxicity are still unclear, and especially the role of dietary modulators and nutritional factors in the control of these molecular pathways is largely unknown. An increase in liver inflammation and decrease of vitamin A stores were found after exposing rats for 7-d to PCB 77 (Azais-Braesco et al., 1997). Similar results were observed in young female mink after a 21-week exposure to Aroclor 1242, with a significant decrease in plasma vitamin A levels (Kakela et al., 2003). We demonstrated that vitamin E and alpha-naphthoflavone (an AhR antagonist and inhibitor of CYP1A1) can markedly reduce PCB 77–mediated oxidative stress, activation of NF-κB and production of inflammatory cytokines (e.g., IL-6), as well as PCB-mediated endothelial barrier dysfunction (Slim et al., 1999). Similar mechanistic phenomena of coplanar PCBs were recently reported in PCB 126–induced toxicity using the chick embryo as a model (Jin et al., 2001). Our studies demonstrate that vitamin E may have potent antiinflammatory properties against PCB insults within vascular tissues.

Dietary Fatty Acids, PCBs, and Endothelial Cell Dysfunction

There is considerable evidence that exposure to PCBs can lead to lipid changes in plasma and tissues and that this may be linked to lipophylic properties of PCBs and their interaction with lipids and, especially, with fatty acids. For example, exposure to Aroclor 1242 modified adipose tissue fatty acids with a decrease of highly unsaturated fatty acids and an increase in monounsaturated fatty acids in membrane phospholipids (Kakela and Hyvarinen, 1999). Interestingly, the decrease in highly unsaturated fatty acids such as docosahexaenoic acid was less in minks, which consumed a marine fish diet rich in fat-soluble vitamins (vitamins E and A). In another study it was found that, compared to a diet high in soybean oil and added PCBs, substitution with highly hydrogenated soybean oil caused decreases in the amounts of monounsaturated and omega-3 polyunsaturated fatty acids in rat livers (Kamei et al., 1996). PCBs also may affect lipid metabolism by interfering with glucogeneic and lipogenic enzymes or pathways (Boll et al., 1998). On the other hand, PCB intestinal bioavailability may be linked to micelle composition (Doi et al., 2000), an event highly dependent on the types of dietary fat being consumed. Micelles of linoleic acid solubilized more PCB 77 than mixed micelles formulated from equal amounts of myristic, palmitic, stearic, and linoleic acids. Furthermore, systemic bioavailability of PCB 77 from an *in situ* perfused intestinal preparation was 2.2-fold greater when delivered to the intestine in linoleic acid micelles as compared to the mixed micelle preparation (Doi et al., 2000). These findings suggest that elevated levels of linoleic acid may facilitate the cellular availability of PCBs. In stranded dolphins, there was a negative correlation between tissue PCBs and arachidonic acid levels (Guitart et al., 1996), suggesting that PCBs can lower tissue arachidonic acid levels significantly enough to affect eicosanoid metabolism. A depressed level of arachidonic acid may be explained in part by reports that coplanar PCBs can reduce the synthesis of physiologically essential long-chain fatty acids, such as arachidonic acid, in rat liver by suppressing delta 5 and delta 6 desaturase activities (Matsusue et al., 1999). Most importantly, a decrease in arachidonic acid in both maternal and cord serum was observed relative to increasing PCB levels, suggesting that inhibition of desaturase activity by PCBs may affect the maintenance of arachidonic acid status during fetal development (Grandjean and Weihe, 2003). Our own preliminary data from plasma and livers of low-density lipoprotein (LDL) receptor-deficient mice support the hypothesis that linoleic acid–rich dietary oils, such as corn oil, facilitate clearance of linoleic acid from plasma into vascular tissues, an event that could exacerbate fatty acid- or PCB-induced oxidative stress and a vascular inflammatory response.

Little is known about the interaction of dietary fats and PCBs in the pathology of atherosclerosis. We hypothesized that selected dietary lipids may increase the atherogenicity of environmental chemicals such as PCBs by cross-amplifying mechanisms leading to dysfunction of the vascular endothelium. To investigate this hypothesis, we treated cultured endothelial cells with linoleic acid, followed by either PCB 77 or PCB 153 (Hennig et al., 1999). PCB 153 had little or no effect on endothelial barrier function. In contrast, PCB 77 disrupted endothelial barrier function, allowing an increase in albumin transfer across endothelial monolayers. We also found that preenrichment of cells with linoleic acid increased the PCB-mediated induction of CYP1A. Our results suggest that certain unsaturated fatty acids can potentiate endothelial cell dysfunction caused by specific PCBs, and that oxidative stress and the CYP1A subfamily may be, in part, responsible for these metabolic events. The synergistic toxicity of linoleic acid and PCBs to endothelial cells may be mediated by cytotoxic epoxide metabolites. In support of this hypothesis we demonstrated that inhibition of cytosolic epoxide hydrolase blocked both linoleic acid–induced cytotoxicity and the additive toxicity of linoleic acid plus PCB 77 to endothelial cells (Slim et al., 2001). Interestingly, cellular uptake and accumulation of linoleic acid in vascular endothelial cells were markedly enhanced in the presence of PCB 77, a metabolic event that could contribute to the added toxicity of both PCBs and fatty acids.

These findings may contribute to a better understanding of the interactive mechanisms of dietary fats and environmental contaminants as mediators of vascular endothelial cell dysfunction (Figure 4.8). Because of the current high intake of linoleic acid in the form of vegetable oils in the average American diet, some caution is warranted if its oxidative metabolites predispose individuals to inflammatory diseases. Thus, this issue of dietary fat may be especially relevant for populations near sites of excessive exposure to environmental pollutants such as PCBs.

PREMETASTATIC AND ATHEROGENIC EFFECTS OF ORTHO-SUBSTITUTED, NONCOPLANAR PCB CONGENERS

It is widely recognized that different PCB structures interact specifically with different cellular targets. Highly ortho-chlorine-substituted, noncoplanar PCB congeners are not typical AhR or constitutive androstane receptor (CAR) agonists. The biological effects of these PCBs include neurotoxicity, estrogenicity, and insulin release, as well as altered regulation of intracellular calcium and signal transduction mechanisms (Fischer et al., 1998; Kodavanti and Tilson, 2000). In addition, it was demonstrated that they can induce apoptotic cell death through a caspase-mediated mechanism (Lee et al., 2003b). Our research focused on the vascular and prometastatic effects of PCB 104, a representative of ortho-chlorine-substituted, noncoplanar PCB congeners.

INVOLVEMENT OF THE VASCULAR ENDOTHELIUM IN BLOOD-BORNE CANCER METASTASIS

The role of endothelial cell dysfunction and activation in the development of atherosclerosis is well known and was described in the preceding text. Therefore, in this

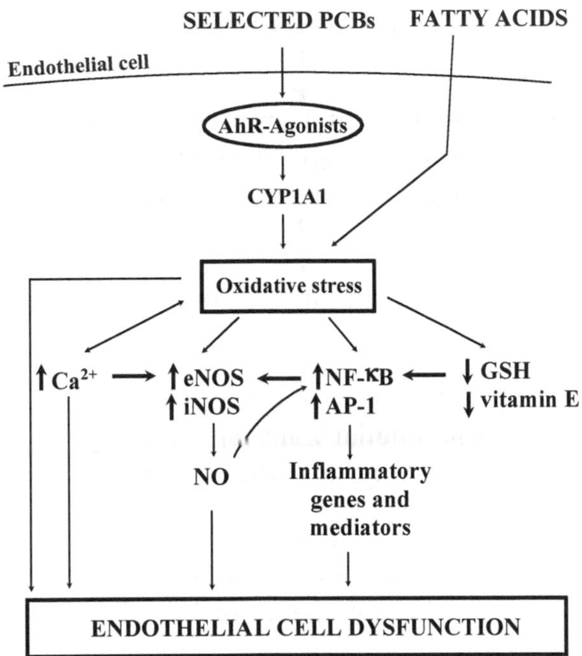

FIGURE 4.8 Schematic diagram of potential mechanisms of interaction of PCBs and fatty acids in endothelial cell dysfunction. CYPs, cytochrome P450s; eNOS, endothelial nitric oxide synthase; iNOS, inducible nitric oxide synthase; NF-κB, nuclear factor-kappa B; AP-1, activator protein-1; GSH, glutathione. (Modified from Hennig, B., Hammock, B.D., Slim, R., Toborek, M., Saraswathi, V., and Robertson, L.W. (2002a), PCB-induced oxidative stress in endothelial cells: modulation by nutrients, *International Journal of Hygiene and Environmental Health*, 205: 95–102.)

section the involvement of the vascular endothelium in the formation of cancer metastasis is briefly discussed (Figure 4.9).

Hematogenous metastasis is a complex process that requires multiple steps. It has been proposed that the extravasation of tumor cells from the circulation to extravascular tissues is one of the most critical events in this process. The extravasation process allows the circulating tumor cells to pass through endothelial barriers to reach selective metastatic sites. Evidence indicates that this step in the metastatic cascade can be regulated by the microvasculature environment and the integrity of the vascular endothelium. For example, endothelial damage induced by oxidative stress promotes the localization and metastasis of circulating cancer cells to the lung in an animal model of pulmonary microvascular injury, and this effect is attenuated after endothelial repair. Free radical–mediated endothelial cell damage also can facilitate the metastasis of pancreatic tumor cells. In addition, several soluble factors derived from tumor cells during their interactions with endothelial cells can induce endothelial cell retraction, which in turn may play a crucial role in the establishment of cancer metastasis.

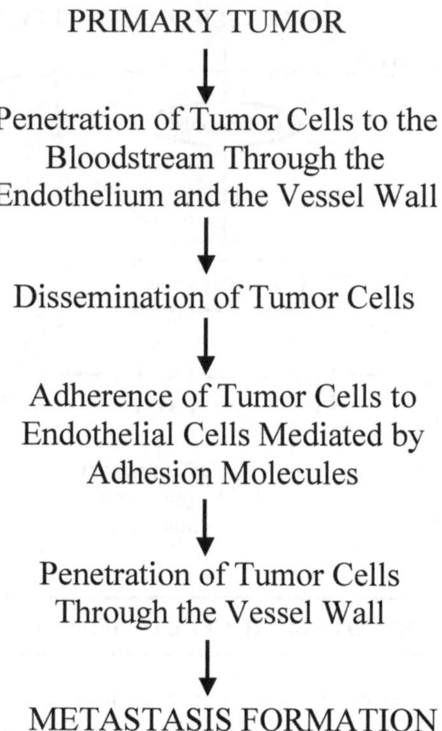

FIGURE 4.9 Schematic diagram of endothelial cell involvement in metastasis formation.

Role of ortho-Chlorine-Substituted, Noncoplanar PCB Congeners in Cancer Metastasis and Atherogenesis

A growing body of evidence indicates that the direct adhesive interaction between tumor cells and endothelial cells is the critical event in metastasis formation. This process requires the binding of tumor cells to specific adhesion molecules on the surface of endothelial cells, followed by migration of tumor cells through the endothelium into underlying tissues. Our research (Choi et al., 2003) indicated that exposure of vascular endothelial cells to PCB 104 can induce overexpression of monocyte chemoattractant protein-1 (MCP-1, a chemokine), and E-selectin and intercellular adhesion molecule-1 (ICAM-1) (both adhesion molecules). Upregulation of chemokines and adhesion molecules plays a critical role in both atherosclerosis and cancer metastasis (Figure 4.10).

MCP-1 is a member of the CC chemokine family and stimulates chemotaxis and transmigration of monocytes, lymphocytes, and granulocytes. Increased production of MCP-1 is associated with a variety of processes, including cancer metastasis (Amann et al., 1998; Hefler et al., 1999) and early stages of atherosclerosis (Boring et al., 1998). At least two distinct mechanisms may be involved in prometastatic effects of MCP-1. First, MCP-1 can exert direct chemotactic effects on tumor cells, as was shown using MCF-7 cells, a cell line obtained from human breast carcinoma.

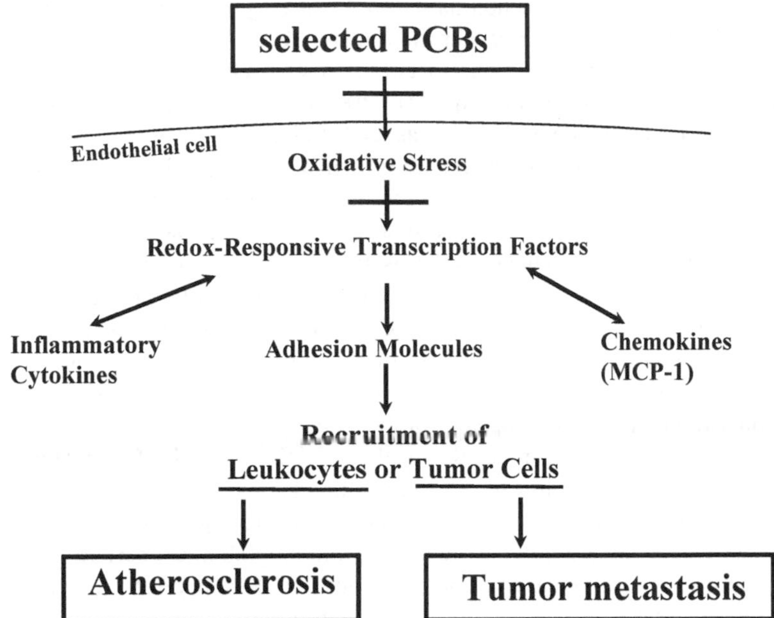

FIGURE 4.10 Schematic diagram of vascular responses associated with the development of atherosclerosis and cancer metastasis induced by specific PCBs.

This chemotactic influence of MCP-1 on tumor cells is mediated by a receptor-stimulated signaling pathway. Thus, it appears that MCP-1 can directly attract tumor cells and induce tumor cell migration across the vascular endothelium with the subsequent generation of tumor metastasis. A second mechanism by which MCP-1 may stimulate the development of cancer metastasis may be related to its chemotactic effects towards leukocytes. Activated leukocytes can migrate across the endothelium and degrade extracellular matrix proteins, which separate the endothelium from the underlying layers of the vascular wall. Such a process can markedly facilitate invasion of tumor cells, a process associated with the development of metastasis. The role of MCP-1 in tumor metastasis has been supported by the observations that the levels of this chemokine were elevated in the serum of ovarian cancer patients (Hefler et al., 1999) and in the urine of patients with bladder cancer (Amann et al., 1998). In addition, the urinary MCP-1 levels were strongly correlated with tumor stage, grade, and distant metastasis (Amann et al., 1998).

Evidence indicates that MCP-1 also can play an important role in atherogenesis. To support its role in the initiation and development of atherosclerosis, it was shown that MCP-1 deficiency significantly reduced atherosclerosis in low-density lipoprotein (LDL) receptor–deficient mice fed a high cholesterol diet (Gu et al., 1998). In a similar study, the selective absence of CCR2, the receptor for MCP-1, markedly decreased atherosclerotic lesion formation in apolipoprotein (apo) E–deficient mice (Boring et al., 1998).

PCB 104–induced overexpression of adhesion molecules may also play a role in vascular pathologies associated with cancer metastasis and atherosclerosis. For example, convincing experimental data have been generated on the involvement of E-selectin in breast and colon cancer metastasis. Several glycoprotein ligands have also been identified on the surface of colon cancer cells, which serve as specific receptors for E-selectin (Tomlinson et al., 2000). In addition, circulating levels of this adhesion molecule were identified as useful clinical markers of tumor progression and metastasis (Alexiou et al., 2001). In addition, inhibition of E-selectin-mediated cancer cell adhesion may be an efficient strategy to inhibit cancer metastasis.

In addition to its role in cancer metastasis, overexpression of E-selectin is associated with the development of atherosclerosis. During atherogenesis, migration of leukocytes through the vascular endothelium initially involves relatively transient adherence of leukocytes to endothelial cells, which results in leukocyte "rolling" over the endothelium. This process is followed by firm leukocyte adhesion and transmigration across the vascular endothelium. Leukocyte rolling is mediated by overexpression of adhesion molecules of the selectin family, such as E-selectin. The importance of the adhesion molecules of the selectin family in the development of atherosclerosis has been confirmed in studies that demonstrated the presence of both E- and P-selectin on the surface of endothelial cells overlying atherosclerotic plaques.

ICAM-1 is an adhesion molecule of the immunoglobulin superfamily critically involved both in cancer metastasis and atherogenesis. To support the role of ICAM-1 in cancer metastasis, it was demonstrated that serum levels of soluble ICAM-1 (sICAM-1) were elevated in patients with non-small-cell lung cancer and correlated with tissue expression of ICAM-1 and tumor stage (Grothey et al., 1998). In addition, metastatic lung cancer was associated with higher sICAM-1 as compared to localized tumors (Grothey et al., 1998), and the highest levels of sICAM-1 were observed in patients with liver metastasis (Sprenger et al., 1997). ICAM-1 expression also correlated with progression of malignant melanoma or renal cell carcinoma (Tanabe et al., 1997). Finally, antisense ICAM-1 oligonucleotides decreased metastasis of malignant melanoma by approximately 50%.

In the development of atherosclerosis, ICAM-1 stimulates firm adhesion of leukocytes to the vascular endothelium. ICAM-1 is expressed at low levels on the surface of nonstimulated endothelial cells. In addition, stimuli such as tumor necrosis factor (TNF)-alpha, IL-1, interferon-gamma, or shear stress can markedly induce expression of this adhesion molecule. The stimulatory involvement of hemodynamic stress in upregulation of ICAM-1 may play an important role in the development of atherosclerosis in hypertension. ICAM-1 is markedly expressed in early stages of atherosclerosis and it stimulates adhesion of monocytes and T lymphocytes. The significance of this adhesion molecule in atherosclerosis was confirmed in clinical studies, which determined elevated levels of sICAM-1 in asymptomatic patients who are prone to developing cardiovascular disease (Ridker et al., 1998).

Apart from upregulation of chemokines and adhesion molecules, several other vascular mechanisms can be responsible for facilitation of extravasation and dissemination of tumor cells. For example, transmigration of cancer cells can be augmented by disruption of cell junctions (Lee et al., 2003a). In addition, several soluble factors such as vascular endothelial growth factor (VEGF) can induce the disruption of endothelial integrity, which in turn may directly enhance penetration of tumor cells and facilitate establishment of cancer metastasis (Lee et al., 2003a). Microvascular hyperpermeability near a metastatic tumor also can accelerate plasma protein extravasation to stimulate tumor growth. Our recent results indicate that exposure to PCB 104 can induce endothelial hyperpermeability and adhesion and transmigration of breast tumor cells across the vascular endothelium (Eum et al., 2004). Most importantly, PCB 104–mediated effects on tumor cell adhesion could be protected by coexposure to antioxidants, pyrrolidine dithiocarbamate (PDTC), and epigallocatechin-3-gallate (EGCG).

DIET AND PROTECTION AGAINST PCB-INDUCED CYTOTOXICITY

ANTIOXIDANTS

Oxidative stress–induced cell injury also includes a general imbalance in the cellular oxidative stress or antioxidant status. PCB uptake and clearance studies suggest that an increase in oxidative stress as a result of redox cycling can markedly alter the activity of critical antioxidant enzymes such as catalase or superoxide dismutase (Rodriguez-Ariza et al., 2003). Both glutathione reductase and glutathione transferase activities were increased significantly in both male and female rats receiving PCB 77 or a combination of PCB 77 and PCB 153 (Twaroski et al., 2001a and 2001b). Furthermore, PCB 77 exposure lowered hepatic total selenium levels and suppressed the critical antioxidant enzyme selenium-dependent glutathione peroxidase (Twaroski et al., 2001a and 2001b). In another study, increases in lipid peroxidation by PCB 126 were associated with significant decreases in glutathione peroxidase activity (Jin et al., 2001). In addition to selenium, cellular levels of other antioxidant nutrients, such as vitamin E, are depressed after exposure to PCBs (Saito, 1990; Hennig et al., 1999). Interestingly, there was a significant increase in alpha-tocopheryl quinone (oxidized vitamin E) within the lipid fraction of rat livers upon treatment with PCBs (Twaroski et al., 2001a). This suggests that alpha-tocopheryl quinone may be a sensitive marker for PCB exposure.

As mentioned earlier, our laboratory has shown that in vascular endothelial cells PCB 77, but not PCB 153, contributed markedly to cellular oxidative stress, manifested by both increased activity and content of CYP1A and decreased levels of vitamin E in the culture medium (Hennig et al., 1999). We also demonstrated that vitamin E and alpha-naphthoflavone can markedly reduce PCB 77–mediated oxidative stress, as well as PCB-mediated endothelial barrier dysfunction (Slim et al., 1999). Furthermore, we studied the cellular glutathione redox status as a modulator of the endothelial defense against PCB toxicity (Slim et al., 2000), and demonstrated that PCB 77 can induce a cellular stress response that is reflected by the activation of stress-activated protein kinases. These results suggest that AhR ligands, such as PCB 77, cause vascular endothelial cell activation and dysfunction by modulating intracellular glutathione, which subsequently leads to induction of stress-specific kinases. Our data, demonstrating that vitamin E protects against PCB-induced vascular endothelial cell dysfunction (Slim et al., 1999), have critical health implications and suggest unique protective properties of this nutrient against toxic effects of certain environmental contaminants. Vitamin E is the only significant lipid-soluble, chain-breaking type of antioxidant present in human blood and all cellular membranes. The primary functions of vitamin E are to terminate lipid peroxidation chain reactions generated by free radicals, particularly in membranes that are rich in polyunsaturated lipids, to regulate cell proliferation and to stabilize cell membranes. These protective actions of vitamin E could have major implications in preventing vessel wall injury and atherosclerotic lesion formation.

Dietary Flavonoids

Epidemiological evidence of a protective role of fruits and vegetables in disease prevention is substantial. Both vitamins and minerals, such as vitamins E and C, zinc, and selenium, and plant-based compounds including phenols, flavonoids, isoflavones, and terpenes, are reported to have antioxidant, antiatherogenic, and anticarcinogenic properties. Little is known about the effect of antioxidant and antiinflammatory plant-derived nutrients on environmental contaminant-induced cytotoxicity. Recent studies, however, suggest that potential protective mechanisms of nutrients against cytotoxicity induced by environmental chemicals, and especially polycyclic aromatic hydrocarbons, involve inhibition of AhR activation and CYP1A1 activity (Henry et al., 1999; Quadri et al., 2000; Shertzer et al., 1999; Casper et al., 1999). Each of these studies supports cytoprotective properties of plant-derived antioxidants.

Dietary flavonoids, such as catechins and quercetin, rich in vegetables, fruits, berries, and beverages (wine and fruit juices) possess antioxidant and antiinflammatory properties. We provide strong evidence that flavonoids can modify PCB-mediated endothelial cytotoxicity (Ramadass et al., 2003). Endothelial cells were treated with epigallocatechin-3-gallate (EGCG) or quercetin with or without PCB 77. EGCG and quercetin strongly and in a concentration-dependent manner inhibited oxidative stress induced by PCB 77 as measured by dichlorofluorescein (DCF) fluorescence. The role of CYP1A1 in the PCB-induced toxicity also was investigated. Both EGCG and quercetin markedly inhibited CYP1A1 mRNA levels and enzyme activity. Furthermore, EGCG or quercetin downregulated the PCB 77–mediated increase in AhR DNA binding activity. These data suggest that protective effects of EGCG and quercetin are initiated upstream from CYP1A1 and that these flavonoids may be of value for inhibiting the toxic effects of PCBs on vascular endothelial cells (Ramadass et al., 2003).

CYP1A1 activity can be inhibited by catechins. Simultaneous treatment of cells with green tea extracts and TCDD inhibited the induced transcription of the CYP1A1-luciferase gene in a concentration-dependent manner, and this correlated with a decrease in CYP1A1 mRNA and protein levels (Williams et al., 2000). It was further demonstrated in HepG2 cells that green tea extracts inhibited TCDD-induced binding of the AhR to DNA and subsequent CYP1A transcription. The authors concluded that the inhibition of TCDD-induced CYP1A1 expression was most likely due to the ability of components of green tea extracts to interact directly with the receptor and function as antagonists of the AhR (Williams et al., 2000).

To demonstrate that the protective effects of EGCG and quercetin are initiated upstream from CYP1A1, i.e., at the functional level of AhR, ethoxyresorufin-*O*-deethylase (EROD) activity and CYP1A1 expression were investigated with the use of alpha-naphthoflavone, a CYP1A1 inhibitor and AhR antagonist (Ramadass et al., 2003). Treatment of cells with alpha-naphthoflavone alone or with alpha-naphthoflavone plus PCB 77 inhibited the EROD activity and CYP1A1 expression. Similar effects were observed when cells were treated with EGCG plus PCB 77 and quercetin plus PCB 77. Beta-naphthoflavone upregulated the CYP1A1 expression and activity, proving its agonist activity towards AhR, whereas in combination with EGCG, it showed a downregulation of CYP1A1 induction. These data strongly suggest that flavonoids act as antioxidants through inhibition of AhR function. Our data provide strong evidence that nutrition, e.g., flavonoids found in foods, can protect against environmental contaminants such as PCB 77 at both functional levels of AhR and CYP1A1. We hypothesize that antioxidant nutrients and related bioactive compounds common in fruits and vegetables protect against environmental toxic insult to the vascular endothelium by downregulation of signaling pathways involved in inflammatory responses and atherosclerosis.

SUMMARY

Polychlorinated biphenyls (PCBs) are polyhalogenated aromatic hydrocarbons that are persistent and widely dispersed in the environment. The molecular structure of individual PCB congeners is governed by the halogens present on the biphenyl rings. Congeners without ortho-chlorine substitutions, known as coplanar PCBs, favor a more coplanar structure and are good AhR agonists. Increasing ortho-chlorine substitution diminishes this binding avidity, but these ortho-substituted PCBs may bind avidly to other cellular receptors. The toxicity of PCBs may be mediated by signal transduction following receptor binding and the myriad effects that follow. Thus the chemical structure of the individual PCB congener may influence its toxicity. PCBs may also be substrates for metabolic attack or activation. Two electrophiles have been identified, one of which, namely PCB-derived quinones, may participate in reactions producing reactive oxygen species and diminishing cellular antioxidants.

The diet is a major route of exposure to PCBs. Because PCBs are fat soluble, fatty foods, especially fish and meat, usually contain higher levels of PCBs than vegetable matter. Once absorbed, PCBs distribute themselves to tissues, especially adipose, where they are in dynamic equilibrium with the blood. PCBs may be transferred to the fetus through the placenta and to a nursing baby via breast milk.

PCBs have broad adverse effects, including cancer induction, immune system suppression, nervous system effects, disruption of normal hormone function, and developmental and reproductive abnormalities, as well as increased cardiovascular and inflammatory diseases. Evidence is increasing that PCBs exhibit toxicity and disease potential via oxidative stress–related mechanisms. This is of particular interest because many common diseases, including cancer, and cardiovascular and neurological diseases, are believed to be initiated through imbalances of the body's oxidative stress or antioxidant status. The contribution of environmental toxins to these processes and the possibility that their initiation and progression could be manipulated by appropriate dietary interventions are of great public health interest.

In studies with cells in culture and with animals, PCBs have been shown to alter oxidative stress via multiple mechanisms (1) release of ROS through the persistent induction of CYP; (2) ROS production by PCB metabolites; (3) altered exogenous and endogenous metabolism through enzyme induction or repression; (4) changes in the cellular redox potential by depletion of GSH and NADPH; (5) a reduction of antioxidant status through a reduction in antioxidant enzymes; and (6) an imbalance of vitamin and mineral metabolism. The biologic and toxic consequences of these changes are diminished antioxidant capacity, resulting in lipid peroxidation, alkylated proteins and DNA and DNA strand breaks, and changes in cell signaling pathways. These changes could be the cause of the adverse health effects of PCBs, including cancer induction and atherosclerosis.

The paradigm of nutrition being able to modify PCB toxicity is of interest to populations at risk, i.e., populations residing near Superfund sites or areas of contamination and populations that rely heavily on fish (and meat) for their nutrition. Proper nutrition counseling should be considered by health officials and the medical community to reduce the overall risk for PCB toxicity and disease development. Very little is known about the interaction of diets and cytotoxicity of environmental contaminants such as PCBs (Figure 4.11). Our data clearly show that nutrition can modulate PCB toxicity and that specific fatty acids rich in plant oils, such as linoleic acid (the parent omega-6 fatty acid), can amplify PCB toxicity in vascular endothelial cells. Preliminary data from studies with mice that lack the LDL receptor gene support our hypothesis that diets high in omega-6 fatty acids can amplify PCB toxicity. Our data also suggest that antioxidants, and especially vitamin E, can protect against PCB toxicity (Figure 4.11). More research is needed to confirm our observed interactions of PCB toxicity with nutritional interventions.

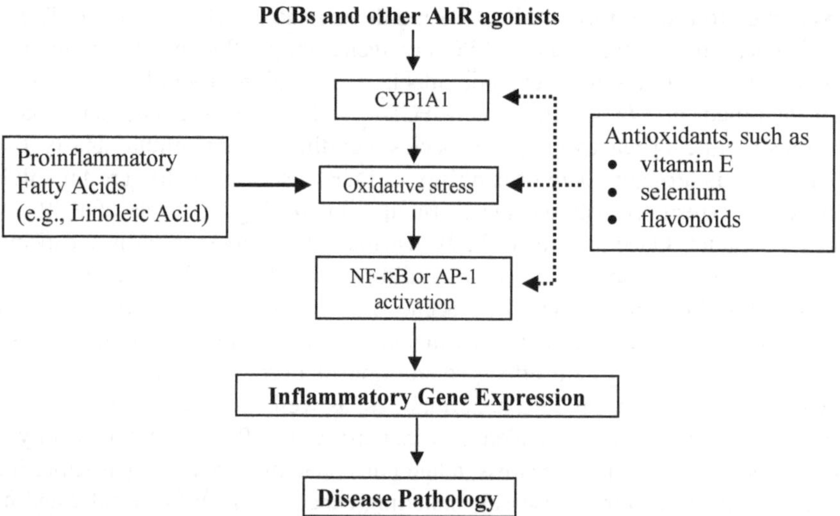

FIGURE 4.11 Potential mechanisms of nutrient-mediated protection against PCB-induced endothelial cell inflammatory response and disease pathology. (Modified from Hennig, B., Hammock, B.D., Slim, R., Toborek, M., Saraswathi, V., and Robertson, L.W. (2002a), PCB-induced oxidative stress in endothelial cells: modulation by nutrients, *International Journal of Hygiene and Environmental Health*, 205: 95–102.)

ACKNOWLEDGMENTS

The authors' research summarized in this review was supported by grant number P42 ES 07380 from the National Institute of Environmental Health Sciences, DAMD17-02-1-0241 from the Department of Defense, and by the Kentucky Agricultural Experiment Station. Its contents are solely the responsibility of the authors and do not necessarily represent the official views of the DOD or the NIH.

ABBREVIATIONS

AhR — aryl hydrocarbon receptor
AP-1 — activator protein-1
ATSDR — Agency for Toxic Substances and Disease Registry
CYP — cytochrome P-450
DCF — dichlorofluorescein
EGCG — epigallocatechin-3-gallate
EPA — Environmental Protection Agency
EROD — ethoxyresorufin-*O*-deethylase
FDA — Food and Drug Administration
GPx — glutathione peroxidase
GR — glutathione reductase
GSH — glutathione
GST — glutathione transferase
HRP — horseradish peroxidase
ICAM-1 — intercellular adhesion molecule-1
IL-6 — interleukin-6
LDL — low density lipoprotein
MCP-1 — monocyte chemoattractant protein-1
NF-κB — nuclear factor-kappaB
OSHA — Occupational Safety and Health Administration
PCB — polychlorinated biphenyl
PCB 38 — 3,4,5-trichlorobiphenyl
PCB 77 — 3,3′,4,4′-tetrachlorobiphenyl
PCB 104 — 2,2′,4,6,6′-pentachlorobiphenyl
PCB 126 — 3,3′,4,4′,5-pentachlorobiphenyl
PCB 153 — 2,2′,4,4′,5,5′-hexachlorobiphenyl
PCB 169 — 3,3′,4,4′,5,5′-hexachlorobiphenyl
ROS — reactive oxygen species
SOD — superoxide dismutase
TCDD — 2,3,7,8-tetrachlorodibenzo-p-dioxin
TNF — tumor necrosis factor
VCAM-1 — vascular cell adhesion molecule-1
VEGF — vascular endothelial growth factor

References

Alexiou, D., Karayiannakis, A.J., Syrigos, K.N., Zbar, A., Kremmyda, A., Bramis, I., and Tsigris, C. (2001), Serum levels of E-selectin, ICAM-1 and VCAM-1 in colorectal cancer patients: correlations with clinicopathological features, patient survival, and tumour surgery, *European Journal of Cancer*, 37: 2392–7.

Alsharif, N.Z., Lawson, T., and Stohs, S.J. (1994), Oxidative stress induced by 2,3,7,8-tetrachlorodibenzo-p-dioxin is mediated by the aryl hydrocarbon (Ah) receptor complex, *Toxicology*, 92: 39–51.

Amann, B., Perabo, F.G., Wirger, A., Hugenschmidt, H., and Schultze-Seemann, W. (1998), Urinary levels of monocyte chemo-attractant protein-1 correlate with tumour stage and grade in patients with bladder cancer, *European Journal of Cancer*, 82: 118–21.

Amaro, A.R., Oakley, G.G., Bauer, U., Spielmann, H.P., and Robertson, L.W. (1996), Metabolic activation of PCBs to quinones: Reactivity toward nitrogen and sulfur nucleophiles and influence of superoxide dismutase, *Chemical Research in Toxicology*, 9: 623–629.

ATSDR. (2000), Toxicological Profile for Polychlorinated Biphenyls (update, November 2000), U.S. Department of Health and Human Service, Public Health Service, Agency for Toxic Substances and Disease Registry, pp. 11–12. Online. Available HTTP: http://www.atsdr.cdc.gov/toxprofiles/tp17.html (accessed October 30, 2003).

Azais-Braesco, V., Hautekeete, M.L., Dodeman, I., and Geerts, A. (1997), Morphology of liver stellate cells and liver vitamin A content in 3,4,3′,4′-tetrachlorobiphenyl-treated rats, *Journal of Hepatology*, 27: 545–53.

Azzi, A., Montecucco, C., and Richter, C. (1975), The use of acetylated ferricytochrome c for the detection of superoxide radicals produced in biological membranes, *Biochemical Biophysical Research Communications*, 65: 597–603.

Bandiera, S., Safe, S., and Okey, A.B. (1982), Binding of Polychlorinated Biphenyls classified as either phenobarbitone-, 3-methylcholanthrene- or mixed-type inducers to cytosolic Ah receptor, *Chemico-Biological Interactions*, 39: 259–277.

Bertazzi, P.A., Bernucci, I., Brambilla, G., Consonni, D., and Pesatori, A.C. (1998), The Seveso studies on early and long-term effects of dioxin exposure: a review, *Environmental Health Perspectives*, 106 Suppl. 2: 625–33.

Bjerregaard, P. and Hansen, J.C. (2000), Organochlorines and heavy metals in pregnant women from the Disko Bay area in Greenland, *The Science of The Total Environment*, 245: 195–202.

Boll, M., Weber, L.W., Messner, B., and Stampfl, A. (1998), Polychlorinated biphenyls affect the activities of gluconeogenic and lipogenic enzymes in rat liver: Is there an interference with regulatory hormone actions?, *Xenobiotica*, 28: 479–92.

Boring, L., Gosling, J., Cleary, M., and Charo, I.F. (1998), Decreased lesion formation in CCR2-/- mice reveals a role for chemokines in the initiation of atherosclerosis, *Nature*, 394: 894–7.

Borlak, J. and Thum, T. (2002), PCBs alter gene expression of nuclear transcription factors and other heart-specific genes in cultures of primary cardiomyocytes: possible implications for cardiotoxicity, *Xenobiotica*, 32: 1173–83.

Butterfield, D., Castegna, A., Pocernich, C., Drake, J., Scapagnini, G., and Calabrese, V. (2002), Nutritional approaches to combat oxidative stress in Alzheimer's disease, *Journal of Nutritional Biochemistry*, 13: 444–61.

Casper, R.F., Quesne, M., Rogers, I.M., Shirota, T., Jolivet, A., Milgrom, E., and Savouret, J.F. (1999), Resveratrol has antagonist activity on the aryl hydrocarbon receptor: implications for prevention of dioxin toxicity, *Molecular Pharmacology*, 56: 784–90.

Chen, K., Thomas, S.R., and Keaney, J.F., Jr. (2003), Beyond LDL oxidation: ROS in vascular signal transduction, *Free Radical Biology and Medicine*, 35: 117–32.

Choi, W., Eum, S.Y., Lee, Y.W., Hennig, B., Robertson, L.W., and Toborek, M. (2003), PCB 104-induced proinflammatory reactions in human vascular endothelial cells: relationship to cancer metastasis and atherogenesis, *Toxicological Sciences*, 75: 47–56.

Clark, L.C., Dalkin, B., Krongrad, A., Combs, G.F., Jr., Turnbull, B.W., Slate, E.H., Witherington, R., Herlong, J.H., Janosko, E., Carpenter, D., Borosso, C., Falk, S., and Rounder, J. (1998), Decreased incidence of prostate cancer with selenium supplementation: Results of a double-blind cancer prevention trial, *British Journal of Urology*, 81: 730–4.

Denomme, M.A., Bandiera, S., Lambert, I., Copp, L., Safe, L., and Safe, S. (1983), Polychlorinated biphenyls as phenobarbitone-type inducers of microsomal enzymes: Structure-activity relationships for a series of 2,4-dichloro-substituted congeners, *Biochemical Pharmacology*, 32: 2955–2963.

Devanaboyina, U. and Gupta, R.C. (1996), Sensitive detection of 8-hydroxy-2-deoxyguanosine in DNA by 32P-postlabeling assay and the basal levels in rat tissues, *Carcinogenesis*, 17: 917–924.

Dogra, S., Filser, J.G., Cojocel, C., Greim, H., Regel, U., Oesch, F., and Robertson, L.W. (1988), Long-term effects of commercial and congeneric polychlorinated biphenyls on ethane production and malondialdehyde levels, indicators of *in vivo* lipid peroxidation, *Archives of Toxicology*, 62: 369–374.

Doi, A.M., Lou, Z., Holmes, E., Li, C., Venugopal, C.S., James, M.O., and Kleinow, K.M. (2000), Effect of micelle fatty acid composition and 3,4,3', 4'-tetrachlorobiphenyl (TCB) exposure on intestinal [(14)C]-TCB bioavailability and biotransformation in channel catfish in situ preparations, *Toxicological Sciences*, 55: 85–96.

Eum, S.Y., Lee, Y.W., Hennig, B., and Toborek, M., VEGF regulates PCB 104-mediated stimulation of permeability and transmigration of breast cancer cells in human microvascular endothelial cells, *Experimental Cell Research,* 296: 231–244, 2004.

Fischer, L.J., Seegal, R.F., Ganey, P.E., Pessah, I.N., and Kodavanti, P.R. (1998), Symposium overview: toxicity of non-coplanar PCBs, *Toxicological Sciences*, 41: 49–61.

Gierthy, J.F., Arcaro, K.F., and Floyd, M. (1995), Assessment and implications of PCB estrogenicity, *Organohalogen Compounds*, 25: 419–423.

Grandjean, P., Weihe, P., Needham, L.L., Burse, V.W., Patterson, D.G., Jr., Sampson, E.J., Jorgensen, P.J., and Vahter, M. (1995), Relation of a seafood diet to mercury, selenium, arsenic, and polychlorinated biphenyl and other organochlorine concentrations in human milk, *Environmental Research*, 71: 29–38.

Grandjean, P. and Weihe, P. (2003), Arachidonic acid status during pregnancy is associated with polychlorinated biphenyl exposure, *The American Journal of Clinical Nutrition*, 77: 715–9.

Grothey, A., Heistermann, P., Philippou, S., and Voigtmann, R. (1998), Serum levels of soluble intercellular adhesion molecule-1 (ICAM-1, CD54) in patients with non-small-cell lung cancer: correlation with histological expression of ICAM-1 and tumour stage, *British Journal of Cancer*, 77: 801–7.

Gu, L., Okada, Y., Clinton, S.K., Gerard, C., Sukhova, G.K., Libby, P., and Rollins, B.J. (1998), Absence of monocyte chemoattractant protein-1 reduces atherosclerosis in low density lipoprotein receptor-deficient mice, *Molecular Cell*, 2: 275–81.

Guitart, R., Guerrero, X., Silvestre, A.M., Gutierrez, J.M., and Mateo, R. (1996), Organochlorine residues in tissues of striped dolphins affected by the 1990 Mediterranean epizootic: relationships with the fatty acid composition, *Archives of Environmental Contamination and Toxicology*, 30: 79–83.

Gustavsson, P. and Hogstedt, C. (1997), A cohort study of Swedish capacitor manufacturing workers exposed to polychlorinated biphenyls (PCBs), *American Journal of Industrial Medicine*, 32: 234–9.

Hansen, L.G. (1999), *In The ortho Side of PCBs: Occurrence and Disposition*, Boston: Kluwer Academic.

Hassoun, E.A., Wang, H., Abushaban, A., and Stohs, S.J. (2002), Induction of oxidative stress in the tissues of rats after chronic exposure to TCDD, 2,3,4,7,8-pentachlorodibenzofuran, and 3,3,4,4,5 pentachlorobiphenyl, *Journal of Toxicology and Environmental Health. Part A*, 65: 825–42.

Hay, A. and Tarrel, J. (1997), Mortality of power workers exposed to phenoxy herbicides and polychlorinated biphenyls in waste transformer oil, *Annals of the New York Academy of Sciences*, 837: 138–56.

Hefler, L., Tempfer, C., Heinze, G., Mayerhofer, K., Breitenecker, G., Leodolter, S., Reinthaller, A., and Kainz, C. (1999), Monocyte chemoattractant protein-1 serum levels in ovarian cancer patients, *British Journal of Cancer*, 81: 855–9.

Hennig, B., Slim, R., Toborek, M., and Robertson, L.W. (1999), Linoleic acid amplifies polychlorinated biphenyl-mediated dysfunction of endothelial cells, *Journal of Biochemical and Molecular Toxicology*, 13: 83–91.

Hennig, B., Hammock, B.D., Slim, R., Toborek, M., Saraswathi, V., and Robertson, L.W. (2002a), PCB-induced oxidative stress in endothelial cells: modulation by nutrients, *International Journal of Hygiene and Environmental Health*, 205: 95–102.

Hennig, B., Meerarani, P., Slim, R., Toborek, M., Daugherty, A., Silverstone, A.E., and Robertson, L.W. (2002b), Proinflammatory properties of coplanar PCBs: *in vitro* and *in vivo* evidence, *Toxicology and Applied Pharmacology*, 181: 174–83.

Henry, E.C., Kende, A.S., Rucci, G., Totleben, M.J., Willey, J.J., Dertinger, S.D., Pollenz, R.S., Jones, J.P., and Gasiewicz, T.A. (1999), Flavone antagonists bind competitively with 2,3,7, 8-tetrachlorodibenzo-p-dioxin (TCDD) to the aryl hydrocarbon receptor but inhibit nuclear uptake and transformation, *Molecular Pharmacology*, 55: 716–25.

Hori, M., Kondo, H., Ariyoshi, N., Yamada, H., and Oguri, K. (1997), Species-specific alteration of hepatic glucose 6-phosphate dehydrogenase activity with coplanar polychlorinated biphenyl: evidence for an Ah-receptor-linked mechanism, *Chemosphere*, 35: 951–8.

Ishii, Y., Hatsumura, M., Ishida, T., Ariyoshi, N., and Oguri, K. (1996), Significant induction of a 54-kDa protein in rat liver with homologous alignment to mouse selenium binding protein by a coplanar polychlorinated biphenyl, 3,4,5,3',4'-pentachlorobiphenyl and 3-methylcholanthrene, *Toxicology Letters*, 87: 1–9.

Janousek, V., Krijt, J., Malbohan, M., Cibula, D., Lukas, W., Zejda, J.E., Lammers, W., Huisman, M., Boersma, E.R., van der Paauw, C.G. et al. (1994), Cord blood levels of potentially neurotoxic pollutants (polychlorinated biphenyls, lead and cadmium) in the areas of Prague (Czech Republic) and Katowice (Poland). Comparison with reference values in The Netherlands. The Czech/Polish/Dutch/German Research Team, *Central European Journal of Public Health*, 2: 73–6.

Jin, X., Kennedy, S.W., Di Muccio, T., and Moon, T.W. (2001), Role of oxidative stress and antioxidant defense in 3,3',4,4',5-pentachlorobiphenyl-induced toxicity and species-differential sensitivity in chicken and duck embryos, *Toxicology and Applied Pharmacology*, 172: 241–8.

Kakela, R. and Hyvarinen, H. (1999), Fatty acid alterations caused by PCBs (Aroclor 1242) and copper in adipose tissue around lymph nodes of mink, *Comparative Biochemistry and Physiology Part C: Pharmacology, Toxicology and Endocrinology*, 122: 45–53.

Kakela, A., Kakela, R., Hyvarinen, H., and Nieminen, P. (2003), Effects of Aroclor 1242 and different fish-based diets on vitamins A1 (retinol) and A2 (3,4-didehydroretinol), and their fatty acyl esters in mink plasma, *Environmental Research*, 91: 104–12.

Kamei, M., Ohgaki, S., Kanbe, T., Shimizu, M., Morita, S., Niiya, I., Matsui-Yuasa, I., and Otani, S. (1996), Highly hydrogenated dietary soybean oil modifies the responses to polychlorinated biphenyls in rats, *Lipids*, 31: 1151–1156.

Kamohara, K., Yagi, N., and Itokawa, Y. (1984), Mechanism of lipid peroxide formation in polychlorinated biphenyls (PCB) and dichlorodiphenyltrichloroethane (DDT)-poisoned rats, *Environmental Research*, 34: 18–23.

Kodavanti, P.R. and Tilson, H.A. (2000), Neurochemical effects of environmental chemicals: *in vitro* and *in vivo* correlations on second messenger pathways, *Annals of the New York Academy of Sciences*, 919: 97–105.

Kwon, O., Lee, E., Moon, T.C., Jung, H., Lin, C.X., Nam, K.S., Baek, S.H., Min, H.K., and Chang, H.W. (2002), Expression of cyclooxygenase-2 and pro-inflammatory cytokines induced by 2,2',4,4',5,5'-hexachlorobiphenyl (PCB 153) in human mast cells requires NF-kappa B activation, *Biological and Pharmaceutical Bulletin*, 25: 1165–1168.

Lee, T.H., Avraham, H.K., Jiang, S., and Avraham, S. (2003a), Vascular endothelial growth factor modulates the transendothelial migration of MDA-MB-231 breast cancer cells through regulation of brain microvascular endothelial cell permeability, *The Journal of Biological Chemistry*, 278: 5277–5284.

Lee, Y.W., Park, H.J., Son, K.W., Hennig, B., Robertson, L.W., and Toborek M. (2003b), 2,2′,4,6,6′-Pentachlorobiphenyl (PCB104) induces apoptosis of human microvascular endothelial cells through the caspase-dependent activation of CREB, *Toxicology and Applied Pharmacology*, 189: 1–10.

Ludewig, G., Srinivasan, A., Espandiari, P., and Robertson, L.W. (1998), Generation of reactive oxygen species (ROS) and genotoxicity by metabolites of polychlorinated biphenyls (PCBs), *Organohalogen Compounds*, 37: 143–146.

Ludewig, G. (2001), Cancer initiation by PCBs, in *PCBs: Recent Advances in Environmental Toxicology and Health Effects*, Robertson, L.W., and Hansen, L.G. Eds., University Press of Kentucky, Lexington, pp. 337–354.

Matsusue, K., Ishii, Y., Ariyoshi, N., and Oguri, K. (1999), A highly toxic coplanar polychlorinated biphenyl compound suppresses Delta5 and Delta6 desaturase activities which play key roles in arachidonic acid synthesis in rat liver, *Chemical Research in Toxicology*, 12: 1158–1165.

McLean, M.R., Bauer, U., Amaro, A.R., and Robertson, L.W. (1996a), Identification of catechol and hydroquinone metabolites of 4-monochlorobiphenyl, *Chemical Research in Toxicology*, 9: 158–164.

McLean, M.R., Robertson, L.W., and Gupta, R.C. (1996b), Detection of PCB-adducts by the ^{32}P-postlabeling technique, *Chemical Research in Toxicology*, 9: 165–171.

McLean, M., Twaroski, T., and Robertson, L.W. (1998), A new mechanism of toxicity for polychlorinated biphenyls (PCBs): Redox cycling and superoxide generation, *Organo-halogen Compounds*, 37: 59–62.

McLean, M.R., Twaroski, T.P., and Robertson, L.W. (2000), Redox cycling of 2-(x′-mono, -di, -trichlorophenyl)- 1, 4-benzoquinones, oxidation products of polychlorinated biphenyls, *Archives of Biochemistry and Biophysics*, 376: 449–55.

Mulvad, G., Pedersen, H.S., Hansen, J.C., Dewailly, E., Jul, E., Pedersen, M., Deguchi, Y., Newman, W.P., Malcom, G.T., Tracy, R.E., Middaugh, J.P., and Bjerregaard, P. (1996), The Inuit diet. Fatty acids and antioxidants, their role in ischemic heart disease, and exposure to organochlorines and heavy metals. An international study, *Arctic Medical Research*, 55 Suppl. 1: 20–24.

Oakley, G.G., Devanaboyina, U.S., Robertson, L.W., and Gupta, R.C. (1996a), Oxidative DNA damage induced by activation of Polychlorinated Biphenyls (PCBs): Implications for PCB-induced oxidative stress in breast cancer, *Chemical Research in Toxicology*, 9: 1285–1292.

Oakley, G.G., Robertson, L.W., and Gupta, R.C. (1996b), Analysis of polychlorinated biphenyl-DNA adducts by 32P-postlabeling, *Carcinogenesis*, 17: 109–114.

Parkinson, A., Safe, S.H., Robertson, L.W., Thomas, P.E., Ryan, D.E., Reik, L.M., and Levin, W. (1983), Immunochemical quantitation of cytochrome P-450 isozymes and epoxide hydrolase in liver microsomes from polychlorinated or polybrominated biphenyl-treated rats: A study of structure-activity relationships, *The Journal of Biological Chemistry*, 258: 5967–5976.

Pelissier, M.A., Boisset, M., Atteba, S., and Albrecht, R. (1990), Lipid peroxidation of rat liver microsomes membranes related to a protein deficiency and/or a PCB treatment, *Food Additives and Contaminants*, 7 Suppl. 1: S172–177.

Puga, A., Barnes, S.J., Chang, C., Zhu, H., Nephew, K.P., Khan, S.A., and Shertzer, H.G. (2000), Activation of transcription factors activator protein-1 and nuclear factor-kappaB by 2,3,7,8-tetrachlorodibenzo-p-dioxin, *Biochemical Pharmacology*, 59: 997–1005.

Püttmann, M., Arand, M., Oesch, F., Mannschreck, A., and Robertson, L.W. (1990), Chirality and the induction of xenobiotic-metabolizing enzymes: Effects of the atropisomers of the polychlorinated biphenyl 2,2',3,4,4',6-hexachlorobiphenyl, in Holmstedt, Frank, and Testa (Eds.), *Chirality and Biological Activity*, New York: Alan R. Liss, pp177–184.

Quadri, S.A., Qadri, A.N., Hahn, M.E., Mann, K.K., and Sherr, D.H. (2000), The bioflavonoid galangin blocks aryl hydrocarbon receptor activation and polycyclic aromatic hydrocarbon-induced pre-B cell apoptosis, *Molecular Pharmacology*, 58: 515–525.

Ramadass, P., Meerarani, P., Toborek, M., Robertson, L.W., and Hennig, B. (2003), Dietary flavonoids modulate PCB-induced oxidative stress, CYP1A1 induction and AhR DNA binding activity in vascular endothelial cells, *Toxicological Sciences*, 76: 212–219.

Ridker, P.M., Hennekens, C.H., Roitman-Johnson, B., Stampfer, M.J., and Allen, J. (1998), Plasma concentration of soluble intercellular adhesion molecule 1 and risks of future myocardial infarction in apparently healthy men, *Lancet*, 351: 88–92.

Robertson, L.W. and Hansen, L.G. (Eds.), (2001), *PCBs: Recent Advances in Environmental Toxicology and Health Effects*, University Press of Kentucky, Lexington.

Rodriguez-Ariza, A., Rodriguez-Ortega, M.J., Marenco, J.L., Amezcua, O., Alhama, J., and Lopez-Barea, J. (2003), Uptake and clearance of PCB congeners in Chamaelea gallina: response of oxidative stress biomarkers, *Comparative Biochemistry and Physiology. Toxicology & Pharmacology: CBP*, 134: 57–67.

Safe, S. and Krishnan, V. (1995), Cellular and molecular biology of aryl hydrocarbon (Ah) receptor-mediated gene expression, *Archives of Toxicology. Supplement*, 17: 99–115.

Saito, M. (1990), Polychlorinated biphenyls-induced lipid peroxidation as measured by thiobarbituric acid-reactive substances in liver subcellular fractions of rats, *Biochimica et Biophysica Acta*, 1046: 301–308.

Schaldach, C.M., Riby, J., and Bjeldanes, L.F. (1999), Lipoxin A4: a new class of ligand for the Ah receptor, *Biochemistry*, 38: 7594–7600.

Schlezinger, J.J., White, R.D., and Stegeman, J.J. (1999), Oxidative inactivation of cytochrome P-450 1A (CYP1A) stimulated by 3,3',4,4'-tetrachlorobiphenyl: production of reactive oxygen by vertebrate CYP1As, *Molecular Pharmacology*, 56: 588–597.

Schlezinger, J.J., Blickarz, C.E., Mann, K.K., Doerre, S., and Stegeman, J.J. (2000), Identification of NF-kappaB in the marine fish Stenotomus chrysops and examination of its activation by aryl hydrocarbon receptor agonists, *Chemico-Biological Interactions*, 126: 137–157.

Schramm, H., Robertson, L.W., and Oesch, F. (1985), Differential regulation of hepatic glutathione transferase and glutathione peroxidase activites in the rat, *Biochemical Pharmacology*, 34: 3735–3739.

Seegal, R.F. (1995), Neurochemical effects of PCBs are structure and recipient age-dependent, *Organohalogen Compounds*, 25: 425–430.

Shertzer, H.G., Puga, A., Chang, C., Smith, P., Nebert, D.W., Setchell, K.D., and Dalton, T.P. (1999), Inhibition of CYP1A1 enzyme activity in mouse hepatoma cell culture by soybean isoflavones, *Chemico-Biological Interactions*, 123: 31–49.

Shimada, T. and Sawabe, Y. (1983), Activation of 3,4,3',4'-tetrachlorobiphenyl to protein-bound metabolites by rat liver microsomal cytochrome P-448-containing monooxygenase system, *Toxicology and Applied Pharmacology*, 70: 486–493.

Silberhorn, E.M., Glauert, H.P., and Robertson, L.W. (1990), Carcinogenicity of polyhalogenated biphenyls: PCBs and PBBs, *Critical Reviews in Toxicology*, 20: 439–496.

Slim, R., Hammock, B.D., Toborek, M., Robertson, L.W., Newman, J.W., Morisseau, C.H., Watkins, B.A., Saraswathi, V., and Hennig, B. (2001), The role of methyl-linoleic acid epoxide and diol metabolites in the amplified toxicity of linoleic acid and polychlorinated biphenyls to vascular endothelial cells, *Toxicology and Applied Pharmacology*, 171: 184–193.

Slim, R., Toborek, M., Robertson, L.W., and Hennig, B. (1999), Antioxidant protection against PCB-mediated endothelial cell activation, *Toxicological Sciences*, 52: 232–239.

Slim, R., Toborek, M., Robertson, L.W., Lehmler, H.J., and Hennig, B. (2000), Cellular glutathione status modulates polychlorinated biphenyl-induced stress response and apoptosis in vascular endothelial cells, *Toxicology and Applied Pharmacology*, 166: 36–42.

Sprenger, A., Schardt, C., Rotsch, M., Zehrer, M., Wolf, M., Havemann, K., and Heymanns, J. (1997), Soluble intercellular adhesion molecule-1 in patients with lung cancer and benign lung diseases, *Journal of Cancer Research and Clinical Oncology*, 123: 632–638.

Srinivasan, A., Lehmler, H.-J., Robertson, L.W., and Ludewig, G. (2001), Production of DNA strand breaks *in vitro* and Reactive Oxygen Species *in vitro* and in HL-60 cells by PCB Metabolites, *Toxicological Sciences*, 60: 92–102.

Srinivasan, A., Robertson, L.W., and Ludewig, G. (2002), Sulfhydryl binding and topoisomerase inhibition by PCB Metabolites, *Chemical Research in Toxicology*, 15: 497–505.

Sugihara, K., Kitamura, S., Yamada, T., Ohta, S., Yamashita, K., Yasuda, M., and Fujii-Kuriyama, Y. (2001), Aryl hydrocarbon receptor (AhR)-mediated induction of xanthine oxidase/xanthine dehydrogenase activity by 2,3,7,8-tetrachlorodibenzo-p-dioxin, *Biochemical and Biophysical Research Communications*, 281: 1093–1099.

Tanabe, K., Campbell, S.C., Alexander, J.P., Steinbach, F., Edinger, M.G., Tubbs, R.R., Novick, A.C., and Klein, E.A. (1997), Molecular regulation of intercellular adhesion molecule 1 (ICAM-1) expression in renal cell carcinoma, *Urological Research*, 25: 231–238.

Thurmond, T.S., Silverstone, A.E., Baggs, R.B., Quimby, F.W., Staples, J.E., and Gasiewicz, T.A. (1999), A chimeric aryl hydrocarbon receptor knockout mouse model indicates that aryl hydrocarbon receptor activation in hematopoietic cells contributes to the hepatic lesions induced by 2,3,7, 8-tetrachlorodibenzo-p-dioxin, *Toxicology and Applied Pharmacology*, 158: 33–40.

Toborek, M., Barger, S.W., Mattson, M.P., Espandiari, P., Robertson, L.W., and Hennig, B. (1995), Exposure to polychlorinated biphenyls causes endothelial cell dysfunction, *Journal of Biochemical Toxicology*, 10: 219–226.

Tokunaga, S., Hirota, Y., and Kataoka, K. (1999), Association between the results of blood test and blood PCB level of chronic Yusho patients twenty five years after the outbreak, *Fukuoka Igaku Zasshi*, 90: 157–161.

Tomlinson, J., Wang, J.L., Barsky, S.H., Lee, M.C., Bischoff, J., and Nguyen, M. (2000), Human colon cancer cells express multiple glycoprotein ligands for E- selectin, *International Journal of Oncology*, 16: 347–353.

Twaroski, T.P., O'Brien, M.L., Larmonier, N., Glauert, H.P., and Robertson, L.W. (2001a), Polychlorinated biphenyl-induced effects on metabolic enzymes, AP-1 binding, vitamin E, and oxidative stress in the rat liver, *Toxicology and Applied Pharmacology*, 171: 85–93.

Twaroski, T.P., O'Brien, M.L., and Robertson, L.W. (2001b), Effects of selected polychlorinated biphenyl (PCB) congeners on hepatic glutathione, glutathione-related enzymes, and selenium status: implications for oxidative stress, *Biochemical Pharmacology*, 62: 273–281.

Williams, S.N., Shih, H., Guenette, D.K., Brackney, W., Denison, M.S., Pickwell, G.V., and Quattrochi, L.C. (2000), Comparative studies on the effects of green tea extracts and individual tea catechins on human CYP1A gene expression, *Chemico-Biological Interactions*, 128: 211–229.

Wong, P.W., Brackney, W.R., and Pessah, I.N. (1997), ortho-Substituted polychlorinated biphenyls alter microsomal calcium transport by direct interaction with ryanodine receptors of mammalian brain, *The Journal of Biological Chemistry*, 272: 15145–15153.

Yoshizawa, K., Willett, W.C., Morris, S.J., Stampfer, M.J., Spiegelman, D., Rimm, E.B., and Giovannucci, E. (1998), Study of prediagnostic selenium level in toenails and the risk of advanced prostate cancer, *Journal of the National Cancer Institute*, 90: 1219–1224.

5 Oxidized Products of Cholesterol: Toxic Effects

Gabriella Leonarduzzi, Barbara Sottero, Veronica Verde, and Giuseppe Poli

CONTENTS

ABSTRACT

Experimental studies, epidemiological surveys, and clinical trials consistently show dietary cholesterol to have a significant effect on plasma cholesterol levels. Increased plasma cholesterol levels are in turn known to promote atherosclerosis and stimulate its progression. Increasing experimental evidence points to cholesterol oxidation products (oxysterols) as being actually responsible for the proatherogenic action of cholesterol, which per se is rather an unreactive molecule. In particular, specific oxysterols of pathophysiological significance exercise strong proapoptotic and proinflammatory action. The most recent findings on oxysterols' toxic and proatherosclerotic effects

0-8493-3516-7/05/$0.00+$1.50

are described in this chapter. To ensure a comprehensive approach to their potential contribution to the atherosclerosis process, the occurrence of cholesterol oxidation products in foodstuffs, methods for their identification and measurement, and the use of antioxidants supplementation to prevent their generation in foods are reviewed analytically.

INTRODUCTION

The relationship between dietary cholesterol, plasma cholesterol, and atherosclerosis has long been the subject of intense research and debate (Hopkins, 1992; McNamara, 2000; Libby et al., 2000; Kritchvsky, 2001). Four lines evidence this relationship: animal feeding studies (Verlangieri et al., 1985; Coutard and Osborne-Pellegrin, 1986), human studies (Rifkind, 1986; Ross, 1986), epidemiological surveys (Pietinen et al., 1997; Toeller et al., 1999), and clinical trials (Shepherd et al., 1995; Sacks et al., 1996). All these studies show that dietary cholesterol has a significant effect on plasma cholesterol levels. In some animal species, dietary cholesterol induces hypercholesterolemia and atherosclerotic lesions, and metabolic experiments have shown a high cholesterol intake to increase plasma cholesterol levels.

Cholesterol is an essential component of cell membranes but, when present in excess in the circulation, it may be deposited in the artery wall, leading to the formation of atherosclerotic plaque (Kroon, 1997). The mechanisms by which cholesterol contributes to the initiation and, in particular, to the progression of atherosclerotic lesions is still unclear and debated. The main plasma carrier of cholesterol, low-density lipoprotein (LDL), is removed from circulation by LDL receptors in the liver. Excess low density lipoprotein may over time accumulate in the artery wall. Increased binding to proteoglycan is a probable principal mechanism for the retention of LDL particles in the arterial intima (Boren et al., 1998; Camejo et al., 1998). Particularly when associated with such extracellular matrix macromolecules, LDL may undergo modifications, some of which may contribute to the atherogenicity of such lipoproteins. The unique structure and composition of LDL make it particularly susceptible to modification by oxidative reactions (Jurgens et al., 1987; Hazen et al., 1999; Heinecke, 1999a, 1999b).

BIOCHEMICAL BACKGROUND

Among the oxidation products of the lipid moiety of LDL, cholesterol oxidation products called oxysterols are of great interest as possible reactive mediators of structural and functional changes of the vascular wall, which are affected by the atherosclerotic process (Smith, 1996; Smith and Johnson, 1989; Luu and Moog, 1991; Ross, 1993; Guardiola et al., 1996, 2002; Brown and Jessup, 1999; Schroepfer, Jr., 2000; Björkhem and Diczfalusy, 2002). Oxysterols are well known for their toxicity. Indeed, the notion that pure cholesterol is reactive per se appears remote, whereas oxysterols should be viewed as toxic because they cause dysfunction of vascular endothelial cells (Peng et al., 1985, 1991) and play an active role in atherosclerotic lesion development (Carpenter et al., 1993, 1995; Breuer et al., 1996).

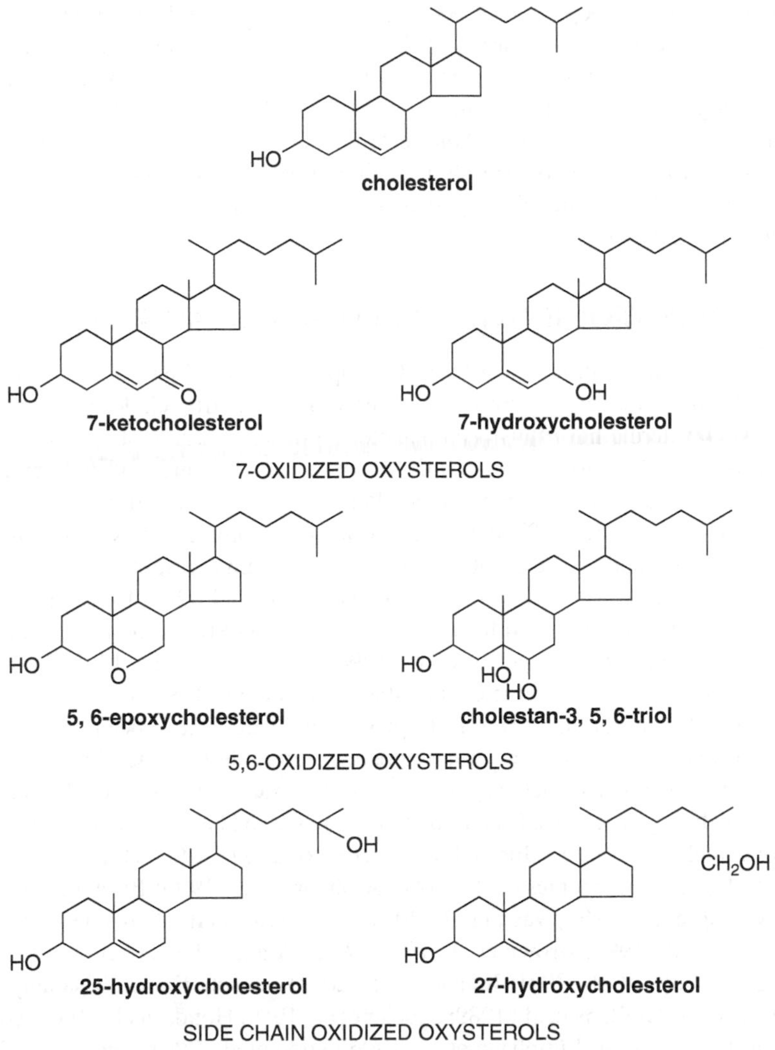

HO
cholesterol

HO O
7-ketocholesterol

HO OH
7-hydroxycholesterol

7-OXIDIZED OXYSTEROLS

HO O
5, 6-epoxycholesterol

HO HO HO
cholestan-3, 5, 6-triol

5,6-OXIDIZED OXYSTEROLS

OH
HO
25-hydroxycholesterol

CH_2OH
HO
27-hydroxycholesterol

SIDE CHAIN OXIDIZED OXYSTEROLS

FIGURE 5.1 Chemical structures of oxysterols of pathophysiological interest.

Oxysterols are 27-carbon products of cholesterol oxidation (Figure 5.1). Introduction of an oxygen function, such as a hydroxyl group, a ketone group, or an epoxide group, onto the sterol nucleus or onto the side chain of the molecule increases the rate at which cholesterol is degraded to more polar compounds, including bile acids. These cholesterol derivatives, in particular those with additional oxygen function on the steroid side chain, can easily be transported out of the cells and thus facilitate elimination of cholesterol from extrahepatic sources (Björkhem et al., 1999). In addition to being intermediates or end products in cholesterol degradation, oxysterols have been implicated in many cellular processes. They have a broad spectrum of biological effects (Brown and Jessup, 1999; Edwards and

Ericsson, 1999; Schroepfer, Jr., 2000; Björkhem and Diczfalusy, 2002; Guardiola et al., 2002), including modulation of the activity of key proteins involved in cholesterol homeostasis (Accad and Farese, 1998; Björkhem, 2002). However, their role in physiology is still controversial. Some of their effects on the regulation of key enzymes are similar to those of cholesterol but much more potent. Notably, one of the critical properties of oxysterols is their ability to pass lipophilic membranes rapidly due to the additional oxygen function (Smondyrev and Berkowitz, 2001; Meaney et al., 2002).

OXYSTEROL IDENTIFICATION AND MEASUREMENT

Several analytical methods have been developed (Smith, 1981, 1996; Schroepfer, Jr., 2000; Guardiola et al., 2002) to identify and quantify cholesterol oxidation products. Oxysterols have thus been detected in different kinds of specimens from foodstuffs (see Guardiola et al., 2002 for an exhaustive and updated review) to biological and cell samples (Pie and Seillan, 1992; Mattsson Hultén et al., 1996; Maor et al., 2000). Great efforts have been made to optimize oxysterol analysis in biological fluids and tissues, both of human and animal origins, because of the potential involvement of oxysterols in atherosclerosis and other disease processes.

Blood sample analysis and hematic oxysterol content have received great attention (Hodis et al., 1991; Sevanian et al., 1994; Dzeletovic et al., 1995; Mol et al., 1997; Salonen et al., 1997; Vine et al., 1998; Zieden et al., 1999; Porkkala-Sarataho et al., 2000; Iuliano et al., 2003). Both plasma and serum have been employed for oxysterol quantitation, but the former appears preferable because chelating agents added as anticoagulants, such as ethylenediaminetetraacetic acid (EDTA), may also prevent oxidizing processes. Oxysterol content has also been reported in lipoprotein fractions, including those subjected to *in vitro* oxidation (Addis et al., 1989; Hodis et al., 1991, 1994; Carpenter et al., 1994; Sevanian et al., 1995; Brown et al., 1996, 1997; Chang et al., 1997; Vaya et al., 2001), in arteries and atherosclerotic lesions (Carpenter et al., 1995; Brown et al., 1997; Rong et al., 1998; Garcia-Cruset et al., 1999, 2001; Vaya et al., 2001; Iuliano et al., 2003), muscle tissues (Csallany et al., 1989), the liver (Csallany et al., 1989; Saucier et al., 1989; Honda et al., 1995; Ariyoshi et al., 2002), the brain (Lütjohann et al., 1996; Kudo et al., 2001; Vega et al., 2003), and other biological materials such as oxidized brain synaptosomes (Vatassery et al., 1997a), erythrocytes (Kucuk et al., 1992; Teng and Smith, 1995), bile, and gallstones (Haigh and Lee, 2001). Although direct comparison among such diverse results is not possible because of the varying analytical methods employed, a common trend may be glimpsed, the following oxysterols often being reported as present: 7-ketocholesterol, 7-hydroxycholesterols, 5,6-epoxycholesterols, and cholestan-3β,5α,6β-triol. The oxysterol 25-hydroxycholesterol is found only in trace amounts; 24-hydroxycholesterol is also found in plasma, though it is predominant in the brain; and 27-hydroxycholesterol is the most abundant oxysterol in atherosclerotic plaque.

Of all the analytical approaches for oxysterol determination, chromatographic techniques are surely those most frequently employed because of their suitability and precision (Smith, 1996; Shan et al., 2003).

Thin-layer chromatography (TLC) has been used extensively to characterize cholesterol oxidation products (Smith, 1981; Ansari and Smith, 1994) by means of silica plates incorporating a fluorescent indicator that detects substances which absorb short-wavelength ultraviolet (UV) light. Of the oxysterols, only 7-ketocholesterol has such a quenching property, and spraying with specific reagents under appropriate conditions is necessary to reveal other components. Identification is easily achieved by comparing color and position (R_f value) of the spots with those of reference standards.

Oxysterols have been quantified based on spot area using a digital planimeter (Finocchiaro et al., 1983) or a scanning densitometer (Jacobson, 1987), but the accuracy of such methods does not appear to be sufficiently reliable. Interestingly, Lebovics (Lebovics et al., 1996) reported a TLC separation followed by enzymatic quantification of oxysterol content in food, but similar approaches have not yet been attempted in blood analysis. However, the principal use of TLC is for qualitative or preparative purposes (Björkhem, 1986; Maerker and Bunick, 1986; Pie et al., 1990; Ansari and Smith, 1994; Dzeletovic et al., 1995), and high-performance liquid chromatography (HPLC) and gas chromatography (GC) are now the most popular techniques to characterize oxidized derivatives of cholesterol.

Both normal- and reversed-phase HPLC can be utilized to separate complex mixture of oxysterols (Shan et al., 2003). When normal-phase HPLC is used, silicic acid columns are usually chosen, and the analysis is performed in isocratic mode by means of a binary mobile phase (Ansari and Smith, 1979; Tsai and Hudson, 1981; Csallany et al., 1989). Alternatively, reversed-phase HPLC columns can be employed (Kritharides et al., 1993; Brown et al., 1996; Ariyoshi et al., 2002), their advantage being the shorter elution time for the more polar oxysterols, e.g., cholestan-3β,5α,6β-triol. Analysis can be optimized using coupled columns in series (Sevanian and McLeod, 1987) — of course, with increasing cost and time.

Several detection systems can be fitted with HPLC instruments; the most common are the UV detector (Kritharides et al., 1993) and the refractive index (RI) detector (Tsai and Hudson, 1981; Sevanian and McLeod, 1987; Chen and Chen, 1994). With regard to the UV detector, it should be said that some of the main oxysterols (e.g., epoxycholesterol epimers, and triol) are not UV-light responsive. Other compounds are also impossible to recognize accurately with a single-wavelength UV light because of their different maximum UV-absorption wavelengths. For these reasons, over the last few years alternative systems have been tested — in particular, the diode array detector (DAD) (Chen and Chen, 1994; Kermasha et al., 1994; Osada et al., 1999), the evaporative light-scattering detector (ELSD) (Kermasha et al., 1994; Osada et al., 1999), and mass spectrometry (MS) (Sevanian et al., 1994).

Despite these and other developments [(e.g., the formation of acetate derivatives (Kudo et al., 1989) or Δ^4-3-ketone derivatives (Teng and Smith, 1995; Zhang et al., 2001) to improve oxysterol characterization by HPLC)], GC is still the most successful technique to measure oxysterols. Figure 5.2 shows a representative GC chromatogram of plasma oxysterols from both normocholesterolemic and hypercholesterolemic subjects.

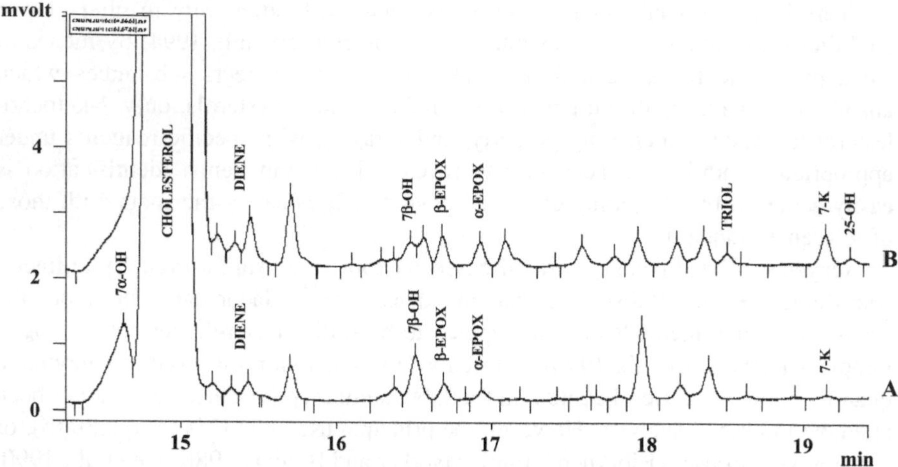

FIGURE 5.2 A representative GC-FID chromatogram of cholesterol and major oxysterols isolated from human plasma (1 ml) of (A) normocholesterolemic and (B) hypercholesterolemic individuals. Peaks: 7α-OH: 7α-hydroxycholesterol; DIENE: cholesta-3,5-diene-7-one; 7β-OH: 7β-hydroxycholesterol; α-EPOX: 5α,6α-epoxycholesterol; β-EPOX: 5β,6β-epoxycholesterol; TRIOL: cholestan-3β,5α,6β-triol; 7-K: 7-ketocholesterol; 25-OH: 25-hydroxycholesterol.

Currently, oxysterols are separated as trimethylsilyl (TMS) ethers by means of nonpolar or low-polar columns. The derivatization step is crucial for reliable analysis of oxysterols: unsuitable silylating agents or conditions can lead to incomplete derivatization of some compounds and consequently to the appearance of multiple peaks or, alternatively, can cause degradation and loss of material (Park and Addis, 1985a; Shan et al., 2003). Detection is usually achieved by the flame ionization detector (FID), which is cheap and simple to use. Sensitivity is also very good: the detection limits for silanized oxysterols range from 1 to 3 μg/ml with split injection techniques, and from 0.1 to 0.5 μg/ml with splitless, on-column, or programmable temperature vaporizing injection modes (Guardiola et al., 2002). However, given the complexity of biological oxysterol samples, confirmation of peak identities by MS is appropriate (Sevanian et al., 1994). Thus GC-MS is probably the most powerful technique applicable to oxysterol quantification. Even better results can be obtained with isotope dilution MS, whereby the use of multiple deuterium-labeled internal standards provides a more precise correction for each compound (Dzeletovic et al., 1995). Nevertheless, determination of cholesterol oxidation products in biological materials is still a difficult task, whatever method is employed, and controversial results have been reported (Brown and Jessup, 1999). Oxysterols have similar structures and chemical behaviors, and are only present in small amounts in a highly complex matrix. For these reasons, some compounds are hardly detectable.

Optimized sample preparation is always necessary before chromatographic analysis to remove interfering substances and enrich trace compound content (Figure 5.3). The first step is usually lipid extraction with suitable solvent mixtures.

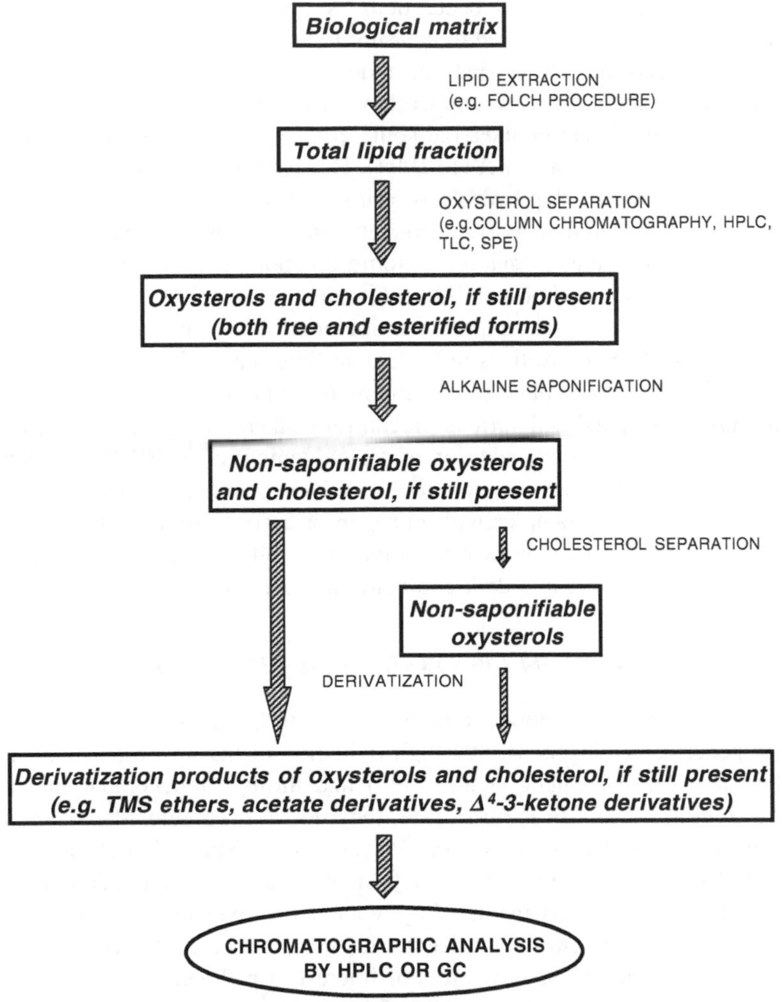

FIGURE 5.3 Scheme of the main steps for isolation and purification of oxysterols from biological samples before chromatographic analysis.

The classic Folch procedure (Folch et al., 1957) is frequently employed, but alternative systems have also been used (Kritharides et al., 1993; Boselli et al., 2001).

Subsequent isolation of oxysterols from other components of the crude lipid extract, including nonoxidized cholesterol, can be achieved by conventional column chromatography (Park and Addis, 1985b), medium-pressure liquid chromatography (MPLC) (Kudo et al., 1989), flash chromatography (Saucier et al., 1989), TLC, and solid phase extraction (SPE) columns (Ruiz-Gutiérrez and Pérez-Camino, 2000). Whatever device is employed, recoveries may differ for each oxysterol with consequent problems in final quantification (Guardiola et al., 1995a). Nevertheless, this risk can be minimized by using several internal standards (Dzeletovic et al., 1995).

Further, because oxysterols occur both as free and as esterified compounds, alkaline saponification is required to measure their exact total content. This process is very critical and must be carried out under strictly controlled conditions (alkali solution concentration, time, and temperature). Of note, hot hydrolysis ensures more efficient breaking of ester bounds, but some compounds decompose — 7-ketocholesterol, for instance (Sevanian et al., 1994). Hence, cold saponification in the dark should be preferred. To stop hydrolysis, mild acids (e.g., acetic acid, phosphorous acid, citric acid) are preferable over stronger mineral acids (e.g., hydrochloric acid) to avoid undesirable transformation of some compounds into others (e.g., epoxycholesterol epimers into cholestan-3β,5α,6β-triol) (Smith, 1981).

Finally, cholesterol's susceptibility to oxidation in the air may lead to overestimation of its oxidized derivatives in biological lipid materials. Thus, analysis must be performed on freshly collected specimens or, alternatively, proper storage conditions must be ensured and various precautions taken during sample preparation, which include the immediate addition of antioxidants such as EDTA and butylated hydroxytoluene (BHT), use of peroxide-free solvents, and operation in oxygen-free conditions (by flushing samples with nitrogen or argon). Additional recommendations are elimination of cholesterol from the analyzed material, and extraction and purification of oxysterols in a dark and cold environment.

ORIGIN OF OXYSTEROLS *IN VIVO*

Oxysterols may enter the blood circulation through the diet (Tai et al., 2000); they are also generated endogenously through oxidation of the lipoprotein lipid moiety (Hughes et al., 1994; Chang et al., 1997) and through intracellular metabolism (Björkhem et al., 1994). In the first case, they derive from autoxidation (nonenzymatic oxidation) of cholesterol present in various foodstuffs (Tai et al., 2000). In the second case, plasma oxysterols may arise either from autoxidation of LDL cholesterol, which is mainly mediated by reactive pro-oxidant species (Hodis et al., 1991, 1992) or specific enzymatic reactions (Björkhem et al., 1994).

The vast majority of the oxysterols of interest in pathophysiology are found in food and food derivatives: in particular, 7-ketocholesterol, 7α-hydroxycholesterol, 7β-hydroxycholesterol, 5α,6α-epoxycholesterol, 5β,6β-epoxycholesterol, cholestan-3β,5α,6β-triol, and 25-hydroxycholesterol. They may be present in fresh or raw foodstuffs containing cholesterol and also in seasonings (Table 5.1). However, more important, several factors are known to accelerate the oxidation of food cholesterol: in particular, γ- and UV-radiation, photo-oxidation, heat, presence of oxygen, presence of pro-oxidant agents, and storage conditions (Dionisi et al., 1998).

This argument will be considered in the following text; however, more nutritionally driven and more comprehensive data have very recently been collected and revised by Guardiola et al. (2002).

Regarding the possible presence of cholesterol oxidation products in fresh or raw foods, oxysterols have been found in trace amounts (though not consistently) in raw beef (Pie et al., 1991), minced pork and veal (Pie et al., 1991), fish (Osada et al., 1993), raw bacon rind (Nourooz-Zadeh and Appelqvist, 1989), fresh cow's milk (Chan et al., 1993), and fresh eggs (Nourooz-Zadeh and Appelqvist, 1987).

TABLE 5.1
Oxysterol Generation in Different Food Products

Food	Treatment	Oxysterol Produced	Amount (ppm)
		Dairy Products	
Skim milk powder	Fresh	7α-OH, 7β-OH, α-EPOX, β-EPOX, 7-K	Trace–2.6
Skim milk powder	Storage (13–37 months)	7α-OH, 7β-OH, α-EPOX, β-EPOX, 7-K, 20α-OH, 25-OH, TRIOL	Trace–23.3
Whole milk powder	Fresh	7α-OH, 7β-OH, α-EPOX, β-EPOX, 7-K	Trace–1.4
Whole milk powder	Storage (12 months)	7α-OH, 7β-OH, α-EPOX, β-EPOX, 7-K, TRIOL	Trace–9.2
Milk powder	Commercial	7α-OH, 7β-OH, 7-K, 25-OH, TRIOL	Trace–22.42
Butter	Fresh	7-K	Trace
Butter	Storage (20°C, 3 or 6 months)	7α-OH, β-EPOX, 7-K	0.22–1.52
Butter oil	Bleaching, storage	7α-OH, 7β-OH, 5,6-EPOX	20–90
Butter powder	Commercial	7β-OH, α-EPOX, β-EPOX, 7-K	3–26
Cheese	Bleaching	7α-OH, 7β-OH, 5,6-EPOX, TRIOL	4–110
		Egg Products	
Dried powdered egg mix	By direct or indirect heating source	7α-OH, 7β-OH, α-EPOX, β-EPOX, 25-OH, 7-K, TRIOL	1.4–50.0
Fresh egg	Commercial	7-K	Trace
Dehydrated egg yolk	Storage (2 months–8 yr)	7α-OH, 7β-OH, α-EPOX, β-EPOX, 25-OH, 20α-OH, 7-K, TRIOL	Trace–46.8
Egg yolk	Spray-dried	7α-OH, 7β-OH, α-EPOX, β-EPOX, 7-K	4.93–9.13
		Fish Products	
Anchovy	Salted-dried	7α-OH, α-EPOX, α-EPOX, 25-OH, 7-K, TRIOL	0.2–48.8
Japanese whiting	Salted-dried	7α-OH, α-EPOX, α-EPOX, 25-OH, 7-K, TRIOL	3.4–24.9
Pacific herring	Salted-dried	7α-OH, α-EPOX, α-EPOX, 25-OH, 7-K, TRIOL	3.5–8.4
		Meat Products	
Minced beef	Raw	7α-OH, 7β-OH, α-EPOX, β-EPOX, 7-K, 20-OH, 25-OH	0.14–1.06
Minced beef	Cooked	7α-OH, 7β-OH, α-EPOX, β-EPOX, 7-K, 20-OH, 25-OH	0.23–2.11
Beef	Irradiated	α-EPOX, β-EPOX, 7-K, 4-cholesten-3-one, 4,6-cholestadien-3-one, 4-cholestene-3,6-dione	2.3–443
Beef	Raw	α-EPOX, β-EPOX, 7-K	13.4–82.9
Freeze-dried beef	Commercial	7α-OH, 7β-OH, α-EPOX, β-EPOX, 7-K, TRIOL	1–27
Veal	Raw	α-EPOX, β-EPOX, 7-K	3.3–22

TABLE 5.1 (continued)
Oxysterol Generation in Different Food Products

Food	Treatment	Oxysterol Produced	Amount (ppm)
Veal	Irradiated	α-EPOX, β-EPOX, 7-K, 4-cholesten-3-one, 4,6-cholestadien-3-one, 4-cholestene-3,6-dione	2.7–183
Minced veal	Raw	7α-OH, 7β-OH, α-EPOX, β-EPOX, 7-K, 20-OH, 25-OH	0.04–0.71
Pork	Raw	α-EPOX, β-EPOX, 7-K	2.3–6.3
Freeze-dried pork	Stored at 22°C with air (3 yr)	7α-OH, 7β-OH, α-EPOX, 7-K, TRIOL	12.5–259.8
Pork	Irradiated	α-EPOX, β-EPOX, 7-K, 4-cholesten-3-one, 4,6-cholestadien-3-one, 4-cholestene-3,6-dione	Trace–108
Minced pork	Raw	7α-OH, 7β-OH, α-EPOX, β-EPOX, 7-K, 25-OH, TRIOL	0.04–0.92
Chicken	Raw	α-EPOX, β-EPOX, 7-K	5.8–12.9
Bacon	Fried	7α-OH, 7β-OH, 7-K, α-EPOX, 25-OH	0.2–0.5
Tallow	Heating (135–180°C)	α-EPOX, 7-K	9.1–43.7
Lard	Refined, deodorized	7α-OH, 7-K, α-EPOX, 20α-OH, 25-OH	Trace–0.4
		Other Products	
French fries	Commercial	7α-OH, 7β-OH, α-EPOX, β-EPOX, 25-OH, 7-K, TRIOL	1–17

Source: Modified from Tai, C.-Y., Chen, Y.C., and Chen, B.H. (2000), Analysis, formation and inhibition of cholesterol oxidation products in foods: an overview (part II), *Journal of Food and Drug Analysis*, 8: 1–15.

Very low levels of oxysterols have been detected in freshly processed meats (Park and Addis, 1987; Monahan et al., 1992), spray-dried eggs (Tsai and Hudson, 1984), and milk powder (Nourooz-Zadeh and Appelqvist, 1988). The inconsistency with which oxysterols have been detected in these materials depends, at least in part, on the different sensitivities of the methods employed.

Unlike fresh or raw foods, those foods undergoing preparation or storage procedures that favor the autoxidation of cholesterol may show decidedly high levels of cholesterol oxides. Still, one should take into account that different cholesterol-rich foods having the same cholesterol content may yield significantly different amounts of oxysterols. For instance, in spite of their very high cholesterol content, eggs appear relatively well protected from oxidation because the lipids are chiefly located in the yolk, which is rich in natural antioxidants (tocopherols, carotenoids) and in metal ion chelators (phosvitin), and is surrounded by the albumen and shell, which provide further protection against atmospheric oxygen (Galobart and Guardiola, 2002).

Seafoods such as fish roes, herring, squid, and prawn are reported to be very rich in cholesterol. In addition, seafoods contain a wide variety of polyunsaturated fatty acids, making them particularly susceptible to oxidation even under mild environmental pro-oxidant conditions. This is particularly true of seafoods with high fat content. Cholesterol oxidation is greatly accelerated by the oxidation of coexisting highly unsaturated lipids (Kim and Nawar, 1991; Galobart et al., 2002). In the case of meat, mincing favors the incorporation of oxygen in the tissue and the disruption of cell barriers, enabling the interaction of membrane lipids with cytosolic pro-oxidants (e.g., oxymyoglobin) and eventually causing generation of hydrogen peroxide, superoxide anion, and hydroxyl radicals (Kerry et al., 2002). Of note, heat treatment of dairy products increases their resistance to oxidative damage by activating the β-lactoglobulin sulphydryl groups (Early, 1992). Because of the different susceptibilities of cholesterol-rich foods to oxidation, the methods employed in food processing may be crucial.

There is no doubt that exposure to high-energy radiation such as UV light and γ-radiation markedly increases oxysterol content in foods through intense generation of various reactive oxygen species (Galobart and Guardiola, 2002). Irradiation is in particularly used to kill naturally occurring microorganisms in shell egg and egg products. For instance, γ-irradiation of spray-dried eggs has been found to increase cholesterol oxide formation in direct correlation with the radiation dosage employed (Du and Ahn, 2000). Regarding UV-irradiation, this has been found to increase oxysterol content of egg yolk powder by 40 to 50 times (van de Bovenkamp et al., 1988). Furthermore, high levels of 7-ketocholesterol, one of the most toxic oxysterols, have been found in Italian salami, mortadella, and cooked ham after UV light treatment (Savage et al., 2002).

Another procedure that greatly accelerates cholesterol oxidation in foodstuffs is heating. Thermal degradation of cholesterol causes all the major oxysterols of pathophysiological interest to form, via mechanisms that may or may not be mediated by free radicals (Lerker and Rodriguez-Estrada, 2002). Of note, heating, as well as UV-irradiation, particularly favor the production of 7-ketocholesterol. Different cooking methods for beef hamburgers were recently compared with regard to cholesterol oxidation. The highest 7-ketocholesterol production was found with combined roasting and microwave heating; this was followed by microwaving alone, barbecuing, frying, and roasting. Boiling was relatively the safest procedure (Rodriguez-Estrada et al., 1997). A peculiar condition may occur during frying if the oil is repeatedly exposed to high temperature for a large number of cycles. If this procedure is used, besides enhancing the oxidation of any cholesterol present in foods, cholesterol-free foods are also supplemented with considerable amounts of oxysterols (Guardiola et al., 1995b; Tai et al., 2000).

The oxidation rate may also be facilitated in the presence of light. The qualitative and quantitative light-dependent production of oxidant species depends on light source and intensity, and illumination time and mode (Maerker, 1987). Photooxidation of cholesterol mainly involves singlet oxygen and is catalyzed by natural pigments such as chlorophylls, flavin, and myoglobin (Lerker and Rodriguez-Estrada, 2002).

The pro-oxidant species hydrogen peroxide is utilized to avoid browning of whole eggs or dried yolk (Guardiola et al., 2002) and, together with benzoyl peroxide, to improve the color of milk products (bleaching) (Nielsen et al., 1996). Food supplementation with hydroperoxides simply increases the quantitative yield of reactive oxidants deriving from heat-dependent breakdown of peroxides (Halliwell et al., 1995).

Food storage most likely represents the crucial step in foodstuff preparation. Inappropriate conservation of cholesterol-containing foods may lead to lipid oxidation and oxysterol generation, even in large quantities. Although few studies have addressed oxysterol production in stored food, there are some indirect indications that control of temperature, water content, and partial oxygen pressure, as well as light restriction, are critical and essential procedures (Paniangvait et al., 1995; Rodriguez-Estrada et al., 1997; Tai et al., 2000).

In conclusion, cholesterol oxidation products may be originally present in foods, but they are primarily formed during the various steps from the selection of fresh food to its final preparation for consumption. Prevention of cholesterol oxidation in foods thus becomes important and could be achieved in various ways. One is by feeding animals a diet containing minimal amounts of oxysterols and supplementing it with antioxidants (Galvin et al., 1998; Decker and Xu, 1998; Morrissey et al., 1998). A complementary approach is to add antioxidants to foods — of course, only when permitted, before processing and storage. A vitamin E-supplemented diet given to rainbow trout (Akhtar et al., 1998), poultry (Li et al., 1996; Galvin et al., 1997, 1998; Maraschiello et al., 1999), pigs (Monahan et al., 1992), and cattle (Buckley et al., 1995) has been shown to prolong the shelf-life of the derived food products. The protective effects of antioxidants against food cholesterol oxidation, as induced by oxidants such as hydrogen peroxide and nitric oxide, or by pro-oxidants such as metal ions, have been variously reported (Huber et al., 1995; Madhavi et al., 1995; Guardiola et al., 1997; Shozen et al., 1997).

Some other additives, not true antioxidants, have been proposed to reduce the cholesterol oxidative impact in food technology. Salt can be used as an effective inhibitor of oxysterols in butteroil (Sander et al., 1989) because it reduces water content and thus slows cholesterol oxidation. Osada and colleagues (2000) reported that cholesterol oxidation in sausages was inhibited by the addition of sodium nitrite in a dose-dependent manner, which inactivates superoxide anion and stabilizes polyunsaturated acids. Storage at low temperature and in the dark, and the use of specific packaging materials or procedures (vacuum packaging) may further contribute to minimize oxysterol formation during storage (Du and Ahn, 2000; Savage et al., 2002).

PROAPOPTOTIC EFFECT OF OXYSTEROLS

Oxysterols, as with the other oxidation products of the lipid moiety of plasma LDL, are consistently found within the walls of major arteries, mainly in the characteristic lesions of atherosclerosis. For this reason, most toxicology data about oxysterols currently available focus on their effects on vascular cells. This does not mean that oxysterols only accumulate in major arteries; recent evidence shows increased formation of cholesterol oxides and cholesterol hydroperoxides in the liver of rats under chronic ethanol treatment (Ariyoshi et al., 2002). It has also been shown that patients with nonalcoholic steatohepatitis (NASH) often have hypercholesterolemia (Reid, 2001). It thus appears likely that accumulation of oxysterols in the liver and possibly in other organs may occur through phagocytosis by local macrophages. This review, therefore, briefly covers the extravascular toxicity of cholesterol oxides. It is already apparent that only in the case of fibroblasts, hepatocytes, and colonic cells do oxysterols appear to kill cells mainly by a necrogenic mechanism, whereas for all other cell types considered thus far, apoptosis appears to be the prevalent mechanism of death, at least for oxysterol concentrations of pathophysiological interest. Furthermore, reports now available are delucidating the toxic effects of the main oxysterols.

While reviewing the recent literature on the cytotoxicity of oxygenated derivatives of cholesterol it became clear that practically all toxicologic studies have been carried out on the major oxysterols taken individually, whereas in biology they always occur in mixtures. In addition, nonoxidized cholesterol had seldom been used for comparative analysis. The review, therefore, also addresses the question of whether oxysterols in mixture may have a different toxicological impact than the individual compounds; particular attention was paid to reports on oxysterols, which also included comparative analyses using the parent compound.

Effect on Vascular Cells

Over the last decade, reliable *in vitro* studies have characterized the potential proapoptotic effect of the major oxysterols with regard to vascular cells, namely smooth muscle cells (Nishio and Watanabe, 1996; Lizard et al., 1999), endothelial cells (Lemaire et al., 1998; Lizard et al., 1999), fibroblasts, and monocyte-macrophages (Christ et al., 1993; Aupeix et al., 1995; Lizard et al., 1999; O'Callaghan et al., 1999).

Nishio and Watanabe (1996) showed that both 7-ketocholesterol and 25-hydroxycholesterol (30 *μM*) induce apoptosis in rabbit cultured smooth muscle cells (SMCs), likely through down regulation of the antiapoptotic factor Bcl-2. Lizard and colleagues confirmed the effect of 7-ketocholesterol, using cultured SMCs obtained from human artery, and also tested 7β-hydroxycholesterol. They found 7β-hydroxycholesterol to be toxic at final concentrations of 20 to 30 *μM* or above (25 to 30% of cells were apoptotic after 48 h incubation), whereas 7-ketocholesterol showed clear proapoptotic effect at 40 to 60 *μM* final concentration (20 to 25% of cells were apoptotic after 48 h incubation) (Lizard et al., 1999). When added to confluent cultures of human umbilical vein endothelial cells (HUVECs), the toxicity of the two compounds was confirmed, as where the concentrations (Lizard et al., 1999). Of note, the same group previously demonstrated that other oxysterols of pathobiological interest induced apoptosis in HUVECs, namely 25-hydroxycholesterol and 5α,6α-epoxycholesterol (Lizard et al., 1996; Lemaire et al., 1998). As far as cells of the macrophage linage are concerned, Aupeix and colleagues showed that apoptosis occurred shortly after treatment of human monocytic cell lines with either 25-hydroxycholesterol or 7β-hydroxycholesterol, at final concentrations of 20 to

30 μM or above (Aupeix et al., 1995). These and other cholesterol oxides, namely, 7α-hydroxycholesterol, 7β-hydroxycholesterol, and 27-hydroxycholesterol, undoubtedly become necrogenic, at least towards cultivated macrophages, when added at final concentrations above 50 μM (Clare et al., 1995). Our own findings, treating J774.A1 murine macrophagic cells in culture with either 7-ketocholesterol or a biologically representative oxysterol mixture (Leonarduzzi et al., 2001), also indicate that specific concentrations of oxysterols lead vascular cells to apoptosis, whereas relatively higher amounts induce straight necrosis. The only apparent exception was the behavior of fibroblasts of vascular origin, which also underwent necrosis in the presence of oxysterol concentrations (20 to 50 μM) exerting proapoptotic effect on the other cell types present in the arterial wall (Lizard et al., 1999).

Effect on Nonvascular Cells

Now, moving on to the potential toxicity of oxysterols in tissues other than the arterial wall, the hypothesis, first proposed by Schaffner's group in 1990, that oxidative cholesterol metabolism and oxysterols may be important factors in the development of alcoholic liver disease is of interest (Ryzlak et al., 1990). Cholesterol oxidation products may actually contribute to the progression of chronic liver disease, including that not of alcoholic origin, providing sustained proinflammatory stimuli through their proapoptotic and pronecrogenic effects, exactly as occurs in the case of the central necrotic region of the atherosclerotic plaque. Of note in chronically ethanol-fed rats, an increased steady-state concentration of cholesterol oxides was recently reported not only in the liver but also in extrahepatic tissues. In a study into the potential generation of cardiotoxic products of cholesterol oxidation in the rat heart following chronic alcohol feeding (6 weeks), Preedy's group demonstrated a marked increase of 7β-hydroxycholesterol in addition to 7α- and 7β-hydroperoxycholesterol in treated vs. untreated animals, whereas 7α-hydroxycholesterol and 7-ketocholesterol did not differ significantly (Adachi et al., 2001). The same group subsequently measured cholesterol hydroperoxides and oxides in the skeletal muscles of chronically ethanol-fed rats (4 to 6 weeks), showing a significant ethanol-induced increase of 7β-hydroxycholesterol both in the soleus and in the plantaris muscles, whereas 7α-hydroxycholesterol and 7-ketocholesterol were increased only in the soleus (Fujita et al., 2002).

The levels of the three 7-oxidized oxysterols were also found to be increased in synaptosomes and mitochondria isolated from rat brain, and challenged with Fe^{2+} and ascorbate as pro-oxidant stimulus. This suggests that oxidation of cholesterol could contribute to brain tissue damage under a condition of oxidative stress (Vatassery et al., 1997b). Further support for the hypothesis that oxysterols are involved in the pathogenesis of brain diseases came from the observation of neurotoxic effect of 24-hydroxycholesterol on human neuroblastoma cells; increased generation of this compound appears to be associated to neuronal cell death (Kolsch et al., 1999). A promising area for further research would appear to be the role of cholesterol oxides in mitochondrial aging and dysfunction, as well as in the pathogenesis of neurodegenerative disorders such as Alzheimer's disease. In brain tissue, on the contrary, increased levels of oxysterols are more probably the consequence of *in situ* enhanced generation (e.g., under oxidative stress) rather than of hematic derivation. Deposition of oxysterols in certain other organs may also rely on hypercholesterolemia. This definitely appears to be the case for accumulation of oxysterols in the retina, gallbladder, and liver. Hypercholesterolemia is considered a primary risk factor for retinal degenerative diseases. Initial experimental evidence has just been obtained on both rat and human retinal cell lines challenged with either 7-ketocholesterol or 25-hydroxycholesterol (final concentration of about 60 μM). The former was more toxic, with a clear proapoptotic effect associated to a marked generation of reactive oxygen species (Ong et al., 2003). The toxicity of 7-ketocholesterol has, on the contrary, been shown to be relatively low in a cultured dog gallbladder epithelial cell experimental model, in which cholestan-3β,5α,6β-triol exerted overt cytotoxicity mainly through apoptosis and markedly depressed mucine secretion (Yoshida et al., 2000). Of interest, both toxicity and functional impairment induced by the triol were significantly reduced when the hydrophilic bile acid tauroursodeoxycholic acid was added to the culture medium (Yoshida et al., 2001). As far as the liver is concerned, besides the recent evidence of increased oxysterol levels in rats chronically fed with ethanol (Ariyoshi et al., 2002), few data are available about the cytotoxicity of specific cholesterol oxides on human transformed, but still differentiated, hepatocytes in culture (HepG2 cell line). O'Brien and colleagues first showed that 30 μM 7β-hydroxycholesterol or 25-hydroxycholesterol induced HepG2 cell death mainly through a necrogenic mechanism (O'Callaghan et al., 1999). They then enlarged the spectrum of oxysterols considered and extended their observations to human colonic adenocarcinoma cells (Caco-2), and also found evidence of cytotoxicity, in terms of neutral red uptake assay, for 5β,6β-epoxycholesterol and 5α,6α-epoxycholesterol (similar concentrations) (O'Sullivan et al., 2003). Finally, little evidence is thus far available regarding the toxicity of oxysterols on T-lymphocytes. Christ and colleagues showed that 25-hydroxycholesterol triggered the apoptotic program in RDM4 murine lymphoma cells but also in normal murine thymocytes (Christ et al., 1993). The inhibition of T-cells cytolytic activity by 25-hydroxycholesterol was confirmed in a different mouse model a year later (Kucuk et al., 1994). It has recently been reported that oxidized ghee, a butter product common in the Indian diet, induced impairment of proliferation both of splenocytes and of peripheral blood lymphocytes from challenged rats, most likely because of its high oxysterol content (Niranjan and Krishnakantha, 2001).

Mechanism of Induction of Apoptosis by Oxysterols

As recently reviewed by our group (Leonarduzzi et al., 2002), the current knowledge of the pathomechanisms through which oxysterols may exert proapoptotic effect still appears incomplete. The downregulation of the antiapoptotic protein Bcl-2 that is induced by 7β-hydroxycholesterol, 25-hydroxycholesterol, and 7-ketocholesterol in different cell models, with consequent activation of the mitochondrial pathway of programed cell death, has been variously reported (Nishio and Watanabe, 1996; Lizard et al., 1997; Harada-Shiba et al., 1998). In addition, evidence has also been provided of the overexpression of programmed death mediators Fas and Fas ligand in vascular smooth muscle cells incubated with 25-hydroxycholesterol or 7β-hydroxycholesterol (Lee and Chau, 2001). In parallel, we found that amounts of 7-ketocholesterol actually detectable *in vivo* upregulated p21 expression and synthesis in J774.A1 cells (Biasi et al., 2004). However, the upstream reactions leading to altered modulation of these and other pro- and antiapoptotic factors have yet to be clarified.

It is now possible to make a reasonable reconstruction of a hypothetical oxysterol-dependent signaling pathway to apoptosis on the basis of the available data. First, it is now recognized that various oxysterols of pathophysiological interest upregulate intracellular steady-state levels of reactive oxygen species (ROS), essentially upregulating NADPH oxidase (Rosenblat et al., 1999; Rosenblat and Aviram, 2002; Biasi et al., 2004). Indirect confirmation of a primary role of oxidative imbalance of the cell redox equilibrium in oxysterol-induced apoptosis comes from the significant protection against this event that occurs when cells are pretreated with antioxidants (Lee and Chau, 2001) or with selective NADPH oxidase inhibitors (Biasi et al., 2004). The steps between ROS overproduction and changes in the expression of genes related to apoptosis remain to be clarified. A significant advance on this matter has been made by the demonstration that the mitogen-activated protein kinases (MAPKs) pathway may be involved. Ares and colleagues (2000) showed that in human aortic smooth muscle cells treated with 7β-hydroxycholesterol extracellular signal-regulated kinase 1 (ERK1) and 2 (ERK2) were rapidly activated, after triggering intracellular homeostasis by perturbation of Ca^{2+}. On the contrary, Jun N-terminal kinases (JNKs) were not activated by 7β-hydroxycholesterol.

The Cytotoxicity of Oxysterols Is Quenched When They Are in a Mixture

A very interesting point has recently been addressed, namely, the relatively lower cytotoxicity of oxysterols when given to cell models as a mixture rather than as individual compounds. Naturally, oxysterols are always present as a mixture, whether in foods, oxidized LDL, or in the core region of atherosclerotic plaque, and molecular interactions often occur among mixed compounds. Using cells of the macrophage lineage, Leonarduzzi and colleagues analyzed the profibrogenic effect both of 7-ketocholesterol and of a biologically representative mixture of oxysterols in a percentage composition consistent with that found in the plasma LDL of hypercholesterolemic patients. Within a concentration range of pathophysiological interest (10 to 30 μM), the oxysterol mixture was markedly profibrogenic, but an equimolar amount of 7-ketocholesterol was not (Leonarduzzi et al., 2001). The reason for such different behavior turned out to be correlated to the much higher toxicity, measured in terms of oligonucleosomal DNA fragments, displayed by 7-ketocholesterol when given individually. Starting from this evidence, we analyzed the effect of 7-ketocholesterol on murine J774.A1 macrophages, in the presence or in the absence of equimolar concentrations of 7β-hydroxycholesterol, in terms of ROS generation, cytochrome c release from mitochondria, caspase-3 activation, p21 upregulation, and morphological appearance of condensed nuclei and apoptotic bodies. All events along the

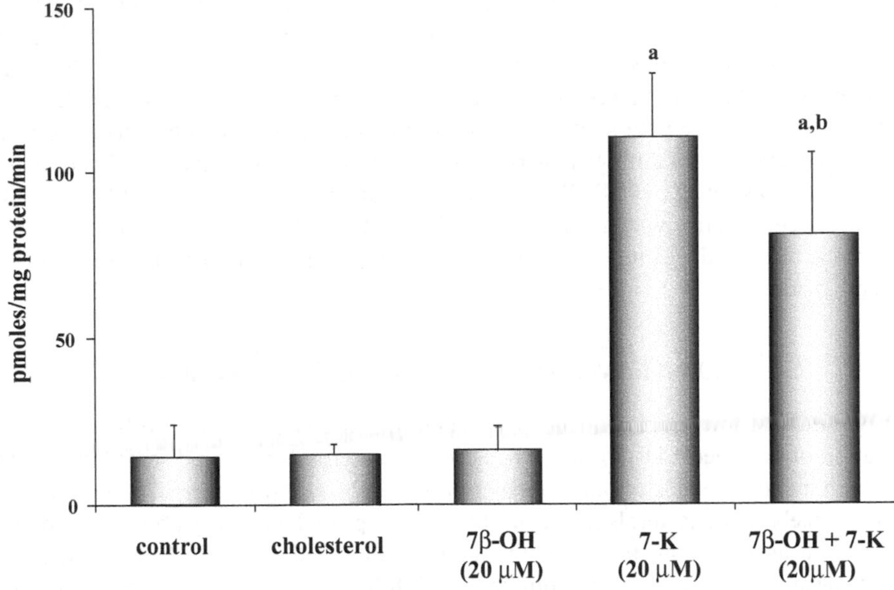

FIGURE 5.4 Caspase-3 activity in J774.A1 macrophages treated for 24 h with 20 *μM* cholesterol, 20 *μM* 7-ketocholesterol (7-K), 20 *μM* 7β-hydroxycholesterol (7β-OH), or 7-ketocholesterol + 7β-hydroxycholesterol at equimolar concentration of 20 *μM*. Caspase activity is expressed as pmoles of fluorescent 7-amino-4-methylcumarin (AMC) released/mg cell protein/min. Data given are means of five experiments ± SD a: significantly different vs. control ($p < .05$); b: significantly different vs. 7-ketocholesterol ($p < .05$).

mitochondrial apoptotic pathway triggered by 7-ketocholesterol were significantly quenched when cells were cotreated with identical amounts of the second sterol oxide (Biasi et al., 2004). Figure 5.4 shows a typical experiment on J774.A1 macrophages challenged with nonoxidized cholesterol, 7β-hydroxycholesterol, or 7-ketocholesterol alone, and 7β-hydroxycholesterol plus 7-ketocholesterol, considering the activity of caspase-3, the executing protease in the apoptotic cascade. Although neither the parent sterol nor 7β-hydroxycholesterol enhanced the constitutive activity of the enzyme, this was strongly potentiated when cells were incubated with 7-ketocholesterol alone. The simultaneous cotreatment with 7β-hydroxycholesterol led to a significant reduction of caspase-3 activity.

As regards the quenching effect occurring among oxysterols in mixture, as far as we are aware, it has only been reported by Aupeix and colleagues (1995), who compared the cytotoxic effect of 7β-hydroxycholesterol and 25-hydroxycholesterol in the U937 human promonocytic cell line. Concentrations above those used in our model of differentiated macrophages (30 to 40 *μM* 7β-hydroxycholesterol) induced apoptosis in the promonocytic cell culture, but the concomitant addition of identical amounts of 25-hydroxycholesterol significantly reduced DNA fragmentation. However, no explanation for this striking effect was given. In our experiments, we also found that the simultaneous addition of oxysterols, namely 7-ketocholesterol and 7β-hydroxycholesterol, appears not to interfere with the uptake of the two oxysterols by cells in which the already constitutively expressed CD36 scavenger receptor is probably involved (Han et al., 1997). In our opinion, the quenching effect on 7-ketocholesterol-induced apoptosis that we observed occurred chiefly at the level of ROS production. Indeed, the concomitant addition of 7β-hydroxycholesterol strongly inhibited a pathobiological rise of intracellular ROS that had been induced by 7-ketocholesterol through a marked upregulation of constitutive NADPH oxidase activity (Biasi et al., 2004). We hypothesize that the concomitant addition of 7β-hydroxycholesterol, which also binds to NADPH oxidase but apparently less efficiently, likely reduces the concentration of free enzyme available for 7-ketocholesterol binding.

PROINFLAMMATORY EFFECT OF OXYSTEROLS

Over the last few years, evidence has accumulated concerning the probable contribution of oxidized LDL to the progression of atherosclerosis, a disease process in which inflammation is certainly a major promoting factor (Steinberg, 2002). It seems quite likely that oxysterols may contribute to the proinflammatory effect of oxidized LDL because of both their quantitative relevance and their biochemical activity. Increasing numbers of reports point to the modulation of proinflammatory molecules by cholesterol oxidation products accumulating within human fibrotic plaques.

With regard to SMCs, 25-hydroxycholesterol was first shown to increase both mRNA levels and synthesis of basic fibroblast growth factor (bFGF), a cytokine with potent mitogenic and fibrogenic activities (Kraemer et al., 1993). It was then shown to upregulate prostaglandin G/H synthase-2 and, consequently, prostaglandin production (Wohlfeil and Campbell, 1999). The same compound strongly induced expression and synthesis of interleukin-1β (IL-1β) in human monocyte-derived macrophages at a final concentration of about 5 *μM* (Rosklint et al., 2002), as well as expression of the group IIA secretory phospholipase A2 gene, again in SMCs (Antonio et al., 2003).

In relation to 7-oxidized oxysterols, the treatment of HUVECs with 7-ketocholesterol, 7α-hydroxycholesterol, and especially with 7β-hydroxycholesterol induced a 10- to 20-fold increase of IL-1β secretion. Further, all three oxysterols induced the expression of vascular cell adhesion molecule-1 (VCAM-1), intercellular adhesion molecule-1 (ICAM-1), and endothelial-selectin (E-selectin) in HUVECs (Lemaire et al., 1998). Another very interesting report on the proinflammatory action of oxysterols showed that many of the cholesterol oxides that can be demonstrated in atherosclerotic human aorta upregulate the production of the key chemokine interleukin-8 (IL-8) by adherent human blood monocytes; IL-8 is active on the migration of neutrophils, lymphocytes, and SMCs. In order of decreasing potency, 25-hydroxycholesterol, cholestan-3β,5α,6β-triol, 24-hydroxycholesterol, 7β-hydroxycholesterol, and 7-ketocholesterol, all significantly enhanced IL-8 synthesis by cells challenged with the same concentration (approximately 10 to 12 *μM*) for 20 h. On the contrary, nonoxidized cholesterol as well as 5α,6α-epoxycholesterol did not show any significant effect (Liu et al., 1997).

Recently, our laboratory has also become involved in investigating the potential proinflammatory action of oxysterols found in human oxidized LDL. As reported in the preceding text, we focused on the expression of the inflammatory cytokine transforming growth factor β1 (TGFβ1), the cytokine having the strongest profibrogenic effect (Ignotz and Massagué, 1986). An oxysterol mixture compatible with that detectable in human hypercholesterolemic plasma, unlike equimolar concentrations of 7-ketocholesterol, markedly upregulated TGFβ1 expression and synthesis in cells of the macrophage lineage (Leonarduzzi et al., 2001). Subsequently, using a human promonocytic cell line (U937) challenged with an oxysterol mixture of pathophysiological interest, pattern-specific gene expression profiling was performed. Preliminary findings appear to indicate that the same oxysterol mixture that was effective with TGFβ1 also upregulates the steady-state levels of messenger RNA specific for a small number of chemokines, in particular that of monocyte chemotactic protein 1 (MCP-1). Of note, knocking out the MCP-1 gene reduces lesion size in murine models of atherosclerosis (Gu et al., 1998), pointing to a leading role for monocyte recruitment in relatively early phases of atherosclerosis. We then performed immunoenzymatic analysis to confirm the gene array data regarding MCP-1 upregulation by the oxysterol mixture. The amount of chemokine actually synthesized by macrophages *in vitro* challenged with the mixture was indeed found to be significantly higher than in control cells. On the contrary, MCP-1 levels in the cell samples treated with equimolar concentrations of 7-ketocholesterol or nonoxidized cholesterol did not show any variation compared to untreated cells (Figure 5.5). Another crucial effect induced by the oxysterol mixture on the monocyte model employed was the overexpression of specific molecules involved in macrophage

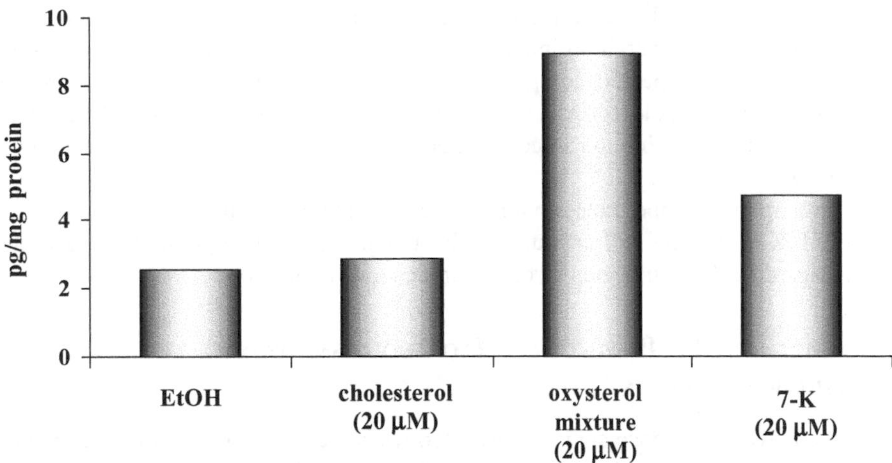

FIGURE 5.5 MCP-1 synthesis in U937 human promonocytic cells. Cells were incubated for 6 h with cholesterol, oxysterol mixture, or 7-ketocholesterol (7-K) at a final concentration of 20 *µM*. Chemokine levels were measured by enzyme-linked immunoassay (ELISA).

adhesion and trapping within the subendothelial space, namely β1-integrin and CD36 scavenger receptor (preliminary data not reported here). A lower but significant overexpression of these two genes was also evident when mononuclear cells were treated with 7-ketocholesterol (not shown). Husemann and colleagues (2001) recently demonstrated that integrins promote adhesion of mononuclear cells to matrixes containing oxidized LDL, also favoring the production of ROS by these adherent cells. Further, CD36 is only expressed when monocytic cells, already differentiated into macrophages, become activated. CD36 expression can undoubtedly be considered as a selective marker of macrophage differentiation and activation.

Thus, the apparent ability of oxysterols to favor the migration of monocytic cells, their differentiation into macrophages, and their adhesion to vascular cells points to a probably significant contribution of these products to the promotion and progression of atherosclerotic lesions within arteries.

Mechanisms Underlying the Proinflammatory Effect of Oxysterols

Very little evidence is available thus far to elucidate the mechanisms by which cholesterol oxides contribute to the progression of atherosclerotic lesions. The only established step is the ability of various oxysterols to upregulate steady-state levels of ROS in target cells by stimulating NADPH oxidase activity (Rosenblat et al., 1999; Rosenblat and Aviram, 2002). Very recent data from our laboratory has demonstrated the inhibition of oxysterol-induced MCP-1 overexpression when macrophages were pretreated with inhibitors of classic protein kinase C isoform. These isoenzymes are recognized to enhance cellular ROS production through NADPH oxidase upregulation. However, there is as yet no data to explain how oxysterol-dependent ROS production may induce cells to express proinflammatory rather than proapoptotic stimuli. In our opinion, the natural occurrence of oxysterols as a mixture would first favor an overall proinflammatory action, partly because of the quenching of the toxicity of specific components, as reported above. In the long run, the concentration threshold for toxic compounds such as 7-ketocholesterol might be overcome, probably through selective metabolism, and then suicide or necrogenic signals would prevail.

Investigation of the transcription pathways effectively stimulated by oxysterol-induced ROS increase will undoubtedly be a major research target in the near future with regard to the pathogenetic role of cholesterol in atherosclerosis.

Antioxidants May Prevent Both Exogenous and Endogenous Oxysterol Generation

A large body of evidence supports the hypothesis that free radical–mediated oxidative processes, particularly cholesterol oxidation and the consequent production of oxysterols, play an important role in atherogenesis. It would, therefore, be useful to inhibit the exogenous generation of oxysterols in cholesterol-containing food by supplementation with antioxidants. Park and Addis (1986) studied the inhibitory effect of vitamin E and ascorbyl palmitate on cholesterol oxidation during tallow heating. Both antioxidants were able to inhibit cholesterol oxidation up to 135°C, but not at higher temperatures, because they underwent degradation. Several other studies on different model systems of food handling and treatment have used various antioxidants as preventive agents (Morgan and Armstrong, 1987; Rankin and Pike, 1993; Huber et al., 1995; Madhavi et al., 1995; Guardiola et al., 1997). The overall finding was that inhibition of oxysterol generation is good, in particular when α-tocopherol, BHT, and propyl gallate are added to foods (Morgan and Armstrong, 1987; Rankin and Pike, 1993; Huber et al., 1995; Madhavi et al., 1995; Guardiola et al., 1997; Zanardi, 2000). Further, the natural presence of antioxidants in foods and dietary ingredients, such as red wine, olive oil, green tea, and licorice, might provide the necessary antioxidant resources for the body to control endogenous oxidation reactions, thus preventing or quenching the generation of products deleterious for the vascular system (Visioli et al., 1995; Hayek et al., 1997; Nicolaiew et al., 1998; Weisburger, 1999). Regarding the endogenous formation of oxysterols, it appears likely that nonenzymatic routes, essentially involving ROS, play a significant role in the body's production of oxysterols. There is increasing experimental evidence that the endogenous formation of oxysterols is influenced by the type of dietary fat and the ingestion of antioxidants (Hodis et al., 1992; Addis et al., 1995; Mahfouz et al., 1997; Mol et al., 1997; Porkkala-Sarataho et al., 2000; Grau et al., 2001; Brandsch et al., 2002). For example, in rats fed with salmon oil, insufficient vitamin E in the diet led to increased generation of 7β-hydroxycholesterol; an adequate vitamin supply, on the contrary, reduced the concentration of LDL oxysterols (Ringeseis and Eder, 2002). Plasma and aortic oxysterol levels (5α,6α-epoxycholesterol, 5β,6β-epoxycholesterol, 7α-hydroxycholesterol, and 7β-hydroxycholesterol) were reduced, independently of plasma cholesterol, in hypercholesterolemic (Hodis et al., 1992) or hypertensive (Hodis et al., 2000) rabbits when they received a 1% probucol supplemented diet for 9 or 12 weeks, respectively.

In vitro experiments support the *in vivo* beneficial effects of antioxidant administration. Oxysterol concentrations, analyzed by gas–liquid chromatography, were lower when bovine aortic smooth muscle cells and human promonocytic cell line U937 were maintained in a vitamin E-enriched medium (Pie and Seillan, 1992). Lizard's group (Lemaire et al., 1998) has observed a decrease in glutathione (GSH), a cellular antioxidant present in abundance, during 7-ketocholesterol-induced apoptosis in U937 cells. Addition of GSH or its precursor *N*-acetylcysteine (NAC) to the culture medium protected the cells against 7-ketocholesterol-induced ROS increase and programmed cell death. Vitamin E has been reported to prevent 7-ketocholesterol-induced apoptosis, using U937 cells (Lizard et al., 2000), smooth muscle cells (Lee and Chau, 2001), and endothelial cells (Uemura et al., 2002).

Further, Chang and colleagues (1998) investigated the effect of 25-hydroxycholesterol (25 μM) in neuronal PC12 cells and demonstrated a protective action by vitamin E, but not by vitamin C. In another study, physiological concentrations of vitamin E (50 to 100 μM) or estradiol-17-β (1 to 100 nM) partially prevented the proapoptotic effect of 24-hydroxycholesterol on differentiated human neuroblastoma cell line, SH-SY5Y, but again vitamin C was not effective (Kolsch et al., 2001). This observation could be extremely important because 24-hydroxycholesterol, the main cholesterol elimination product of the brain, has been shown to be increased in the serum of Alzheimer's patients (Kolsch et al., 2001).

The effect of certain antioxidants — vitamin E, in particular — has been studied in human clinical trials (Heinecke, 2001). The evidence is controversial. It may be that animal studies of antioxidants have focused on early events in atherosclerosis, whereas the human clinical trials involved individuals with advanced disease. However, Mol and colleagues (1997) found that cholestan-3β,5α,6β-triol, 7-ketocholesterol, and 7β-hydroxycholesterol were decreased in diabetic subjects and that 5α,6α-epoxycholesterol was decreased in smokers, both when supplemented with vitamin E (600 IU/d for 4 weeks). Such findings are closely correlated with two secondary prevention studies: one by Stephens and colleagues (1996) in which an average daily supplementation of 600 IU of vitamin E significantly reduced the risk of cardiovascular complications; the other by Boaz and colleagues (2000) in which hemodialysis patients with preexisting cardiovascular disease, randomized to receive 800 IU/d vitamin E, showed a significant reduction of cardiovascular disease end points and myocardial infarction. However, other recent randomized, placebo-controlled clinical trials have not substantiated a beneficial role for vitamin E consumption in retarding atherosclerotic events (Anonymous, 1999; Yusuf et al., 2000). The cause of these conflicting results is not clear. Many factors might play key roles — for instance, the fact that studies were limited to diseased individuals, the daily dose of vitamin used, or the interference of food intake with the bioavailability of vitamin E, a highly lipophilic molecule (Iuliano et al., 2001).

Prevention studies including subjects initially free of clinical signs of disease are now under way. Contrary to clinical trials, this type of investigation has a better chance of demonstrating a beneficial effect of antioxidants, given individually or in mixture, against both the onset and the progression of atherosclerosis.

In an already published prevention study, Salonen's group (Porkkala-Sarataho et al., 2000) observed the long-term effects (12 and 36 months) of vitamins E and C, in individual or combined supplementation, on serum cholesterol oxidation products and on *in vivo* and *in vitro* lipid oxidation resistance in nondepleted men. Supplementation with moderate doses of vitamin E reduced the oxidation of isolated atherogenic lipoproteins. Vitamin C had no effect, a plausible explanation being that, during the lipoprotein separation process, the vitamin is eliminated. The combined supplementation with both vitamins increased the resistance to oxidation of total serum lipids more efficiently than supplementation with vitamin E or with vitamin C alone. The serum level of 7β-hydroxycholesterol was significantly reduced by a combination supplement with vitamins E and C. Moreover, vitamin C supplementation reduced 5α,6α-epoxycholesterol and 5β,6β-epoxycholesterol.

These data show that this combination of vitamins E and C may be beneficial because of possible synergistic interactions. For example, vitamin C can regenerate vitamin E from the α-tocopheroxyl radical.

CONCLUSIONS

The *in vivo* generation of cholesterol oxidation products is both frequent and simple. Because of the several biochemical activities that these compounds may exert, the presence of oxysterols in foodstuffs is now receiving great attention. A further matter for consideration stems from the increasing experimental evidence in support of a primary contribution by oxysterols to the development of atherosclerosis. Oxidation appears essential for the cholesterol molecule to become able to express its proatherogenic potential.

In contrast to the so-called stable atherosclerotic plaque, the unstable variety may become enlarged and lead to vascular complications such as further stenosis of the lumen or plaque rupture and thrombosis. Unstable plaque is characterized by the presence of a sustained inflammatory stimulus, which in turn is maintained by sustained partial cell death. Oxysterols, consistently present in significant amounts in unstable atherosclerotic plaque, most likely provide a primary contribution to the whole process through their pronounced proinflammatory and proapoptotic effects (Figure 5.6).

The proatherogenic effects of oxysterols are difficult if not impossible to quench once these compounds circulate in LDL particles or accumulate in the arterial wall. A promising window of protection from such deleterious events may be envisaged at the level of food preparation and intake.

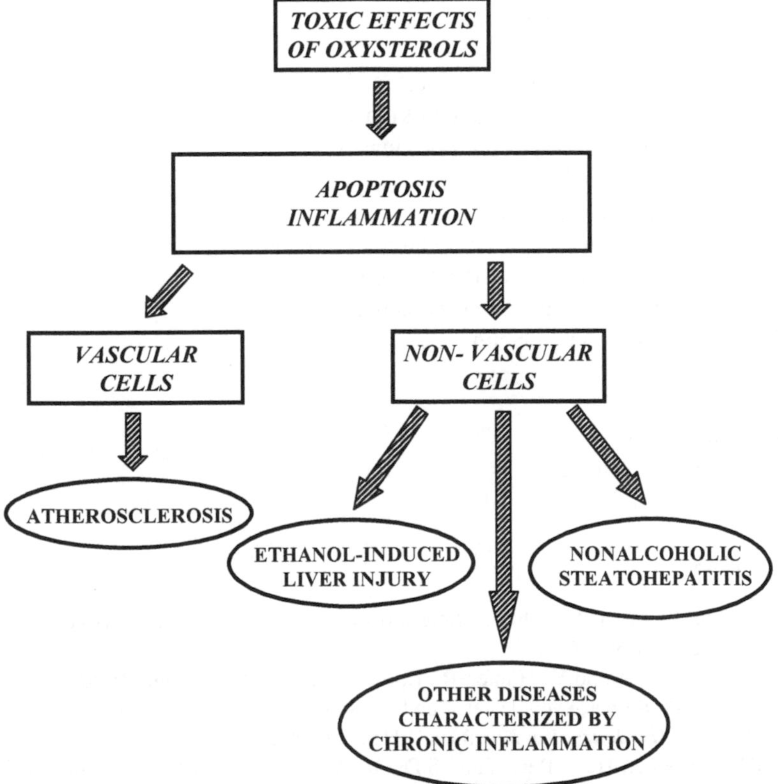

FIGURE 5.6 Oxysterols of pathophysiological interest may contribute to various diseases characterized by chronic inflammation, exerting their toxic effects on both vascular and nonvascular cells.

Finally, oxysterols may, in principle, also exert toxicity toward nonvascular cells (Figure 5.6). They could be involved in the pathogenesis of a number of disease processes characterized and sustained by chronic inflammation. Suitable nutrition may thus be considered in the cotreatment of at least some of these diseases — first of all, in our opinion, nonalcoholic steatohepatitis.

ACKNOWLEDGMENTS

The authors' original data that are reported and discussed in this review were obtained with the support of research grants in Italy from the Ministry for the University, Rome (PRIN, 2002), the Region of Piedmont (finalized projects, 2002; 2003), and the University of Torino.

ABBREVIATIONS

LDL — low density lipoprotein
EDTA — ethylenediaminetetraacetic acid
BHT — butylated hydroxytoluene
ROS — reactive oxygen species
bFGF — basic fibroblast growth factor
ERK — extracellular signal-regulated kinase
MAPKs — mitogen-activated protein kinases
JNKs — Jun N-terminal kinases
SMCs — smooth muscle cells
HUVECs — human umbilical vein endothelial cells
VCAM-1 — vascular cell adhesion molecule-1
ICAM-1 — intercellular adhesion molecule-1
E-selectin — endothelial-selectin
IL-1 — interleukin-1
IL-8 — interleukin-8
GSH — glutathione
NAC — N-acetylcysteine

References

Accad, M. and Farese, R.V. (1998), Cholesterol homeostasis: a role for oxysterols, *Current Biology*, 8: R601–604.

Adachi, J., Kudo, R., Ueno,Y., Hunter, R., Rajendram, R., Want, E., and Preedy, V.R. (2001), Heart 7-hydroperoxycholesterol and oxysterols are elevated in chronically ethanol-fed rats, *The Journal of Nutrition*, 131: 2916–2920.

Addis, P.B., Emanuel, H.A., Bergmann, S.D., and Zavoral, J.H. (1989), Capillary GC quantification of cholesterol oxidation products in plasma lipoproteins of fasted humans, *Free Radical Biology and Medicine*, 7: 179–182.

Addis, P.B., Carr, T.P., Hassel, C.A., Huang, Z.Z., and Warner, G.J. (1995), Atherogenic and antiatherogenic factors in the human diet, *Biochemical Society Symposium*, 61: 259–271.

Akhtar, P., Gray, J.I., Booren, A.M., and Gomaa A. (1998), The effects of dietary α-tocopherol and surface application of oleoresin rosemary on lipid oxidation and cholesterol oxide formation in cooked rainbow trout (*Oncorhynchus mykiss*) muscle, *Journal of Food Lipids*, 5: 59–71.

Anonymous (1999), Dietary supplementation with n-3 polyunsaturated fatty acids vitamin E after myocardial infarction: results of the GISSI-Prevenzione Trial. Gruppo Italiano per lo Studio della Streptochinasi nell'Infarto Miocardico (GISSI), *Lancet*, 354: 447–455.

Ansari, G.A.S. and Smith, L.L. (1979), High-performance liquid chromatography of cholesterol oxidation products, *Journal of Chromatography*, 175: 307–315.

Ansari, G.A.S. and Smith, L.L. (1994), Assay of cholesterol autoxidation, *Methods in Enzymology*, 233: 332–338.

Antonio, V., Janvier, B., Brouillet A., Andreani, M., and Raymondjean, M. (2003), Oxysterol and 9-*cis*-retinoic acid stimulate the group IIA secretory phospholipase A2 gene in rat smooth muscle cells, *The Biochemical Journal*, 376: 351–360.

Ares, M.P., Porn-Ares, M.I., Moses, S., Thyberg, J., Juntti-Berggren, L., Berggren, P., Hultgardh Nilsson, A., Kallin, B., and Nilsson, J. (2000), 7β-hydroxycholesterol induces Ca(2+) oscillations, MAP kinase activation and apoptosis in human aortic smooth muscle cells, *Atherosclerosis*, 153: 23–35.

Ariyoshi, K., Adachi, J., Asano, M., Ueno, Y, Rajendram, R., and Preedy, V.R. (2002), Effect of chronic ethanol feeding on oxysterols in rat liver, *Free Radical Research*, 36: 661–666.

Aupeix, K., Weltin, D., Mejia, J.E., Christ, M., Marchal, J., Freyssinet, J.M., and Bischoff, P. (1995), Oxysterol-induced apoptosis in human monocytic cell lines, *Immunobiology*, 194: 415–428.

Biasi, F., Leonarduzzi, G., Vizio, B., Zanetti, D., Sevanian, A., Sottero, B., Verde, V., Zingaro, B., Chiarpotto, E., and Poli, G. (2004), Oxysterol mixture prevent proapoptotic effects of 7-ketocholesterol in macrophages: implications for proatherogenic gene modulation, *Faseb Journal,* 18: 693–695.

Björkhem, I. (1986), Assay of unesterified 7-oxocholesterol in human serum by isotope dilution-mass spectrometry, *Analytical Biochemistry*, 154: 497–501.

Björkhem, I. (2002), Do oxysterols control cholesterol homeostasis?, *The Journal of Clinical Investigation*, 110: 725–730.

Björkhem, I. and Diczfalusy, U. (2002), Oxysterols: friends, foes, or just fellow passengers?, *Arteriosclerosis, Thrombosis, and Vascular Biology*, 22: 734–742.

Björkhem, I., Andersson, O., Diczfalusy, U., Sevastik, B., Xiu, R.-J., Duan, C., and Lund, E. (1994), Atherosclerosis and Sterol 27-Hydroxylase: Evidence for a Role of This Enzyme in Elimination of Cholesterol from Human Macrophages, *Proceedings of the National Academy of Sciences of the United States of America*, 91: 8592–8596.

Björkhem, I., Diczfalusy, U., and Lütjohann, D. (1999), Removal of cholesterol from extrahepatic sources by oxidative mechanisms, *Current Opinions in Lipidology*, 10: 161–165.

Boaz, M., Smetana, S., Weinstein, T., Matas, Z., Gafter, U., Iaina, A., Knecht, A., Weissgarten, Y., Brunner, D., Fainaru, M., and Green, M.S. (2000), Secondary prevention with antioxidants of cardiovascular disease in endstage renal disease (SPACE): randomised placebo-controlled trial, *Lancet*, 356: 1213–1218.

Boren, J., Olin, K., Lee, I., Chait, A., T.N., Wight, T.N., and Innerarity, T.L. (1998), Identification of the principal proteoglycan-binding site in LDL. A single-point mutation in apo-B100 severely affects proteoglycan interaction without affecting LDL receptor binding, *The Journal of Clinical Investigation*, 101: 2658–2664.

Boselli, E., Velazco, V., Caboni, M.F., and Lercker, G. (2001), Pressurized liquid extraction of lipids for the determination of oxysterols in egg-containing food, *Journal of Chromatography A*, 917: 239–244.

Brandsch, C., Ringseis, R., and Eder, K. (2002), High dietary iron concentrations enhance the formation of cholesterol oxidation products in the liver of adult rats fed salmon oil with minimal effects on antioxidants status, *The Journal of Nutrition*, 132: 2263–2269.

Breuer, O., Dzeletovic, S., Lund, E., and Diczfalusy, U. (1996), The oxysterols cholest-5-ene-3β, 4α-diol, cholest-5-ene-3β, 4β-diol and cholestane-3β, 5α, 6α-triol are formed during *in vitro* oxidation of low density lipoprotein, and are present in human atherosclerotic plaques, *Biochimica et Biophysica Acta*, 1302: 145–152.

Brown, A.J. and Jessup, W. (1999), Oxysterols and atherosclerosis, *Atherosclerosis*, 142: 1–28.

Brown, A.J., Dean, R.T., and Jessup, W. (1996), Free and esterified oxysterols: formation during copper-oxidation of low density lipoprotein and uptake by macrophages, *Journal of Lipid Research*, 37: 320–335.

Brown, A.J., Leong, S., Dean, R.T., and Jessup, W. (1997), 7-Hydroperoxycholesterol and its products in oxidized low density lipoprotein and human atherosclerotic plaque, *Journal of Lipid Research*, 38: 1730–1745.

Buckley, D.T., Morrissey, P.A., and Gray, J.I. (1995), Influence of dietary vitamin E on oxidative stability and qualities of pig meat, *Journal of Food Protection*, 43: 265–267.

Camejo, G., Hurt-Camejo, E., Wiklund, O., and Bondjers, G. (1998), Association of apo B lipoproteins with arterial proteoglycans: pathological significance and molecular basis, *Atherosclerosis*, 139: 205–222.

Carpenter, K.L., Taylor, S.E., Ballantine, J.A., Fussell, B., Halliwell, B., and Mitchinson, M.J. (1993), Lipids and oxidized lipids in human atheroma and normal aorta, *Biochimica et Biophysica Acta*, 1167: 121–130.

Carpenter, K.L., Wilkins, G.M., Fussell, B., Ballantine, J.A., Taylor, S.E., Mitchinson, M.J., and Leake, D.S. (1994), Production of oxidised lipids during modification of low-density lipoprotein by macrophages or copper, *The Biochemical Journal*, 304: 625–633.

Carpenter, K.L., Taylor, S.E., van der Veen, C., Williamson, B.K., Ballantine, J.A., and Mitchinson, M.J. (1995), Lipids and oxidized lipids in human atherosclerotic lesions at different stage of development, *Biochimica et Biophysica Acta*, 1256: 141–150.

Chan, S.H., Gary, J.I., Gomaa, E.A., Harte, B.R., Kelly, P.M., and Buckley, D.J. (1993), Cholesterol oxidation in whole milk powders as influenced by processing and packaging, *Food Chemistry*, 47: 321–328.

Chang, Y.H., Abdalla, D.S., and Sevanian, A. (1997), Characterization of cholesterol oxidation products formed by oxidative modification of low density lipoprotein, *Free Radical Biology and Medicine*, 23: 202–214.

Chang, J.Y., Phelan, K.D., and Liu, L.-Z. (1998), Neurotoxicity of 25-hydroxycholesterol in NGF-differentiated PC12 cells, *Neurochemical Research*, 23: 7–16.

Chen, B.H. and Chen, Y.C. (1994), Evaluation of the analysis of cholesterol oxides by liquid chromatography, *Journal of Chromatography*, 661: 127–36.

Christ, M., Luu, B., Mejia, J.E., Moosbrugger, I., and Bischoff, P. (1993), Apoptosis induced by oxysterols in murine lymphoma cells and in normal thymocytes, *Immunology*, 78: 455–460.

Clare, K., Hardwick, S.J., Carpenter, K.L.H., Weeratunge, N., and Mitchinson, M.J. (1995), Toxicity of oxysterols to human monocyte-macrophages, *Atherosclerosis*, 118: 67–75.

Coutard, M. and Osborne-Pellegrin, M.J. (1986), Spontaneous endothelial injury and lipid accumulation in the rat caudal artery, *American Journal of Pathology*, 122: 120–128.

Csallany, A.S., Kindom, S.E., Addis, P.B., and Lee, J.H. (1989), HPLC method for quantitation of cholesterol and four of its major oxidation products in muscle and liver tissues, *Lipids*, 24: 645–651.

Decker, E.A. and Xu, Z. (1998), Minimizing rancidity in muscle foods, *Food Technology*, 52: 54–59.

Dionisi, F., Golay, P.A., Aeschlimann, J.M., and Fay, L.B. (1998), Determination of cholesterol oxidation products in milk powders: methods comparison and validation, *Journal of Agricultural and Food Chemistry*, 46: 2227–2233.

Du, M. and Ahn, D.U. (2000), Effects of antioxidant and packaging on lipid and cholesterol oxidation and color changes of irradiated egg yolk powder, *Journal of Food Science*, 65: 625–629.

Dzeletovic, S., Breuer, O., Lund, E., and Diczfalusy, U. (1995), Determination of cholesterol oxidation products in human plasma by isotope dilution-mass spectrometry, *Analytical Biochemistry*, 225: 73–80.

Early, R.B. (Ed.) (1992), *The technology of dairy products*, New York: VCH.

Edwards, P.A. and Ericsson, J. (1999), Sterols and isoprenoids: signaling molecules derived from the cholesterol biosynthetic pathway, *Annual Reviews in Biochemistry*, 68: 157–185.

Finocchiaro, E.T., Lee, K., and Richardson, T. (1983), Identification and quantification of cholesterol oxides in grated cheese and bleached butter oil, *Journal of the American Oil Chemists' Society*, 61: 877–883.

Folch, J., Lees, M., and Sloane-Stanley, G.H. (1957), A simple method for the isolation and purification of total lipids from animal tissues, *The Journal of Biological Chemistry*, 226: 497–509.

Fujita, T., Adachi, J., Ueno, Y., Peters, T.J., and Preedy, V.R. (2002), Chronic ethanol feeding increases 7-hydroperoxycholesterol and oxysterols in rat skeletal muscle, *Metabolism*, 51: 737–742.

Galobart, J. and Guardiola, F. (2002), Formation and content of cholesterol oxidation products in egg and egg products, in F. Guardiola, P.C. Dutta, R. Codony, and G.P. Savage (Eds.) *Cholesterol and Phytosterol Oxidation Products: Analysis, Occurrence and Biological Effects,* Champaign, IL: AOCS.

Galobart, J., Guardiola, F., Barroeta A.C., Lopéz-Ferrer, S., and Baucells, M.D. (2002), Influence of dietary supplementation with α-tocopherol acetate and canthaxanthin on cholesterol oxidation in ω-3 and ω-6 fatty acid enriched spray-dried eggs, *Journal of Food Science*, 67: 2460–2466.

Galvin, K., Morrissey, P.A., and Buckley, D.J (1997), Cholesterol oxides in processed chicken muscle as influenced by dietary α-tocopherol supplementation, *Food Chemistry*, 62: 185–190.

Galvin, K., Morrissey, P.A., and Buckley, D.J (1998), Effect of dietary α-tocopherol supplementation and γ-irradiation on α-tocopherol retention and lipid oxidation in cooked minced chicken, *Food Chemistry*, 62: 185–190.

Garcia-Cruset, S., Carpenter, K.L., Guardiola, F., and Mitchinson, M.J. (1999), Oxysterols in cap and core of human advanced atherosclerotic lesions, *Free Radical Research*, 30: 341–350.

Garcia-Cruset, S., Carpenter, K.L., Guardiola, F., Stein, B.K., and Mitchinson, M.J. (2001), Oxysterols profiles of normal human arteries, fatty streaks and advanced lesions, *Free Radical Research*, 35: 31–41.

Grau, A., Codony, R., Grimpa, S., Baucells, M.D., and Guardiola, F. (2001), Cholesterol oxidation in frozen dark chicken meat: influence of dietary fat source, and α-tocopherol and ascorbic acid supplementation, *Meat Science*, 57: 197–208.

Gu, L., Okada, Y., Clinton, S.K., Gerard, C., Sukhova, G.K., Libby, P., and Rollins, B.J. (1998), Absence of monocyte chemoattractant protein-1 reduces atherosclerosis in low density lipoprotein receptor-deficient mice, *Molecular Cell*, 2: 275–281.

Guardiola, F., Codony, R., Rafecas, M., and Boatella, J. (1995a), Comparison of three methods for the determination of oxysterols in spray-dried food, *Journal of Chromatography A*, 705: 289–304.

Guardiola, F., Codony, R., Rafecas, M., and Boatella, J. (1995b), Stability of polyunsaturated fatty acids in egg powder processed and stored under various conditions, *Journal of Agricultural and Food Chemistry*, 43: 2254–2259.

Guardiola, F., Codony, R., Addis, P.B., Rafecas, M., and Boatella, J. (1996), Biological effects of oxysterols: current status, *Food Chemistry and Toxicology*, 34: 193–211.

Guardiola, F., Codony, R., Rafecas, M., Grau, A., Jordan, A., and Boatella, J. (1997), Oxysterols formation in spray-dried egg processed and stored under various conditions: prevention and relationship with other quality parameters, *Journal of Agricultural and Food Chemistry*, 45: 2229–2243.

Guardiola, F., Dutta, P.C., Codony, R., and Savaga, G.P. (Eds.) (2002), *Cholesterol and Phytosterol Oxidation Products: Analysis, Occurrence and Biological Effects*, Champaign, IL: AOCS.

Haigh, W.G. and Lee, S.P. (2001), Identification of oxysterols in human bile and pigment gallstones, *Gastroenterology*, 121: 118–123.

Halliwell, B., Aeschbach, R., Loliger, J., and Aruoma, O.J. (1995), The characterization of antioxidants, *Food Chemistry and Toxicology*, 33: 601–607.

Han, J., Hajjar, D.P., Febbraio, M., and Nicholson, A.C. (1997), Native and modified low density lipoproteins increase the functional expression of the macrophage class B scavenger receptor CD36, *The Journal of Biological Chemistry*, 272: 21654–21659.

Harada-Shiba, M., Kinoshita, M., Kamido, H., and Shimokado, K. (1998), Oxidized low density lipoprotein induces apoptosis in cultured human umbilical vein endothelial cells by common and unique mechanisms, *The Journal of Biological Chemistry*, 273: 9681–9687.

Hayek, T., Fuhrman, B., Vaya, J., Rosenblat, M., Belinky, P., Coleman, R., Elis, A., and Aviram, M. (1997), Reduced progression of atherosclerosis in apolipoprotein E-deficient mice following consumption of red wine, or its polyphenols quercetin or catechin, is associated with reduced susceptibility of LDL to oxidation and aggregation, *Arteriosclerosis, Thrombosis, and Vascular Biology*, 17: 2744–2752.

Hazen, S.L., Hsu, F.F., Gaut, J.P., Crowley, J.W., and Heinecke, J.W. (1999), Modification of proteins and lipids by myeloperoxidase, *Methods in Enzymology*, 300: 88–105.

Heinecke, J.W. (1999a), Mass spectrometric quantification of amino acid oxidation products in proteins: insights into pathways that promote LDL oxidation in the artery wall, *FASEB Journal*, 13: 1113–1120.

Heinecke, J.W. (1999b), Mechanisms of oxidative damage by myeloperoxidase in atherosclerosis and other inflammatory disorders, *The Journal of Laboratory and Clinical Medicine*, 133: 321–325.

Heinecke, J.W. (2001), Is the emperor wearing clothes? Clinical trials of vitamin E and LDL oxidation hypothesis, *Arteriosclerosis, Thrombosis, and Vascular Biology*, 21: 1261–1264.

Hodis, H.N., Crawford, D.W., and Sevanian, A. (1991), Cholesterol feeding increases plasma and aortic tissue cholesterol oxide levels in parallel: further evidence for the role of cholesterol oxidation in atherosclerosis, *Atherosclerosis*, 89: 117–126.

Hodis, H.N., Chauban, A., Hashimoto, S., Crawford, D.W., and Sevanian, A. (1992), Probucol reduced plasma and aortic wall oxysterol levels in cholesterol-fed rabbits independently of its plasma cholesterol-lowering effect, *Atherosclerosis*, 96: 125–134.

Hodis, H.N., Kramsch, D.M., Avogaro, P., Bittolo-Bon, G., Cazzolato, G., Hwang, J., Peterson, H., and Sevanian, A. (1994), Biochemical and cytotoxic characteristics of an *in vivo* circulating oxidized low density lipoprotein (LDL-), *Journal of Lipid Research*, 35: 669–677.

Hodis, H.N., Hashimoto, S., Mack, W.J., and Sevanian, A. (2000), Probucol reduces oxysterol formation in hypertensive rabbits, *Hypertension*, 36: 436–441.

Honda, A., Yoshida, T., Tanaka, N., Matsuzaki, Y., He, B., Shoda, J., and Osuga, T. (1995), Accumulation of 7α-hydroxycholesterol in liver tissue of patients with cholesterol gallstones, *Journal of Gastroenterology*, 30: 651–656.

Hopkins, P.N. (1992), Effects of dietary cholesterol on serum cholesterol: a meta-analysis and review, *American Journal of Clinical Nutrition*, 55: 1060–1070.

Huber, K.C., Pike, O.A., and Huber C.S. (1995), Antioxidant inhibition of cholesterol oxidation in a spray-dried food system during accelerated storage, *Journal of Food Science*, 60: 909–912.

Hughes, H., Mathews, B., Lenz, M.L., and Guyton, J.R. (1994), Cytotoxicity of oxidized LDL to porcine smooth muscle cells is associated with the oxysterols 7-ketocholesterol and 7-hydroxycholesterol, *Arteriosclerosis and Thrombosis,* 14: 1177–1185.

Husemann, J., Obstfeld, A., Febbraio, M., Kodama, T., and Silverstein, S.C. (2001), CD11b/CD18 mediates production of reactive oxygen species by mouse and human macrophages adherent to matrixes containing oxidized LDL, *Arteriosclerosis, Thrombosis, and Vascular Biology*, 21: 1301–1305.

Ignotz, R.A. and Massagué, J. (1986), Transforming growth factor-β stimulates the expression of fibronectin and collagen and their incorporation into the extracellular matrix, *The Journal of Biological Chemistry*, 261: 4337–4345.

Iuliano, L., Micheletta, F., Maranghi, M., Frati, G., Diczfalusy, U., and Violi, F. (2001), Bioavailability of vitamin E as function of food intake in healthy subjects, *Arteriosclerosis, Thrombosis, and Vascular Biology*, 21: 1261–1264.

Iuliano, L., Micheletta, F., Natoli, S., Ginanni Corradini, S., Iappelli, M., Elisei, W., Giovannelli, L., Violi, F., and Diczfalusy, U. (2003), Measurements of oxysterols and α-tocopherol in plasma and tissue samples as indices of oxidant status, *Analytical Biochemistry*, 312: 217–223.

Jacobson, M.S. (1987), Cholesterol oxides in Indian ghee: possible cause of unexplained high risk of atherosclerosis in Indian immigrant population, *Lancet*, 2: 656–658.

Jurgens, G., Hoff, H.F., Chisolm, G.M.I., and Esterbauer, H. (1987), Modification of human serum low density lipoprotein by oxidation — characterization and pathophysiological implications, *Chemistry and Physics of Lipids*, 45: 315–336.

Kermasha, S., Kubow, S., and Goetghebeur, M. (1994), Comparative high-performance liquid chromatography analysis of cholesterol and its oxidation products using diode-array ultraviolet and laser light-scattering detection, *Journal of Chromatography A*, 685: 229–235.

Kerry, J.P., Gilroy, D.A., and O'Brien, N.M. (2002), Formation and content of cholesterol oxidation products in meat and meat products in F. Guardiola, P.C. Dutta, R. Codony, and G.P. Savage (Eds.) *Cholesterol and Phytosterol Oxidation Products: Analysis, Occurrence and Biological Effects*, Champaign, IL: AOCS.

Kim, S.K. and Nawar, W.W. (1991), Oxidative interactions of cholesterol oxides in marine fish product induced by triacylglycerols, *Journal of the American Oil Chemists' Society*, 68: 931–934.

Kolsch, H., Lütjohann, D., Tulke, A., Björkhem, I., and Rao, M.L. (1999), The neurotoxic effect of 24-hydroxycholesterol on SH-SY5Y human neuroblastoma cells, *Brain Research*, 818: 171–175.

Kolsch, H., Ludwig, M., Lütjohann, D., and Rao, M.L. (2001), Neurotoxicity of 24-hydroxycholesterol, an important cholesterol elimination product of the brain, may be prevented by vitamin E and estradiol-17beta, *Journal of Neural Transmission*, 108: 475–488.

Kraemer, R., Pomerantz, K.B., Joseph-Silverstein, J., and Hajjar, D.P. (1993), Induction of basic fibroblast growth factor mRNA and protein synthesis in smooth muscle cells by cholesteryl ester enrichment and 25-hydroxycholesterol, *The Journal of Biological Chemistry*, 268: 8040–8045.

Kritchvsky, D. (2001), Diet and atherosclerosis, *The Journal of Nutrition, Health and Aging*, 5: 155–159.

Kritharides, L., Jessup, W., Gifford, J., and Dean, R.T. (1993), A method for defining the stages of low-density lipoprotein oxidation by the separation of cholesterol- and cholesteryl ester-oxidation products using HPLC, *Analytical Biochemistry*, 213: 79–89.

Kroon, P.A. (1997), Cholesterol and atherosclerosis, *Australian and New Zealand Journal of Medicine*, 27: 492–496.

Kucuk, O., Lis, L.J., Dey, T., Mata, R., Westerman, M.P., Yachnin, S., Szostek, R., Tracy, D., Kauffman, J.W., Gage, D.A., and Sweeley, C.C. (1992), The effects of cholesterol oxidation products in sickle and normal red blood cell membranes, *Biochimica et Biophysica Acta*, 1103: 296–302.

Kucuk, O., Stoner-Picking, J., Yachnin, S., Gordon, L.I., Williams, R.M., Lis, L.J., and Westerman, M.P. (1994), Inhibition of cytolytic T lymphocyte activity by oxysterols, *Lipids*, 29: 657–660.

Kudo, K., Emmons, G.T., Casserly, E.W., Via, D.P., Smith, L.C., St. Pyrek, J., and Schroepfer, G.J. (1989), Inhibitors of sterol synthesis. Chromatography of acetate derivatives of oxygenated sterols, *Journal of Lipid Research*, 30: 1097–1111.

Kudo, R., Adachi, J., Uemura, K., Maekawa, T., Ueno, Y., and Yoshida, K. (2001), Lipid peroxidation in the rat brain after inhalation is temperature dependent, *Free Radical Biology and Medicine*, 31: 1417–1423.

Lebovics, V.K., Antal, M., and Gaál, Ö. (1996), Enzymatic determination of cholesterol oxides, *Journal of the Science of Food and Agriculture*, 71: 22–26.

Lee, T. and Chau, L. (2001), Fas/Fas ligand mediated death pathway is involved in oxLDL-induced apoptosis in vascular smooth muscle cells, *American Journal of Physiology. Cell Physiology*, 280: 131–138.

Lemaire, S., Lizard, G., Monier, S., Miguet, C., Gueldry, S., Volot, F., Gambert, P., and Neel, D. (1998), Different patterns of IL-1beta secretion, adhesion molecule expression, and apoptosis induction of human endothelial cells treated with 7alpha-, 7beta-hydroxycholesterol, or 7-ketocholesterol, *FEBS Letters*, 440: 434–439.

Leonarduzzi, G., Sevanian, A., Sottero, B., Arkan, M.C., Biasi, F., Chiarpotto, E., Basaga, H., and Poli, G. (2001), Up-regulation of the fibrogenic cytokine TGF-β1 by oxysterols: a mechanistic link between cholesterol and atherosclerosis, *FASEB Journal*, 15: 1619–1621.

Leonarduzzi, G., Sottero, B., and Poli, G. (2002), Oxidized products of cholesterol: dietary and metabolic origin, and proatherosclerotic effects (review), *The Journal of Nutritional Biochemistry*, 13: 700–710.

Lerker, G. and Rodriguez-Estrada, M.T. (2002), Cholesterol oxidation mechanisms, in F. Guardiola, P.C. Dutta, R. Codony and G.P. Savage (Eds.) *Cholesterol and Phytosterol Oxidation Products: Analysis, Occurrence and Biological Effects*, Champaign, IL: AOCS.

Li, S.X., Cherian, G., and Sim, J.S. (1996), Cholesterol oxidation in egg yolk powder during storage and heating as affected by dietary oils and tocopherols, *Journal of Food Science*, 61: 721–725.

Libby, P., Aikawa, M., and Schönbeck, U. (2000), Cholesterol and atherosclerosis, *Biochimica et Biophysica Acta*, 1529: 299–309.

Liu, Y., Hulten, L.M., and Wiklund, O. (1997), Macrophages isolated from human atherosclerotic plaques produce IL-8, and oxysterols may have a regulatory function for IL-8 production, *Arteriosclerosis, Thrombosis, and Vascular Biology*, 17: 317–323.

Lizard, G., Deckert, V., Dubrez, L., Moisant, M., Gambert, P., and Lagrost, L. (1996), Induction of apoptosis in endothelial cells treated with cholesterol oxides, *American Journal of Pathology*, 148: 1625–1638.

Lizard, G., Lemaire, S., Monier, S., Gueldry, S., Neel, D., and Gambert, P. (1997), Induction of apoptosis and of interleukin-1β secretion by 7β-hydroxycholesterol and 7-ketocholesterol: partial inhibition by Bcl-2 overexpression, *FEBS Letters*, 419: 276–280.

Lizard, G., Monier, S., Cordelet, C., Gesquiere, L., Deckert, V., Gueldry, S., Lagrost, L., and Gambert, P. (1999), Characterization and comparison of the mode of cell death, apoptosis versus necrosis, induced by 7β-hydroxycholesterol and 7-ketocholesterol in the cells of the vascular wall, *Arteriosclerosis, Thrombosis, and Vascular Biology*, 19: 1190–1200.

Lizard, G., Miguet, C., Bessede, G., Monier, S., Gueldry, S., Neel, D., and Guambert, P. (2000), Impairment with various antioxidant of the loss of mitochondrial transmembrane potential and of the cytosolic release of the cytochrome c occurring during 7-ketocholesterol-induced apoptosis, *Free Radical Biology and Medicine*, 28: 743–753.

Lütjohann, D., Breuer, O., Ahlborg, G., Nennesmo, I., Sidén, A., Diczfalusy, U., and Björkhem, I. (1996), Cholesterol Homeostasis in Human Brain: Evidence for an Age-Dependent Flux of 24*S*-Hydroxycholesterol from the Brain into Circulation, *Proceedings of the National Academy of Sciences of the United States of America*, 93: 9799–9804.

Luu, B. and Moog, C. (1991), Oxysterols: biological activities and physicochemical studies, *Biochimie*, 73: 1317–1320.

McNamara, D.J. (2000), Dietary cholesterol and atherosclerosis, *Biochimica et Biophysica Acta*, 1529: 310–320.

Madhavi, D.L., Deshpande, S.S., and Salunkhe, D.K. (Eds.) (1995), *Food Antioxidants: Technological, Toxicological and Health Perspectives*, New York: Dekker.

Maerker, G. (1987), Cholesterol autoxidation current status, *Journal of the American Oil Chemists' Society*, 64: 388–392.

Maerker, G. and Bunick, F.J. (1986), Cholesterol oxides I. Isolation and determination of some cholesterol oxidations products, *Journal of the American Oil Chemists' Society*, 63: 767–770.

Mahfouz, M.M., Kawano, H., and Kummerow, F.A. (1997), Effect of cholesterol-rich diets with and without added vitamins E and C on the severity of atherosclerosis in rabbits, *The American Journal of Clinical Nutrition*, 66: 1240–1249.

Maor, I., Kaplan, M., Hayek, T., Vaya, J., Hoffman, A., and Aviram, M. (2000), Oxidized monocyte-derived macrophages in aortic atherosclerotic lesion from apolipoprotein E-deficient mice and from human carotid artery contain lipid peroxides and oxysterols, *Biochemical and Biophysical Communications*, 269: 775–780.

Maraschiello, C., Sarraga, C., and Garcia Regueiro, J.A. (1999), Glutathione peroxidase activity, TBARS, and alpha-tocopherol in meat from chickens fed different diets, *Journal of Agricultural and Food Chemistry*, 47: 867–872.

Mattsson Hultén, L., Lindmark, H., Diczfalusy, U., Björkhem, I., Ottosson, M., Liu, Y., Bondjers, G., and Wiklund, O. (1996), Oxysterols present in atherosclerotic tissue decrease the expression of lipoprotein lipase messenger RNA in human monocyte-derived macrophages, *The Journal of Clinical Investigation*, 97: 461–468.

Meaney, S., Bodin, K., Diczfalusy, U., and Björkhem, I. (2002), On the rate of translocation *in vitro* and kinetics *in vivo* of the major oxysterols in human circulation: critical importance of the position of the oxygen function, *Journal of Lipid Research*, 43: 2130–2135.

Mol, M.J.T.M., de Rijke, Y.B., Demacker, P.N.M., and Stalenhoef, A.F.H. (1997), Plasma levels of lipid and cholesterol oxidation products and cytokines in diabetes mellitus and cigarettes smoking: effects of vitamin E treatment, *Atherosclerosis*, 129: 169–176.

Monahan, F.J., Gray, J.I., Moren, A.M., Miller, E.R., Buckley, D.J., Morrisey, P.A., and Gomma, E.A. (1992), Influence of dietary treatment on lipid and cholesterol oxidation in pork, *Journal of Agricultural and Food Chemistry*, 40: 1310–1315.

Morgan, J.N. and Armstrong, D.J. (1987), Formation of cholesterol-5,6-epoxides during spray-drying of egg yolk, *Journal of Food Science*, 52: 1224–1227.

Morrissey, P.A., Sheeny, P.J., Galvin, K., Kerry, J.P., and Buckley, D.J. (1998), Lipid stability in meat and meat products, *Meat Science*, 49: S73–86.

Nicolaiew, N., Lemort, N., Adorni, L., Berra, B., Montorfano, G., Rapelli, S., Cortesi, N., and Jacotot, B. (1998), Comparison between extra virgin oil and oleic acid rich sunflower oil: effects on postprandial lipemia and LDL susceptibility to oxidation, *Annals of Nutrition and Metabolism*, 42: 251–260.

Nielsen, J.H., Olsen, C.E., Lyndon, J., Sensen, J., and Skibsted, L.H. (1996), Cholesterol oxidation in feta cheese produced from high temperature bleached and from non-bleached butteroil from bovine milk, *Journal of Dairy Research*, 63: 615–621.

Niranjan, T.G. and Krishnakantha, T.P. (2001), Effect of dietary ghee: the anhydrous milk fat on lymphocytes in rats, *Molecular and Cellular Biochemistry*, 226: 39–47.

Nishio, E. and Watanabe, Y. (1996), Oxysterols induced apoptosis in cultured smooth muscle cells through CPP32 protease activation and bcl-2 protein downregulation, *Biochemical and Biophysical Research Communications*, 226: 928–934.

Nourooz-Zadeh, J. and Appelqvist, L.-A. (1987), Cholesterol oxides in swedish foods and food ingredients: fresh eggs and dehydrated egg products, *Journal of Food Science*, 52: 57–67.

Nourooz-Zadeh, J. and Appelqvist, L.-A. (1988), Cholesterol oxides in swedish foods and food ingredients: milk powder products, *Journal of Food Science,* 53: 74–87.

Nourooz-Zadeh, J. and Appelqvist, L.-A. (1989), Cholesterol oxides in swedish food and food ingredients: lard and bacon, *Journal of the American Oil Chemists' Society*, 66: 586–592.

O'Callaghan, Y.C., Woods, J.A., and O'Brien, N.M. (1999), Oxysterol-induced cell death in U937 and HepG2 cells at reduced and normal serum concentrations, *European Journal of Nutrition*, 38: 255–262.

Ong, J., Aoki, A., Seigel, G., Sacerio, I., Castellon, R., Nesburn, A.B., and Kenney, M.C. (2003), Oxysterol-induced toxicity in R28 and ARPE-19 cells, *Neurochemistry Research*, 28: 883–891.

Osada, K., Kodama, T., Cui, L., Yamada, K., and Sugano, M. (1993), Levels and formation of oxidized cholesterols in processed foods, *Journal of Agricultural and Food Chemistry*, 41: 1893–1898.

Osada, K., Ravandi, A., and Kuksis, A. (1999), Rapid analysis of oxidized cholesterol derivatives by high-performance liquid chromatography combined with diode-array ultraviolet and evaporative laser light-scattering detection, *Journal of the American Oil Chemists' Society*, 66: 863–871.

Osada, K., Hoshima, S., Nakamura, S., and Sugano, M. (2000), Cholesterol oxidation in meat products and its regulation by supplementation of sodium nitrite and apple polyphenol before processing, *Journal of Agricultural and Food Chemistry*, 48: 3823–3829.

O'Sullivan, A.J., O'Callaghan, Y.C., Woods, J.A., and O'Brien, N.M. (2003), Toxicity of cholesterol oxidation products to Caco-2 and HepG2 cells: modulatory effects of alpha- and gamma-tocopherol, *Journal of Applied Toxicology*, 23: 191–197.

Paniangvait, P., King, A.J., Jones, A.D., and German, B.G. (1995), Cholesterol oxides in foods of animal origin, *Journal of Food Science*, 60: 1159–1174.

Park, S.W. and Addis, P.B. (1985a), Capillary gas-liquid chromatographic resolution of oxidised cholesterol derivatives, *Analytical Biochemistry*, 149: 275–283.

Park, S.W. and Addis, P.B. (1985b), HPLC determination of C-7 oxidised cholesterol derivatives in foods, *Journal of Food Science*, 50: 1437–1444.

Park, S.W. and Addis, P.B. (1986), Identification and quantitative estimation of oxidized cholesterol derivatives in heated tallow, *Journal of Agricultural and Food Chemistry*, 34: 653–659.

Park, S.W. and Addis, P.B. (1987), Cholesterol oxidation products in some muscle foods, *Journal of Food Science*, 52: 1500–1503.

Peng, S.K., Taylor, C.B., Hill, R.J., and Morin, R.J. (1985), Cholesterol oxidation derivatives and arterial endothelial damage, *Atherosclerosis*, 54: 121–133.

Peng, S.K., Hu, B., and Morin, R.J. (1991), Angiotoxicity and atherogenicity of cholesterol oxides, *Journal of Clinical Laboratory Analysis*, 5: 144–152.

Pie, J.E. and Seillan, C. (1992), Oxysterols in cultured bovine aortic smooth muscle cells and in the monocyte-like cells line U937, *Lipids*, 27: 270–274.

Pie, J.E., Spahis, K., and Seillan, C. (1990), Evaluation of oxidative degradation of cholesterol in food and food ingredients: identification and quantification of cholesterol oxides, *Journal of Agriculture and Food Chemistry*, 38: 973–979.

Pie, J.E., Spahis, K., and Seillan, C. (1991), Cholesterol oxidation in meat products during cooking and frozen storage, *Journal of Agricultural and Food Chemistry*, 39: 250–254.

Pietinen, P., Ascherio, A., Korhonen, P., Hartman, A.M., Willett, W.C., Albanes D., and Virtamo, J. (1997), Intake of fatty acids and risk of coronary heart disease in a cohort of Finnish men. The alpha-tocopherol, beta-carotene cancer prevention study, *American Journal of Epidemiology*, 145: 876–887.

Porkkala-Sarataho, E., Salonen, J.T., Nyyssönen, K., Kaikkonen, J., Salonen, R., Ristonmaa, U., Diczfalusy, U., Brigelius-Flohe, R., Loft, S., and Poulsen, H.E. (2000), Long-term effects of vitamin E, vitamin C, and combined supplementation on urinary 7-hydro-8-oxo-2-deoxyguanosine, serum cholesterol oxidation products, and oxidation resistance of lipid in nondepleted men, *Arteriosclerosis, Thrombosis, and Vascular Biology*, 20: 2087–2093.

Rankin, S.A. and Pike, O.A. (1993), Cholesterol autoxidation inhibition varies among several natural antioxidants in an aqueous model system, *Journal of Food Science*, 58: 653–655.

Reid, A.E. (2001), Nonalcoholic steatohepatitis, *Gastroenterology*, 121: 710–723.

Rifkind, B.M. (1986), Diet, plasma, cholesterol and coronary heart disease, *The Journal of Nutrition*, 116: 1578–1580.

Ringeseis, R. and Eder, K. (2002), Insufficient dietary vitamin E increases the concentration of 7β-hydroxycholesterol in tissues of rats fed salmon oil, *The Journal of Nutrition*, 132: 3732–3735.

Rodriguez-Estrada, M.T., Penazzi, G., Caboni, M.F., Bertacco, G., and Lercker, G. (1997), Effect of different cooking methods on some lipid and protein components of hamburgers, *Meat Science*; 45: 365–375.

Rong, J.X., Rangaswamy, S., Shen, L., Dave, R., Chang, Y.H., Peterson, H., Hodis, H.N., Chisolm, G.M., and Sevanian, A. (1998), Arterial injury by cholesterol oxidation products causes endothelial dysfunction and arterial wall cholesterol accumulation, *Arteriosclerosis, Thrombosis, and Vascular Biology*, 18: 1885–1894.

Rosenblat, M., and Aviram, M. (2002), Oxysterol-induced activation of macrophage NADPH-oxidase enhances cell-mediated oxidation of LDL in the atherosclerotic apolipoprotein E deficient mouse: inhibitory role for vitamin E, *Atherosclerosis*, 160: 69–80.

Rosenblat, M., Belinky, P., Vaya, J., Levy, R., Hayek, T., Coleman, R., Merchav, S., and Aviram M. (1999), Macrophage enrichment with the isoflavan glabridin inhibits NADPH oxidase-induced cell-mediated oxidation of low density lipoprotein. A possible role for protein kinase C, *The Journal of Biological Chemistry*, 274: 13790–13799.

Rosklint, T., Ohlsson, B.G., Wiklund, O., Noren, K., and Hulten, L.M. (2002), Oxysterols induce interleukin-1 beta production in human macrophages, *European Journal of Clinical Investigation*, 32: 35–42.

Ross, R. (1986), The pathogenesis of atherosclerosis — an update, *New England Journal of Medicine*, 314: 488–500.

Ross, R. (1993), The pathogenesis of atherosclerosis: a perspective for the 1990s, *Nature*, 362: 801–809.

Ruiz-Gutiérrez, V. and Pérez-Camino, M.C. (2000), Update on solid-phase extraction for the analysis of lipid classes and related compounds, *Journal of Chromatography A*, 885: 321–341.

Ryzlak, M.T., Fales, H.M., Russell, W.L., and Schaffner, C.P. (1990), Oxysterols and alcoholic liver disease, *Alcohol Clinical and Experimental Research*, 14: 490–495.

Sacks, F.M., Pfeffer, M.A., Moye, L.A., Rouleau, J.L., Rutherford, J.D., Cole, T.G., Brown, L., Warnica, J.W., Arnold, J.M., Wun, C.C., Davis, B.R., and Braunwald, E. (1996), The effect of pravastatin on coronary events after myocardial infarction in patients with average cholesterol levels. Cholesterol and Recurrent Events Trial investigators, *New England Journal of Medicine*, 335: 1001–1009.

Salonen, J.T., Nyyssönen, K., Salonen, R., Porkkala-Sarataho, E., Tuomainen, T., Diczfalusy, U., and Björkhem, I. (1997), Lipoprotein oxidation and progression of carotid atherosclerosis, *Circulation*, 95: 840–845.

Sander, B.D., Smith, D.E., Addis, P.B., and Park, S.W. (1989), Effects of prolonged and adverse storage conditions on levels of cholesterol oxidation products in dairy products, *Journal of Food Science,* 54: 874–879.

Saucier, S.E., Kandutsch, A.A., Gayen, A.K., Swahn, D.K., and Spencer, T.A. (1989), Oxysterol regulation of 3-hydroxy-methylglutaryl-CoA reductase in liver. Effect of dietary cholesterol, *The Journal of Biological Chemistry*, 264: 6863–6869.

Savage, G.P., Dutta, P.C., and Rodriguez-Estrada, M.T. (2002), Cholesterol oxides: their occurrence and methods to prevent their generation in foods, *Asian Pacific Journal of Clinical Nutrition*, 11: 72–78.

Schroepfer, G.J., Jr. (2000), Oxysterols: modulators of cholesterol metabolism and other processes, *Physiological Reviews*, 80: 361–554.

Sevanian, A. and McLeod, L.L. (1987), Cholesterol autoxidation in phospholipid membrane bilayers, *Lipids*, 22: 627–636.

Sevanian, A., Seraglia, R., Traldi, P., Rossato, P., Ursini, F., and Hodis, H. (1994), Analysis of plasma cholesterol oxidation products using gas- and high-performance liquid chromatography/mass spectrometry, *Free Radical Biology and Medicine*, 17: 397–409.

Sevanian, A., Hodis, H.N., Hwang, J., McLeod, L.L., and Peterson, H. (1995), Characterization of endothelial cell injury by cholesterol oxidation products found in oxidized LDL, *Journal of Lipid Research*, 36:1971–1986.

Shan, H., Pang, J., Li, S., Chiang, T.B., Wilson, W.K., and Schroepfer Jr, G.J. (2003), Chromatographic behavior of oxygenated derivatives of cholesterol, *Steroids*, 68: 221–233.

Shepherd, J., Cobbe, S.M., Ford, I., Isles, C.G., Lorimer, A.R., MacFarlane, P., McKillop, J.H., and Packard, C.J. (1995), Prevention of coronary heart disease with pravastatin in men with hypercholesterolemia, *New England Journal of Medicine*, 333: 1301–1307.

Shozen, K., Ohshima, T., Ushio, H., Takiguchi, A., and Koizumi, C. (1997), Effects of antioxidants and packing on cholesterol oxidation in processed anchovy during storage, *Lebensmittel-Wissenschaft und Technologie*, 29: 94–99.

Smith, L.L. (Ed.) (1981), *Cholesterol Autoxidation*, New York: Plenum.

Smith, L.L. (1996), Review of progress in sterol oxidations: 1987–1995, *Lipids*, 31: 453–487.

Smith, L.L. and Johnson, B.H. (1989), Biological activities of oxysterols, *Free Radical Biology and Medicine*, 7: 285–332.

Smondyrev, A.M. and Berkowitz, M.L. (2001), Effects of oxygenated sterol on phospholipid bilayer properties: a molecular dynamics simulation, *Chemistry and Physics of Lipids*, 112: 31–39.

Steinberg, D. (2002), Atherogenesis in perspective: hypercholesterolemia and inflammation as partners in crime, *Nature Medicine*, 8: 1211–1217.

Stephens, N.G., Parsons, A., Schofield, P.M., Kelly, F., Cheeseman, K., and Mitchinson, M.J. (1996), Randomised controlled trial of vitamin E in patients with coronary disease: Cambridge Hearth Antioxidant Study (CHAOS), *Lancet*, 347: 781–786.

Tai, C.-Y., Chen, Y.C., and Chen, B.H. (2000), Analysis, formation and inhibition of cholesterol oxidation products in foods: an overview (part II), *Journal of Food and Drug Analysis*, 8: 1–15.

Teng, J.I. and Smith, L.L. (1995), High-performance liquid chromatographic analysis of human erythrocyte oxysterols as delta 4-3-ketone derivatives, *Journal of Chromatography A*, 691: 247–254.

Toeller, M., Buyken, A.E., Heitkamp, G., Scherbaum, W.A., Krans, H.M.J., and Fuller, J.H. (1999), EURODIAB IDDM Complications Group. Associations of fat and cholesterol intake with serum lipid levels and cardiovascular disease: the EURODIAB IDDM Complications Study, *Experimental Clinical Endocrinology*, 107: 512–521.

Tsai, L.S. and Hudson, C.A. (1981), High-performance liquid chromatography of oxigenated cholesterols and related compounds, *Journal of the American Oil Chemists' Society*, 58: 931–934.

Tsai, L.S. and Hudson, C.A. (1984), Cholesterol oxides in commercial dry egg products: isolation and identification, *Journal of Food Science*, 49: 1245–1248.

Uemura, M., Manabe, H., Yoshida, N., Fujita, N., Ochiai, J., Matsumoto, N., Takagi, T., Naito, Y., and Yoshikawa, T. (2002), α-Tocopherol prevents apoptosis of vascular endothelial cells via a mechanism exceeding that of mere antioxidation, *European Journal of Pharmacology*, 456: 29–37.

van de Bovenkamp, P., Kosmeijer-Schuil, T.G., and Katan, M.B. (1988), Quantification of oxysterols in Dutch foods: egg products and mixed diets, *Lipids*, 23: 1079–1085.

Vatassery, G.T., Quach, H.T., Smith, W.E., Krick, T.P., and Ungar, F (1997a), Analysis of hydroxy and keto cholesterols in oxidized brain synaptosome, *Lipids*, 32: 101–107.

Vatassery, G.T., Quach, H.T., Smith, W.E., and Ungar, F. (1997b), Oxidation of cholesterol in synaptosomes and mitochondria isolated from rat brains, *Lipids*, 32: 879–886.

Vaya, J., Aviram, M., Mahmood, S., Hayek, T., Grenadir, E., Hoffman, A., and Milo, S. (2001), Selective distributions of oxysterols in atherosclerotic lesions and human plasma lipoproteins, *Free Radical Research,* 34: 485–497.

Vega, G.L., Weiner, M.F., Lipton, A.M., Von Bergmann, K., Lütjohann, D., Moore, C., and Svetlik, D. (2003), Reduction in levels of 24*S*-hydroxycholesterol by statin treatment in patients with Alzheimer disease, *Archives in Neurology*, 60: 510–515.

Verlangieri, A.J., Cardin, B.A., and Bush, M. (1985), The interaction of aortic glycosaminoglycans and ^{3}H-inulin endothelial permeability in cholesterol induced rabbit atherogenesis, *Research Communications in Chemical Pathology and Pharmacology*, 47: 85–96.

Vine, D.E., Mamo, J.C.L., Beilin, L.J., Mori, T.A., and Croft, K.D. (1998), Dietary oxysterols are incorporated in plasma triglyceride-rich lipoproteins, increase their susceptibility to oxidation and increase aortic cholesterol concentrations of rabbits, *Journal of Lipid Research*, 39: 1995–2004.

Visioli, F., Bellomo, G., Montedoro, G., and Galli, C. (1995), Low density lipoprotein oxidation is inhibited by olive oil constituents, *Atherosclerosis*, 117: 25–32.

Weisburger, J.H. (1999), Mechanisms of action of antioxidant as exemplified in vegetables, tomatoes and tea, *Food and Chemical Toxicity*, 37: 943–948.

Wohlfeil, E.R. and Campbell, W.B. (1999), 25-Hydroxycholesterol increases eicosanoids and alters morphology in cultured pulmonary artery smooth muscle and endothelial cells, *Arteriosclerosis, Thrombosis, and Vascular Biology*, 19: 2901–2908.

Yoshida, T., Klinkspoor, J.H., Kuver, R., Wrenn, S.P., Kaler, E.W., and Lee, S.P. (2000), Cholestan-3β,5α,6β-triol, but not 7-ketocholesterol, suppresses taurocholate-induced mucin secretion by cultured dog gallbladder epithelial cells, *FEBS Letters*, 478: 113–118.

Yoshida, T., Klinkspoor, J.H., Kuver, R., Poot, M., Rabinovitch, P.S., Wrenn, S.P., Kaler, E.W., and Lee, S.P. (2001), Effects of bile salts on cholestan-3β,5α,6β-triol-induced apoptosis in dog gallbladder epithelial cells, *Biochimica et Biophysica Acta*, 1530: 199–208.

Yusuf, S., Dagenais, G., Pogue, J., Bosch, J., and Sleight, P. (2000), Vitamin E supplementation and cardiovascular events in high-risk patients. The heart outcomes prevention evaluation study investigators, *New England Journal of Medicine*, 342: 154–160.

Zanardi, E., Novelli, E., Ghiretti, G.P., and Chizzolini, R. (2000), Oxidative stability of lipids and cholesterol in salame Milano, coppa and Parma Ham: dietary supplementation with vitamin E and oleic acid, *Meat Science*, 55: 169–175.

Zhang, Z., Dansu, L., Blanchard, D.E., Lear, S.R., Erickson, S.K., and Spencer, T.A. (2001), Key regulatory oxysterols in liver: analysis as Δ^4-3-ketone derivatives by HPLC and response to physiological perturbations, *Journal of Lipid Research*, 42: 649–658.

Zieden, B., Kaminskas, A., Kristenson, M., Kucinskienê, Z., Vessby, B., Olsson, A.G., and Diczfalusy, U. (1999), Increased plasma 7β-hydroxycholesterol concentrations in a population with a high risk for cardiovascular disease, *Arteriosclerosis, Thrombosis, and Vascular Biology*, 19: 967–971.

6 Adverse Effects and Toxicity of Nutraceuticals

E. Davies, D. Greenacre, and G.B. Lockwood

CONTENTS

0-8493-3516-7/05/$0.00+$1.50

ABSTRACT

This review covers examples of published safety data on nutraceuticals, general adverse effects, drug interactions, and problems with formulated products, and gives a more detailed investigation of certain important nutraceuticals; carnitine, acetyl-L-carnitine, isoflavones, proanthocyanidins, melatonin, β-carotene, and glucosamine, and examines studies regarding the adverse effects and toxicity of these products. Although the major nutraceuticals included in this article are components of food or are present in normal human metabolism, a large number of adverse effects and toxicities have been reported, and there is evidence of problems with certain formulated products. Increasing use of these compounds by consumers at higher doses and chronic administration may reveal further adverse effects.

INTRODUCTION

The range of dietary supplements available to the consumer has been increasing, and nutraceuticals are an expanding sector of this market. They are being studied not only for use in diet supplementation but also for development as drugs for the treatment of a huge range of major diseases including cancer and Alzheimer's disease.

Stephen De Felice of the Foundation for Innovation in Medicine was the first to use the term, defining a nutraceutical as a "food, or parts of a food, that provide medical or health benefits, including the prevention and treatment of disease" (Rapport and Lockwood, 2002). These nutraceuticals are generally sourced from plants and plant products, e.g., isoflavones from soy, but some are derivatives of naturally occurring substances in the human body, e.g., carnitine.

Due to the ambiguity of the nature of nutraceuticals as foods or medicines and the difficulty of obtaining a patent as a medicine, companies often seek to promote their product as a supplement. This means that little or no research into safety and quality control is required by law, which has potentially serious implications for the safety of the product and the health of the consumer.

This review discusses examples of published safety data on nutraceuticals, general adverse effects, drug interactions, and problems with formulated products, and gives a more detailed investigation of certain important nutraceuticals, carnitine, acetyl-L-carnitine, isoflavones, proanthocyanidins, melatonin, β-carotene, and glucosamine. It also examines studies regarding the adverse effects and toxicity of these products.

SAFETY DATA

Table 6.1 shows LD_{50} values and published doses found to be safe in animal studies. Even allowing for the species differences, these selected nutraceuticals appear very safe at therapeutic levels. The data shown represent single-dose treatment, but chronic long-term dosing may reveal sporadic or more widespread adverse effects.

GENERAL ADVERSE EFFECTS

Problems with large-scale use of unregistered nutraceuticals have been reported by U.S. poison control centers. Wide-ranging symptoms from mild to severe have been

TABLE 6.1
LD_{50} Values and Reported Safe Doses for Selected Nutraceuticals

Nutraceutical	LD_{50} g/kg	Animal	Route[a]	Safe Doses g/kg	Animal	References
Glucosamine	15	Mouse	p.o.	—	—	Lenga, 1988
Chondroitin	1.58	Mouse	i.v.	—	—	Yang et al., 1978
Chondroitin[b]	9.8/>10	Mouse	i.p or s.c	—	—	Takeuchi and Edanaga, 1972
Chondroitin[b]	2.9/3.7	Rat	i.p. or s.c.	—	—	Takeuchi and Edanaga, 1972
Carnitine	19.2	Mouse	p.o.	—	—	Kelly, 1998
Melatonin	>3.2	Mouse	p.o.	—	—	Sugden, 1983
Melatonin	1.2/>>1.6	Mouse	i.p. or s.c.		—	Sugden, 1983
Pycnogenol			p.o.	0.15	Dog	Rohdewald, 2002
GPSE	>5	Rat	p.o.	4	Rat	Yamakoshi et al., 2002
MSM	—	—	g.	2	Rat	Horvath et al., 2002
Lutein	—	—	p.o.	0.26	Rat	Kruger et al., 2002
Policosanol	—	—	p.o.	0.25	Dog	Aleman et al., 2001
Lycopene	—	—	g.	3	Rat	Mellert et al., 2002

[a] p.o. = per os, i.v. = intravenous, i.p. = intraperitoneal, s.c. = subcutaneous, g. = gavage.
[b] = Sodium chondroitin polysulphate.

noted in adult and pediatric cases, and associations are difficult to verify if a product has more than one ingredient or is incompletely labeled (Palmer et al., 2003). Nutraceuticals are frequently involved in basic metabolic pathways in the body and, as such, are closely involved in the metabolism of these nutrients. Availability of one nutrient may impair or enhance the action of another in the immune system, as reported for nutrients such as dietary fatty acids and vitamin A. (Dillard and German, 2000). Decreased absorption of any prescription drug may occur with concomitant dosing with certain nutraceuticals such as flaxseed products (Ly et al., 2002). Recently, an investigation into possible increased antibiotic and antimicrobial resistance caused by certain nutraceuticals was published, and certain products showed interaction with antibiotics (Ward et al., 2002).

Adverse Effects

A number of nutraceuticals have been involved in animal toxicological studies and shown to be safe, namely lutein (Kruger et al., 2002) and MSM (Horvath et al., 2002). Many nutraceuticals have been shown to exhibit only minor side effects such as the reported 2% of animal subjects with gastrointestinal upset after treatment with a standardized glucosamine or chondroitin product (Anderson et al., 1999) and the 1.5% experiencing similar effects after taking Pycnogenol® (Rohdewald, 2002). One report of asthma exacerbation by a glucosamine or chondroitin product has appeared (Tallia and Cardone, 2002). Patients taking *S*-adenosyl methionine also reported mild to moderate gastrointestinal complaints, with the incidence of complaints sometimes as high as 20%. In addition, there is a possible risk of more significant psychiatric and cardiovascular adverse effects (Fetrow and Avila, 2001). Similar gastrointestinal effects are reported to occur with carnitine (Kelly, 1998). α-Lipoic acid has been found to cause allergic skin reactions and possible hypoglycemia in diabetic patients as a consequence of improved glucose utilization associated with high doses (Packer et al., 1995). There is evidence from both mice and human studies that *trans*-10, *cis*-12 conjugated linoleic acid may induce liver hypertrophy and insulin resistance via a redistribution of fat deposition that resembles lipodystrophy, and may decrease fat contents of both human and bovine milk (Larsen et al., 2003). A complex relationship between melatonin and other antioxidants *in vitro* has been reported (Medina-Navarro et al., 1999), in which melatonin was shown to exhibit pro-oxidant qualities, possibly due to formation of secondary oxidation products such as endoperoxides. It has been reported that hepatitis has been induced by use of shark cartilage over a 3-week period, after one patient experienced nausea, vomiting, diarrhea, and anorexia (Ashar and Vargo, 1996). Levels of endogenous metabolites may be adversely affected by administration of certain nutraceuticals; glucose levels may be depressed by coenzyme Q 10, and levels of thyroid hormone depressed by carnitine or soy products (Harkness and Bratman, 2003).

Adverse effects of niche supplements are also being reported. 1,4-Butanediol is converted to gamma-hydroxybutyrate after ingestion, and this is marketed as a dietary supplement to enhance bodybuilding. Monitoring of adverse effects in a number of cases showed a wide range of effects including vomiting, incontinence, agitation, respiratory depression, and death (Zvosec et al., 2001). γ-Hydroxybutyrate itself has been reported to cause serious adverse effects. After taking doses of 1 to 60 cc, a number of patients experienced coma and tonic seizure-like activity (Dyer, 1991). Major breathing and cardiac problems have also been reported with this supplement when in combination with drugs of abuse (Chin et al., 1998).

Drug Interactions

Not surprisingly, interactions between nutraceuticals and prescribed medicines or even other nutraceuticals are being reported. Table 6.2 shows some of these interactions, which occur with either drugs or nutraceuticals.

Interactions between nutraceuticals include those experienced with β-carotene (see the section titled, "Carnitine and Acetyl L-Carnitine"). Two studies on carotenoids have shown interactions between β-carotene and lutein and lycopene (Van den Berg, 1999), and reduction of β-carotene absorption after treatment with lutein in both animal and human studies (Van den Berg, 1998). β-Carotene has been reported to interfere with levels of concomitant administration of α-tocopherol (Simone et al., 2002) and has also been shown to potentiate analgesia produced by morphine by a factor of 100% in animal studies (Penn, 1995). Melatonin was shown to exacerbate methamphetamine-induced neurochemical effects, again in animal studies (Gibb et al., 1997), and policosanol has been shown to stimulate the antiulcer effects of cimetidine in rats (Valdes et al., 2000). Neither of these two interactions would have been expected. Interactions between *S*-adenosyl methionine and clomipramine have

TABLE 6.2
Drug Interactions with Specific Nutraceuticals

Nutraceutical	Nutraceutical or Drug	References
β-Carotene	Morphine	Penn, 1995
β-Carotene	Lutein	van den Berg, 1998
β-Carotene	Lycopene	van den Berg, 1999
β-Carotene	Tocopherol	Simone et al., 2002, Blakely et al., 1990, Mayne et al., 1986
Melatonin	Methamphetamine	Gibb et al., 1997
Policosanol	Aspirin or Warfarin	Harkness and Bratman, 2003
Policosanol	Cimetidine	Valdes et al., 2000
S-Adenosyl methionine	Clomipramine	Iruela et al., 1993
S-Adenosyl methionine	Clomipramine or Opiates	Fetrow and Avila, 2001
Linolenic acid or Omega-3 fatty acids	Indomethacin	Codde et al., 1985
Coenzyme Q 10	Warfarin	Landbo and Almdal, 1998
L-Carnitine	Nicoumalone	Martinez et al., 1993
L-Carnitine	Pentylenetetrazol	Kelly, 1998

been reported (Fetrow and Avila, 2001). This not unexpected interaction between two compounds with antidepressant activity dictates that care should be taken with concomitant treatment with prescription drugs and also with certain opium derivatives, and foods high in tyramine should be avoided. Simultaneous administration of both omega-3 fatty acids and linolenic acid and indomethacin was shown to cause reductions in prostaglandin excretion (Codde et al., 1985). An interaction between warfarin and coenzyme Q10 has been reported in which a 72-year-old showed less responsiveness to the warfarin treatment, possibly because of the structural similarity between coenzyme Q10 and the K-vitamins (Landbo and Almdal, 1998). An increased anticoagulant effect has similarly been reported with policosanol (Harkness and Bratman, 2003). It has been reported that L-carnitine should be used cautiously, if at all, with pentylenetetrazol, a respiratory stimulant because evidence suggests that the combination might enhance the side effects of the pharmaceutical (Kelly, 1998).

In addition to interactions between nutraceuticals and drugs, there appears to be instances where drugs depress levels of nutraceuticals. Levels of coenzyme Q 10 have been depressed by administration of acetohexamide, statins, propranolol, phenothiazine, and tricyclic antidepressants, and both carnitine and acetyl carnitine levels depressed by valproic acid (Harkness and Bratman, 2003).

Problems with Formulated Products

Because nutraceuticals are not classified as medicinal products, their manufacture is often not legally regulated. Nutraceuticals are widely available for sale, and the many suppliers may formulate their products differently. This leads to variations between products of different manufacturers. For effective and safe use, the products must contain the same active ingredient at the same concentration, have known bioavailabilty per delivery system, and state detailed and correct information on packaging or patient leaflet. However, possibly due to purchase from nonpharmaceutical sources such as the Internet, there is often little information or data on efficacy, drug interactions, effect of long-term use or abrupt discontinuation, or potential adverse effects available to patients, as is the case with dihydroepiandrosterone (DHEA) (Vacheron-Trystram et al., 2002). Research in Canada has shown that patient-oriented literature often does not mention what is a subeffective dose of β-carotene (generally when products are not available in suitable doses) (Chang et al., 2003) and often may not give information concerning maximum safe doses and drug interactions (Kava et al., 2002).

Researchers have analyzed 23 glucosamine products; 2 out of 10 products did not contain within ±10% of what was stated on the label and, also, there was wide variation of levels within batches. One out of seven chondroitin products showed a variance of 55% from the level stated on the label (Adebowale et al., 1999). More recently, a survey of 14 glucosamine formulations from Canada revealed wide-ranging levels of active constituents with only one product containing within ±10% of the stated dosage (Russell et al., 2002). In a move expected to improve the quality problems, three regulatory bodies in the U.S. are now producing official monographs with standards for glucosamine, melatonin, and β-carotene (Wechsler, 2003).

Many clinical trials that have been performed on a number of nutraceuticals have given satisfactory results and have shown them to be safe; however, there are indications that the trials have not been carried out in an unbiased way. For example, many of the glucosamine trials to date have had sponsorship from companies that manufacture the supplements (Dieppe and Chard, 2001). Due to this, all information relating to the effects of glucosamine not specifically used in clinical trials needs to be treated with caution. It has also been claimed that some soy products, particularly those of European manufacture, may be devoid of phytoestrogens, and consequently may be unable to elicit the physiological responses expected from them (Sirtori, 2000).

As with all medicines and related substances, serious problems may be caused due to poor quality control; an example of an impurity in one formulated product, L-tryptophan, has been reported to have caused subchronic toxicity due to the presence of 3-phenylamino alanine (Sato et al., 1995).

CARNITINE AND ACETYL-L-CARNITINE

Carnitine (β-carotene-hydroxy-γ-*N*, *N*, *N*-trimethylaminobutyric acid) is a naturally occurring amino acid derivative first discovered in muscle extracts in 1905, with the structure being assigned in 1927. It is synthesized from lysine and methionine in the liver, kidney, and brain (Rapport and Lockwood, 2002). Carnitine is sometimes known as vitamin B_T but is not officially recognized as a vitamin. It is involved in fatty acid metabolism by participating in transesterification reactions where medium to long-chain fatty acids (or short-chain organic acids) are transferred from coenzyme

A to the hydroxyl group of carnitine. In the mitochondria, carnitine is essential for the production of energy from long-chain fatty acids. (Rebouche, 1999);

$$\text{Carnitine} + \text{Acyl CoA} \leftrightarrow \text{Acyl carnitine} + \text{Coenzyme A}$$

Carnitine acyltransferase enzymes catalyse these reactions. In the human body carnitine exists in both esterified and nonesterified forms.

The suggested range of metabolic functions of carnitine is listed below:

- Transfer of long-chain fatty acids across the inner mitochondrial membrane
- Facilitation of branched chain α-ketoacid oxidation
- Shuttling of acyl-CoA products of peroxisomal α-oxidation to mitochondrial matrix of the liver
- Formulation of the acyl-CoA/CoA ratio in mammalian cells
- Esterification of the potentially toxic acyl CoA metabolites that maintain the Krebs and urea cycles and pathways for gluconeogenesis and fatty acid oxidation during acute clinical crises (DeVivo, D.C. et al., 1998)

The predominant ester of carnitine in biological systems is acetylcarnitine. (Rebouche, 1999). It is clear then that carnitine and acetylcarnitine are essential components of metabolism, but it should be noted that only the L-isomer of carnitine and acetylcarnitine is naturally occurring, with the D-isomer (synthetic) having more severe adverse effects and hence the demand for supplementation with L-carnitine and acetyl-L-carnitine.

Although the body normally synthesizes sufficient carnitine for its requirements, dietary supplementation has been tested for use in Alzheimer's disease, HIV, chronic fatigue syndrome, and cardiovascular disease. It has also been suggested that carnitine may improve exercise performance (Mason, 2001a).

Primary carnitine deficiency is of unknown prevalence and is thought to be rare. It is genetic and results from a defect in the carnitine transporters, causing a major reduction in carnitine in affected tissues. Secondary carnitine deficiency is more common and produces a less dramatic decrease in blood or tissue concentrations of carnitine or increase in the ratio of esterified to free carnitine. This disorder can be due to inherited metabolic error, acquired disease, e.g., chronic renal failure, or iatrogenic procedures, i.e., by drug administration (DeVivo et al., 1998). The symptoms of carnitine deficiency include fatigue, thrombocytopenia, muscle cramping, and exercise intolerance (Semeniuk et al., 2000, Rapport and Lockwood, 2002). Acetyl-L carnitine deficiency symptoms may include chronic muscle weakness, recurrent episodes of coma and hypoglycemia, encephalopathy, and cardiomyopathy (Carta et al., 1993).

Drug Interactions

Drug-induced carnitine deficiency has been linked with the drugs sodium valproate, isotretinoin, and pivampicillin (Rapport and Lockwood, 2002). These deficiencies have led to the use of L-carnitine supplementation. For example, sodium valproate interferes with carnitine uptake by:

- Forming valproyl-CoA and valproyl carnitine resulting in direct competitive inhibition of carnitine uptake at the transport site
- Forming valproic acid metabolites with secondary inhibition of β-oxidation and excessive CoA
- Decreasing ATP due to inhibition of β-oxidation inhibition, pyruvate metabolism, oxidative phosphorylation, and gluconeogenesis, all of which contribute to a decrease in efficiency of the carnitine transporters (DeVivo et al., 1998)

Supplementation is very important for at-risk patients, e.g., the very young and those with a family history of neurological problems or established metabolic abnormalities. Although valproate toxicity is not always due to carnitine deficiency, carnitine supplementation is recommended (Rapport and Lockwood, 2002). Intravenous carnitine supplementation is clearly indicated for valproate-induced hepatotoxicity, overdose, and other acute metabolic crises associated with carnitine deficiency, whereas oral supplementation is recommended for nonemergency situations, e.g., infants and young children taking valproate (De Vivo et al., 1998).

A study of L-carnitine supplementation in cystic acne patients on isotretinoin therapy has produced positive results. Some of the adverse effects of isotretinoin therapy are shared with carnitine deficiency, i.e., myalgia, weakness, and hypotension. When 230 patients were asked to report any muscular symptoms, 40 reported symptoms. Of these patients suffering adverse effects, 20 were given L-carnitine supplementation and 20 were given placebo. The L-carnitine-supplemented patients found their symptoms disappeared and carnitine levels returned to normal without isotretinoin discontinuation or reduction. The placebo-treated group continued to complain of myalgias. This study would appear to indicate a clear interaction between isotretinoin and carnitine levels. It has been suggested that this is due to liver dysfunction caused by isotretinoin, as the liver is the organ where carnitine is synthesized from methionine and protein-bound lysine (Georgala et al., 1999; Bieber, 1998).

The pivalic acid present in the pivampicillin structure aids its absorption. In the intestinal mucosa, pivalic acid is cleaved by nonspecific esterases. Pivalic acid forms an ester with carnitine and is excreted in the urine as pivaloyl carnitine. Prolonged treatment with these antibiotics leads to a decrease in circulating carnitine levels as well as tissue carnitine. Carnitine supplementation can help to avoid the reduced energy metabolism which may result from pivampicillin therapy (Rebouche, 1999; Rapport and Lockwood, 2002).

Adverse Effects and Toxicity

In clinical trials, L-carnitine has been found to be relatively nontoxic and side effects are rare and generally minor (Arsenian et al., 1996). Acetyl-L-carnitine has equally been found to result in few side effects (Biagiotti and Cavallini, 2001; Torrioli et al., 1999).

However, adverse events have been reported for the use of carnitine. In a randomized, double-blind, placebo-controlled double crossover study involving children receiving carnitine for treatment of attention-deficit hyperactivity disorder, one child (out of 24) dropped out of the study due to experiencing a pungent skin odor following carnitine administration. This is an established side effect likely to be due to the formation of trimethylamine. A fishy body or urine odor was also noted in a study on the use of carnitine in Rett syndrome (Ellaway et al., 1999).

Gastrointestinal side effects have also been reported from the use of carnitine and acetylcarnitine supplementation. Nausea and vomiting, as well as diarrhea and muscle cramping have been reported (Semeniuk et al., 2000). In one study it was found that these side effects only occurred with oral administration and that intravenous L-carnitine had no side effects, although this is not confirmed by other studies (Singh and Aslam, 1998).

It has been suggested that carnitine consumed orally in large quantities (about 5 g/d by an adult) may cause diarrhea as well as fish-odor syndrome (Rebouche, 1999). In addition, aggression and agitation have been reported in some patients following administration of acetyl-L-carnitine (Furlong, 1996).

Carnitine is promoted as a potential ergogenic aid, which promotes exercise performance. Concern has been raised that excessive consumption of nutritional supplements, including those containing carnitine, may lead to adverse effects. This is partly due to the fact that dosage guidelines are inadequate, and quality control is poor (Beltz and Doering, 1993).

A suggested mechanism by which carnitine improves exercise performance is via its vasodilatory properties.The mechanism for this is unknown, but this potential for it to influence blood flow could impact physical performance and recovery (Kraemer and Volek, 2000). If carnitine has significant vasodilatory properties, there may be potential negative as well as positive clinical effects. Theoretically, the potential exists for carnitine to cause hypotension.

Animal studies have shown transient seizure activity induced by acetylcarnitine. This epileptogenic property was demonstrated in rats following intracerebral injection of acetyl-L-carnitine. This is thought to be due to cholinomimetic and GABA uptake interference. (Fariello et al., 1984). In mice, L-carnitine was shown to increase acute paraquat toxicity. This is thought to be due to the chemical structure of carnitine as opposed to its physiological action (Garcia-Rubio et al., 1996).

D-Carnitine

The L-stereoisomers of carnitine and acetylcarnitine are naturally occurring, and D-isomers are synthetic. Most dietary supplements contain L-carnitine or a D L-carnitine mixture. Compounds containing the D-isomer should be avoided as this isomer can interfere with normal function of the L-isomer, leading to adverse effects (Mason, 2001a). The D-isomer does this by acting as a competitive inhibitor of uptake systems for L-carnitine and hinders mitochondrial fatty acid oxidation and energy formation. D-Carnitine administration leads to a depletion in L-carnitine in cardiac and skeletal muscle, which consequently can cause cardiac arrhythmias and muscle weakness (Fuhrmann, 2000). Serious toxic effects have been observed with the administration of the racemic mixture, and D-carnitine is not safe for human consumption; it has effectively been banned in the U.S. (Fuhrmann, 2000).

ISOFLAVONES

Isoflavones are heterocyclic phenols found in plants, particularly soybeans and soy products (Messina et al., 1994). They are called phytoestrogens and occur naturally in both conjugated and unconjugated forms. For example, tofu (soybean curd), contains high levels of glycoside conjugates daidzin and genistin, which are hydrolyzed in the large intestine to release unconjugated daidzein and genistein. Fermented soybean products such as miso have approximately 90% of isoflavones present as in the unconjugated form, i.e., daidzein and genistein (Cassidy, 1996).

The isoflavones have shown weak estrogenic activity varying between 1/500–1/1000 of the activity of mammalian estrogen, 17-β-estradiol. They have produced typical estrogenic responses when administered to animals (Cassidy, 1996; Juniewicz et al., 1998).

The binding of isoflavones to estrogen receptors seems to depend on the individual and has led to suggestions that isoflavones may be antiestrogenic premenopausally, but may prove to be estrogen receptor agonists postmenopausally. The isoflavone daidzein can be metabolized by bacteria in the large intestine to form the estrogenic equol or the nonestrogenic σ-desmethylangolensin. Genistein is metabolized to the nonestrogenic ρ-ethyl phenol. Individual variation in the ability to metabolize daidzein could consequently influence the effects of isoflavones (Mason, 2001b).

Isoflavones have generated major interest as drugs and nutritional supplements due largely to the fact that the incidence of breast and prostate cancer is significantly lower in Oriental countries, where soy isoflavone intake is highest (Adlercreutz et al., 1995; Cassidy, 1996). The Japanese diet, for example, is high in soy, and an epidemiologic study carried out in Singapore found an inverse relation between the consumption of soybean products and breast cancer risk in premenopausal women (Adlercreutz et al., 1991). This apparent protection against hormonal cancers with the Asian diet is lost by the second generation following migration to Western countries, where soy isoflavone intake is significantly lower, e.g., the U.S. (Bloedon et al., 2002).

Of the isoflavones, genistein has been identified as most likely to exhibit anticarcinogenic properties, despite the similar structure and increased bioavailability of daidzein. (Record et al., 1995).

Isoflavones are being studied for their use in therapy with cardiovascular diseases due to their cholesterol-lowering ability; in cancer, due to their potential protective effect; with osteoporosis, as a result of studies which have shown that soy isoflavones preserve bone mineral density; and with menopausal symptoms as a natural hormone replacement therapy (HRT) supplement (Mason, 2001a).

Mechanisms of Action of Soy Isoflavones

The full range of actions for the soy isoflavones has not yet been elucidated. However, it has been proposed that soy isoflavones do the following:

- Activate estrogen receptors
- Inhibit the activity of growth-promoting steroid hormones by inhibition of the enzymes progesterone 5α-reductase and 17β-hydroxysteroid dehydrogenase
- Inhibit protein-tyrosine kinase-mediated signal transduction
- Inhibit ATP hydrolyzing DNA topoisomerase
- Inhibit transforming growth factor β_1-mediated signal transduction
- Inhibit angiogenesis (Busby et al., 2002)

Despite the fact that soy has been widely consumed as a regular part of the diet in many cultures which benefit from these wide ranging effects, there is likely to be some degree of toxicity and adverse reactions from these otherwise potentially useful compounds which must be assessed. These are discussed in the following text.

Toxicity

Potentially mutagenic, carcinogenic, and teratogenic properties of phytoestrogens have been reported (Sirtori, 2000), and animal studies have shown an influence of phytoestrogens on reproductive physiology. An infertility syndrome involving ewes grazing on pastures of *Trifolium subterraneum* containing the isoflavone precursor formononetin was linked to the isoflavone. The metabolism of this compound leads to the formation of equol, a weak estrogen. Equol is absorbed, achieving high blood concentrations which result in permanent histological damage to the reproductive organs of the ewe (Cassidy, 1996). Infertility and veno-occlusive liver disease have also been attributed to isoflavones in the diet of the captive cheetah. This was explained to be due to the inability of the feline liver to deactivate the unconjugated isoflavones by conjugating with glucuronic acid. As a result, a high level of unconjugated isoflavones is present in the systemic circulation (Cassidy, 1996).

Conversely, developmental studies in rats treated during the prepubertal period did not find significant alterations to fertility, number of male and female offspring, body weight, anogenital distance, vaginal opening, testes descent, estrus cycle, or follicular development in rats tested with the unconjugated isoflavone, genistein. This study suggested that genistein is too weak an estrogen to cause endocrine and reproductive tract alterations in prepubertal exposure (Lamartiniere et al., 1998).

Phytoestrogens present in soybeans inhibit human breast cancer cell proliferation *in vitro* and breast cancer development in animal models. However, in a study involving 47 premenopausal women with histologically normal breasts, a significant increase in the number of epithelial cells in the S phase of the cell cycle was observed when subjects supplemented their normal diet with soy. An upregulation in progesterone receptor expression and an elevation in the number of proliferating epithelial cells occurred. No evidence of estrogen antagonism was found and the findings of the study suggested that dietary soy protein may have estrogen agonist effects on the breast (McMichael-Phillips et al., 1998). If these findings were to be repeated in a subject with breast cancer, the potential exists for the soy to stimulate the tumor cells.

The antiproliferative effects of genistein may not, however, be mediated by the estrogen receptor but may involve tyrosine kinase inhibition, topoisomerase II, angiogenesis, and the transformation of growth factor-β. DNA topoisomerases are essential enzymes that govern DNA topology during fundamental nuclear metabolic processes, and the isoflavone genistein has been shown to inhibit tyrosine kinase activity *in vitro* and stabilize the topoisomerase II-DNA complex, thus functioning as a topoisomerase poison (Martin-Cordero et al., 2000). Genistein is a nonintercalative inhibitor of topoisomerase II and the main lesions induced by both nonintercalative and intercalative inhibitors are double-stranded breaks, which are considered to be the ultimate DNA lesions leading to chromosomal aberrations (Liu, 1989; Kulling et al., 1999).

The ability of isoflavones to inhibit topoisomerase is a mechanistic concern associated with isoflavones due to the possibility of this leading to DNA strand breaks and subsequent mutations. When PTI G-2535 (an investigational soy isoflavone drug product) was subjected to mouse lymphoma cell mutagenesis experiments, it was revealed that, with and without metabolic activation, statistically significant dose-related increases in mutation frequency was measured. As part of the same study, it was found that there were small but statistically significant increases in the frequency of micronucleated polychromatic erythrocytes in male, but not in female, mice receiving 500 to 1000 mg/kg body weight of PTI-G2535 (Misra et al., 2002).

Micronucleus formation is an event associated with genetic damage, possibly resulting from the production of double-stranded DNA breaks due to the inhibition of topoisomerase II (Record et al., 1995). It is suggested that an increase of 1 micronucleated cell/1000 cytokinesis-blocked cells is equivalent to exposure of 1 rad of x-radiation (Record et al., 1995). Exposure of mouse splenocytes to the level of genistein incurred in Record's 1995 study was considered equivalent to greater than 100 rad of x-radiation. Results showed that although *in vitro* studies induced cytogenic damage, sufficient concentrations were not reached *in vivo* to have a clastogenic effect, despite a massive dietary intake (Record et al., 1995).

In a separate study, phytoestrogens, including genistein and daidzein, were investigated to examine their ability to induce chromosomal aberrations in cultured human peripheral blood lymphocytes. This was done by exposing the lymphocytes to differing concentrations of daidzein, genistein, and the lignan coumestrol. It was found that genistein and the coumestrol clearly increased the number of aberrant metaphases in the cell cycle, whereas daidzein did not. As discussed in the preceding text, these genotoxic actions are not mediated by estrogen receptors and are, therefore, independent of the hormonal activity. The observed chromatid aberrations in this study were chromatid gaps, breaks, and chromatid type interchanges (Kulling et al., 1999).

The ability of genistein to inhibit tyrosine kinase has been linked to potential anticarcinogenic activity of the compound as several growth factor receptors and oncogenes are regulated by tyrosine phosphorylation. However, it has also been suggested that tyrosine phosphorylation inhibition plays a significant role in the response of checkpoints in the cell cycle to DNA damage. For example, progress from the G2- to M-phase of the cell cycle is controlled in part by a cyclin-dependent kinase and is regulated by phosphorylation of tyrosine. If this tyrosine kinase is inhibited, the checkpoint may be inactivated resulting in mitosis of damaged cells.

The survival of such damaged cells will result in cells with permanent genetic alterations. (Kulling et al., 1999).

Neuronal apoptosis was found to result from high doses of genistein, thus raising the issue that there may be potential nervous system side effects resulting from the use of genistein in high doses as a therapeutic agent. This neuronal apoptosis was found when rat primary cortical neurons were exposed to genistein 50 μ*M*. Genistein clearly induced a potent cytotoxicity in primary cortical neurons within 24 h of treatment. The dying cells were characterized by fragmented nuclei, appearance of apoptotic bodies, DNA laddering, and capsase-dependent poly (ADP ribose) polymerase cleavage (Linford et al., 2001). Daidzein was found to show no such toxicity at similar concentrations as, despite its similar estrogen agonist properties to genistein, it does not show tyrosine kinase and topoisomerase II inhibition (Linford et al., 2001).

Linford's study (2001) examining neuronal apoptosis showed that genistein is not associated with downregulation of the antiapoptotic proteins Bcl-2 and Bcl-X_L. It was shown that genistein-induced apoptosis does involve the activation of proteases, specifically the caspase family and intranucleosomal cleavage of genomic DNA. In response to genistein, p42/p44 and p38 mitogen-activated protein kinases (MAPK) are converted into their dual phosphorylated active form. In these cells p42/44 MAPK activation is proapoptotic, but p38 does not appear to be critical. Calcium release from intracellular stores is involved in genistein-induced toxicity. By blocking calcium release and p42/4 MAPK activation, toxicity can be reduced (Linford et al., 2001).

The tyrosine kinase and topoisomerase inhibition properties of genistein have been attributed as the primary intracellular targets for genistein apoptosis induction. It is likely that the combination of these as well as direct effects on mitochondria result in the apoptosis observed in primary cortical neurons.

The above study leads to concerns when considering the use of genistein in chemotherapy as the levels of genistein needed to initiate a proapoptotic effect in tumor cells are similar to those which result in toxicity to cultured neurons (Polkowski and Mazurek, 2000; Linford et al., 2001).

Isoflavones have also been linked to infant leukemia (Misra et al., 2002; Strick et al., 2000). Chromosomal translocations involving the MLL gene occur in approximately 80% of infant leukemia, and isoflavones have been identified as substances that cause DNA cleavage in the MLL breakpoint cluster region (BCR) *in vivo*. Again, this is linked to DNA topoisomerase II inhibition. Isoflavones that showed cleavage of MLL equivalent to that caused by 25 μ*M* of the epipodophyllotoxin VP16 were 50 μ*M* genistein and 100 μ*M* genistein. 200 μ*M* Daidzein produced 50% MLL cleavage compared to the VP16, which is a known eukaryotic topoisomerase II inhibitor. Genistein was the most potent topoisomerase II inhibitor of the isoflavones tested and was shown to have topoisomerase II inhibition similar to that of VP16 and the chemotherapeutic drug doxorubicin (Strick et al., 2000).

Liu (1989) proposed a two-stage pathway for processing topoisomerase II-inhibiting substances: Stage 1 involves the stabilization of topoisomerase II cleavable complexes by forming drug (e.g., isoflavone: topoisomerase II: DNA) ternary complexes on chromosomal DNA. This stage is reversible by DNA repair or religation. Stage 2 is irreversible, leads to cell death, and is activated by accumulation of these ternary complexes. Strick et al. (2000) state that illegitimate chromosome translocations, which may result in leukemia, occur only during the first reversible stage of MLL DNA breakage.

The suggestion that infant leukemias could be linked to bioflavonoid intake can also be associated with studies that have shown an increased incidence of infant ALL and AML in several Asian cities, e.g., Hong Kong and Osaka, compared with Western countries. This difference could be attributed to the high dietary intake of bioflavonoids, such as the isoflavones genistein and genistein (Strick et al., 2000). A preliminary epidemiological study demonstrated that a tenfold higher risk of AML occurred following maternal consumption of topoisomerase II inhibitor-containing foods, including isoflavones (Ross et al., 1996).

Adverse Effects

When single dose safety studies of purified unconjugated genistein, daidzein, and glycitein were carried out on 30 healthy men, some adverse events were reported. Four of these, specifically two episodes of loss of appetite, one case of pedal edema, and one case of abdominal tenderness were judged to be related possibly to the isoflavone due to results from animal studies. A further eight events may have been related to isoflavone administration as no other cause was identified. These included two cases of elevated lipase, one case of elevated amylase, one case of leucopenia, and four cases of hypophosphatemia (Busby et al., 2002). A similar study undertaken in women showed two cases of pedal edema, one case of nausea, and one of breast tenderness potentially attributable to the isoflavones. Blood pressure reduction was also reported with one test formulation, as was a decrease in the neutrophil count (Bloedon et al., 2002).

The hypophosphataemia resulting from isoflavone administration may be due to a possible mechanism whereby genistein inhibits phosphate reabsorption and increases phosphorous excretion by the kidney. Genistein may also change phosphorous deposition in the bone, and despite the fact that this degree of hypophosphataemia shows no clinical toxicity, low phosphorous over an extended period of time may inhibit bone formation (Busby et al., 2002). Breast tenderness may have been due to the estrogenic effects of the isoflavones or may have been the consequence of interactions with estrogen replacement therapy as both of the subjects experiencing these symptoms were prescribed HRT.

Changes in lipase levels may also be attributable to the isoflavones as genistein or daidzein might alter the regulation of activity or the distribution of lipoprotein lipase. Genistein inhibits a protein kinase, which mediates the modulation of lipoprotein lipase activity by glucose, as well as inhibiting the cyclic-AMP-mediated release of lipoprotein lipase activity from fat pads (Busby et al., 2002).

It can be seen, then, that isoflavones have a variety of actions and are potentially very useful substances in medicine. However, caution should also be taken with excessive consumption of the isoflavones as their full toxicity to humans is yet to be examined.

PROANTHOCYANIDINS

Proanthocyanidins, as with the isoflavones, are a subgroup of the bioflavonoids. They are also known as vegetable tannins, polyphenols, or condensed tannins. They occur naturally in the flowers, roots, and bark of a number of plants as well as in many fruits, vegetables, and nuts (Rapport and Lockwood, 2002; Bagchi et al., 1998). The health benefits of many fruits and vegetables are, in part, thought to arise from the presence of proanthocyanidins acting as antioxidants in the body. They were first extracted from pine bark in 1951 and from grape seed in 1970 (Fine, 2000).

The proanthocyanidins are polyphenols based on the flavan-3-ol unit. This unit can either be catechin, which has a *trans* arrangement or epicatechin, which has a *cis* arrangement, as shown in Figure 6.1.

These structures may join together to form dimers, trimers, and oligomers, which make up the family of proanthocyanidins (Rapport et al., 2000). The (–)-epicatechin can be isolated from cacao beans, whereas, (–)-epigallocatechin and its 3-gallate are mainly found in green teas, along with (+)-gallocatechin, (+)-catechin, (–)-epicatechin, and its 3-gallate (Chung et al., 1998a).

Proanthocyanidins are being used for their potential benefit in various areas of healthcare. They have antioxidant properties, which means that they appear to act as scavengers of free radicals, including superoxide anions, singlet oxygen, and lipid peroxyl radicals (Mason, 2001a). The presence of proanthocyanidins such as GPSE (grape seed proanthocyanidin extract) in wine may very well contribute to the prevention of atherosclerosis as part of the "French paradox" (German and Walzem, 2000). Proanthocyanidins are also being suggested for various other uses, such as the prevention and treatment of cancer and vascular disorders, as antiviral agents, as natural sweeteners, and even to promote hair growth (Rapport and Lockwood, 2002).

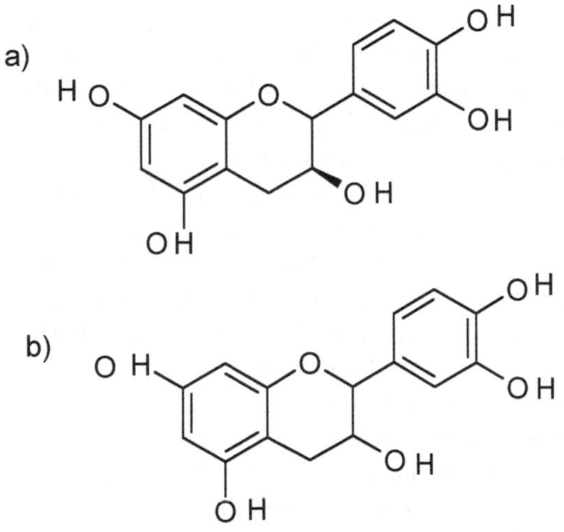

FIGURE 6.1

As they are commonly consumed in a wide range of food and plant products, such as wine and tea, proanthocyanidins are generally thought to be safe. A safety evaluation of proanthocyanidin-rich extract from grape seeds, which subjected the grape seed extract to a series of toxicological tests to determine its safety for use in various foods, found that their studies showed a lack of toxicity (Yamakoshi et al., 2002). However, a review of these studies in the following text shows potential toxicity and adverse effects of these compounds.

Toxicity and Adverse Effects

It has been shown that very high levels of antioxidants in some individuals may cause pro-oxidation, potentially worsening cardiovascular damage and atherosclerosis. It may prove necessary in the long term to screen patients to determine how they would benefit from antioxidants such as proanthocyanidins, although it is thought that the numbers affected would be small (Rapport and Lockwood, 2002).

Incidences of certain cancers such as esophageal cancer have been related to consumption of tannin-rich foods such as betel nuts and herbal teas, suggesting that tannins may in fact be carcinogenic. However, it has also been suggested that this carcinogenic activity may not be related directly to the tannins but to those components with which they associate (Chung et al., 1998b).

Tannins, and particularly the condensed type (i.e., proanthocyanidins), are reported to inhibit most digestive enzymes including pectinase, amylase, lipase, and proteolytic enzymes. They are said to interfere with the digestion and absorption of carbohydrates from sorghum. One mole of tannin is said to bind 12 mol of protein, and, therefore, it is believed that it is the formation of poorly digestible complexes with dietary protein, rather than inhibition of enzymes, which results in adverse effects. High levels of tannins in domestic feeds, including grape seed, are said to be responsible for the poisoning of domestic animals (Chung et al., 1998a, 1998b).

Gastroenteritis and congestion of the intestinal wall in rats were attributed to the ingestion of condensed tannins. This effect was suggested to be due to the reduction in the efficiency with which the digested and absorbed nutrients were converted into new body substances rather than direct inhibition of digestion. It can be seen, then, that consumption of large quantities of proanthocyanidins in the diet is nutritionally undesirable (Chung et al., 1998a).

Despite being investigated for anticarcinogenic effects, tannins in teas, including catechin, have been suspected to cause cancers, although it is thought that this carcinogenic activity may be due to the irritation and cellular damage that they promote as opposed to direct mutation of DNA (Chung et al., 1998a).

A screening program of compounds derived from Chinese herbal medicines, with a view to discovering a lead compound for a male contraceptive, found that, *in vitro*, epicatechin-(4β-8)-catechin and catechin have a strong inhibitory effect on sperm motility, showing a dose–response relationship. It is possible that these effects may not be exerted *in vivo* (Chiang et al., 1992). However, this may lead to potential contraceptive drugs for men and also may raise the question whether excessive consumption or supplementation with these substances may affect male fertility.

In a study of the *in vivo* effects of (+)-catechin on iron-overloaded rats (Das et al., 1988), (+)-catechin was found to be toxic to erythrocytes, hepatocytes, and the kidney in very high dose 100 mg/100 g body weight, whether used alone or in combination with iron loading. In the iron-loaded animals, alone, fatty acid metabolism was also affected by (+)-catechin. This again shows that proanthocyanidins may be toxic in very large doses and also interfere with iron metabolism.

MELATONIN

Melatonin is a hormone produced by the pineal gland in darkness but not in bright light. There are receptors found in the brain within a nucleus situated above the optic chiasma. These react with melatonin, and so this nucleus becomes synchronized to the day–night rhythm by the light and dark effects on the release of melatonin. This hormone is a derivative of serotonin, which is also believed to play a role in sleeping and the body's own day–night orientation (Rapport and Lockwood, 2002), therefore affecting the circadian rhythms of the body.

Natural melatonin has been found in bananas, tomatoes, cucumbers, and beetroots, although only in small amounts. Large amounts of these foods would have to be consumed to dramatically alter the levels within the body. As melatonin is proposed to control the circadian rhythms it has been researched as an aid for jet lag, sleep disorders, aging, antioxidant properties, cancer, and enhanced immunity and reproduction. (Rapport and Lockwood, 2002)

Melatonin is available without a prescription in the U.S., but there is little or no information readily available on its toxicity. The lowest dose available in an over-the-counter preparation (0.3 mg) produces levels that are comparable with natural nocturnal melatonin (Dollins et al., 1993). In Europe, it is classified as a nutrohormone that cannot be sold over the counter. (Guardiola-Lemaitre, 1997). As a hormone, its action is dependant on its timing, concentration, and relationship with other compounds and receptors within the body to serve its purpose. One study suggested that the lengths of pregnant mothers' days directly affected the shyness of their children (Gortmaker et al., 1997). Although the evidence behind this claim is not clear, it does raise the possibility that such a hormone could have disastrous effects if abused by taking at the wrong time of day. During the synthesis of melatonin, which derives from L-tryptophan, the metabolite L-kynurenine, along with others, are found in the metabolic pathway. This compound has been reported to have a convulsant effect, and another metabolite, quinolinic acid, has been reported to be a neurotoxin that has neuroexcitatory activity at the level of the *N*-methyl D-aspartate receptor. (Huether et al., 1992; Heyes et al., 1994) This raises the question as to whether supplements such as melatonin should be so widely available to the general public with little or no regulation of the formulated products.

ADVERSE EFFECTS

There is limited data on the use of melatonin, and little is known as to the pharmacokinetic or pharmacodynamic nature of the compound as a supplement. Side effects that have been demonstrated include inhibition of reproductive function, delayed timing of puberty, and influence (when taken during pregnancy and lactation) on the circadian status of the foetus and neonate, and on its future development (Arendt, 1997).

There have been several recent studies carried out to assess the potential for treatment of insomnia and circadian rhythm disorders. Drugs that increase sleepiness — for example, sedatives and antihistamines — can cause inappropriate effects on neurobehavioral performance and, therefore, can cause side effects and potential risk to users. One study with 16 young healthy volunteers, involved giving 5 mg of oral melatonin at 12.30 h, followed by neurobehavioral performance tasks. This study found significant decrements in performance on the tracking task, as well as on the response and reaction time scores for the visual choice and extended two-choice tasks (Rogers et al., 1998). A similar study observed melatonin supplements for sleep disorders and emotional or behavior disturbances of patients with developmental disorders. In the study, 42 patients out of the 50 experienced improvement of sleep following melatonin; however, 17 experienced side effects. These effects included residual drowsiness the next morning, awakening in the middle of sleep, and excitement after wakening and before going to sleep (Ishizaki et al., 1999). These effects were also seen in a study looking at melatonin for jet lag (Samel, 1999), which also reported that inappropriate timing of intake caused sleep disturbances and unfavourable shifts in the circadian system. A study was conducted on 22 patients with advanced metastatic cutaneous and ocular malignant melanoma, with each patient given high dose oral melatonin. Little significant therapeutic benefit was seen in these patients, and it was concluded that melatonin was safe and well tolerated as no patients withdrew from the study due to adverse effects; however, some patients did report drowsiness (Robinson et al., 1995).

A study of 97 patients with circadian rhythm disorders used a dose of 5 mg melatonin every evening 5 h before endogenous melatonin production occurred, for a period of 2 to 12 months. During the study a total of 35 adverse reactions were reported by 25 of the subjects. These included fever, hyperkinesis, dizziness, gastrointestinal disorders, headaches, hemorrhages, pigmentation, ankle edema, flushing, diplopia, hepatic pain, thrombosis, and hyperglycemia in type 1 diabetes patients receiving treatment (Nagtegaal et al., 1997). Tolerance, fatigue, and other side effects have been reported, and, therefore, it should be avoided on consecutive nights and only the lowest possible dose to produce the desired effect should be used (Kendler, 1997).

In one study that was carried out on healthy male volunteers, 3 mg of oral melatonin was given to each subject, and a series of autonomic tests were carried out. The results concluded that melatonin had a significant hypotensive effect on blood pressure and on the central cardiovascular regulatory mechanism, such as lowering of the baroreflex set point (Kitajima et al., 2001). This finding may cause concern, not only for already hypotensive patients who may be self medicating with melatonin, but also for patients concurrently receiving an antihypotensive, the effect of which may be enhanced by taking this supplement and may push them below a safe blood pressure level. In contrast to this, an *in vitro* experiment has been performed in isolated rat caudal and cerebral arteries as well as isolated pig and human coronary arteries (Mahle et al., 1997). In these experiments, melatonin was shown to produce vasoconstriction of the vessels showing the need for further and possibly more rigorous studies as to the effect that melatonin has on systolic and diastolic blood pressure in human subjects. Melatonin receptors have been identified in many different structures within the body such as the arteries, and although the mechanism by which it exerts its effect is not known, a study has shown a higher risk of coronary heart disease in shift workers (Kawachi et al., 1995). This also provides an interesting line of thought that melatonin secretion decreases with age, and so this factor could be related to the changes to vascular tone that is seen as aging occurs. However, it also highlights the dangers of inappropriate dosing of such a supplement.

The possibility that melatonin could be related to autoimmune hepatitis was suggested when a patient developed the features of autoimmune hepatitis after being given melatonin therapy for insomnia. In this study, a liver biopsy performed on the patient demonstrated histologic features of autoimmune hepatitis. When given steroid immunosuppressive therapy, symptomatic and biochemical improvement was rapidly demonstrated in the patient; however, the hepatitis reoccurred after withdrawal of the therapy. The study raises the possible question that melatonin supplementation may have been a key factor in the pathogenesis of the observed patient's autoimmune disease (Hong and Riegler, 1997).

Some studies as to the effects and toxicity of melatonin have demonstrated toxicity in rats. One such experiment administered 10 mg/kg of melatonin to normal and 30-d streptozotocin-induced diabetic male Sprague–Dawley rats, and then measured activities of phase I and phase II biotransformation enzymes in the liver, kidney, intestine, and spleen. This study concluded after observation of the rats showed that melatonin may induce some phase II enzymes in normal but not diabetic rats (Maritim et al., 2000), although this study was only of short duration and further studies of this nature are needed.

Melatonin has been shown to influence the timing of puberty in species such as Siberian hamsters, sheep, ferrets, and monkeys (all of which are seasonal breeders), possibly as these animals are the most widely studied. The timing of puberty in these animals has been shown to be affected by day lengths experienced in the postnatal period. An attempt was made to create a contraceptive pill that combined melatonin with norethindrone, although the project was not pursued. The major side effects that were experienced by those who took the tablet are shown in Table 6.3 (Cohen et al., 1995). Although there is no evidence to suggest that melatonin supplementation alters the onset of puberty in humans, it raises the possibility that slight alterations in melatonin levels within the body could cause unwanted effects (Weaver et al., 1997).

Arendt's criticism of melatonin being widely used as a contraceptive is that toxicological data is not publicly available (Arendt, 1997). Dopamine and melatonin have been shown to control the daily rhythms of nonmammalian retinas. During the day the cones contract, rods elongate, and pigment granules migrate to the optical villi of retinal pigment epithelium cells, with the opposite happening at night. Melatonin controls the reaction in the dark, and dopamine, the reaction in light (Lemaître, 1997). Although the exact mechanism has not been demonstrated in

TABLE 6.3
The Major Side Effects Experienced by Women Taking Melatonin (75 mg) and Norethindrone (0.5 mg) as a Daily Contraceptive in Different Formulations for a Period of Months to Years

Adverse Effect	Percent Experiencing the Effect
Abnormal bleeding	11.1–36.9
Gastrointestinal complaints	11.1–17.7
Breast complaints	3.17–9.36
Neurosensory problems	1.6–4.8

human subjects, the involvement of the two compounds has been. It could be concluded that any number of eye-related problems could arise from inappropriate administration of melatonin.

Toxic Effects and Drug Interactions

Melatonin has also been found to interact with several prescription drugs. Fluvoxamine, which is a selective serotonin reuptake inhibitor used for depressive illness and obsessive–compulsive disorder, has been found to significantly increase serum melatonin levels. This effect seems to be due to fluvoxamine reducing the metabolism of melatonin by inhibiting human cytochrome P450 enzymes, CYP1A2 or CYP2C9, which metabolize diazepam and haloperidol, two drugs that also have sedative properties. Therefore, these drugs taken concurrently could lead to exaggerated enhancement of their sedative effects (Hartter et al., 2000). More problematic, melatonin has also shown to interact with the antihypertensive drug nifedipine. This appears to be due to melatonin having some form of effect on the cardiovascular regulation and so increasing the blood pressure and heart rate of patients taking the two products concurrently. Studies have shown that melatonin-binding sites exist at the level of the gastrointestinal tract. Human studies have reported that nonsteroidal antiinflammatory drugs, known for their gastrointestinal side effects, produce these effects more readily if given in the morning rather than the evening, which shows a time-dependant effect (Moore and Goo, 1987). As melatonin has a worsening effect on these ulcerations, this highlights the need for careful monitoring of those susceptible to gastric ulcer, especially those taking nonsteroidal antiinflammatory drugs while taking melatonin supplements.

No long-term safety data for melatonin is available. It is probably safe to say that it should not be given to children or pregnant or lactating females, as it crosses the placenta and has shown rhythmic variations in milk in both humans and goats. Furthermore, Lemaître has discussed animal data that show melatonin interfering with glucose metabolism (Lemaître, 1997); therefore monitoring of diabetic patients using this hormone may also be needed. As melatonin is secreted in the urine and some of its metabolites have been shown to have pharmacological effect, it is quite possible that metabolites or indeed melatonin itself could be found in underground water stores (Arendt, 1997). Melatonin could have effects on other animal species or even humans if it is reentering the water supply, thus reinforcing the fact that there is a great need for more research to be done on this compound. It also raises the question of whether such a compound should be made available without prescription with such little available data, coupled with evidence to suggest that it may have toxic side effects for users.

β-CAROTENE

β-carotene is a carotenoid widely available in fruits and vegetables. It can be converted to retinol (Vitamin A) by the action of dioxygenenase. Vitamin A is a fat-soluble vitamin that is largely found in foods of animal origin. It is needed for growth, night vision, and maintenance of soft mucous tissue. As β-carotene is found in foods such as fruits and vegetables, and as observational epidemiological studies indicate that an increase of such foods within the diet decreases the incidence of cancer, high β-carotene levels have been associated with cancer prevention and particularly cancer of the lung. (Bast and Haenen, 2002)

In the U.K. the safe upper limit of β-carotene per day is 7 mg. It is estimated that the average intake for a male is 2.4 mg/d and for a woman 2.1 mg/d (Mason, 2003). The recommended daily intake is 750 μg of retinol; this is equivalent to 4.5 mg of β-carotene. Normal dietary sources supply insufficient β-carotene.

Toxic Effects

β-Carotene has been investigated in recent years to assess whether it could help prevent cancer. The studies have focused on smokers and the incidence on pulmonary cancer and disease. However, two recent trials found some rather unexpected results, and there is evidence appearing that β-carotene induces lung cancer in smokers (Yeh, 2001).

One study looked at 29,133 male smokers aged between 59 and 69 in southwest Finland. The men were given one of the following four regimens:

- α-tocopherol alone (n = 7,286)
- α-tocopherol and β-carotene (n = 7,278)
- β-carotene alone (n = 7,282)
- placebo (n = 7,287)

A total of 14,564 individuals took α-tocophenol at a dose of 50 mg/d and 14,560 took β-carotene at a dose of 20 mg/d. The follow up of the experiment continued for 5 to 8 yr or until death. The study found no evidence of any decrease in the incidence of lung cancer in the supplemented groups or of any benefits compared with placebo. In fact the results showed an increase in the incidence of cardiovascular disease and lung cancer in the group that had been supplemented with β-carotene (The Massachusetts Society, 1994). This effect was also shown in a more recent trial that was stopped because of an increased risk of developing cancer in the supplemented group (Omenn et al., 1996).

TABLE 6.4
Particular CYP Enzymes that Were Found to Increase Significantly in Rat Lung after Supplementation with β-Carotene

Cytochrome P450 (CYP) Changes	Role
CYP1A1/2	Activating aromatic amines, polychlorinated biphenyls, dioxins, and PAHs
CYP3A	Activating aflatoxins, 1-nitropyrene, and PAHs
CYP2B1	Activating olefins and halogenated hydrocarbons
CYP2A	Activating butadiene, esamethyl phosphoramide, and nitrosamines

Another study was conducted on rat lung, and it was found that β-carotene produced a powerful booster effect on phase 1 carcinogen-bioactivating enzymes, including activators of polycyclic aromatic hydrocarbons (PAH). This induction was associated with the generation of oxidative stress. These findings may help to explain why β-carotene may increase the risk of lung cancer in smokers (Paolini et al., 1999). In the same study, significant increases in the carcinogen-metabolizing enzymes (see Table 6.4) were recorded.

If corresponding findings occur in humans, then the individual would be predisposed to an increased cancer risk from the bioactivated tobacco-smoke procarcinogens. Thus, we conclude that β-carotene's cocarcinogenic properties and its ability to cause oxidative stress could have been responsible for the observed mortality in the lung cancer trials (Paolini et al., 1999). It has also been suggested that β-carotene may have cocarcinogenic properties on latent tumors rather than initiating new ones (Hinds et al., 1997).

Drug Interactions

β-carotene has also been found to interact with alcohol, which interferes with the latter's conversion to retinol. Studies on human subjects have shown a correlation between plasma β-carotene and alcohol consumption. Heavy drinkers generally have a low β-carotene plasma concentration due to lack of intake; in one study, alcohol appeared to increase plasma concentration for women drinking as little as two drinks a day (Ahmed et al., 1994). The effect was further studied in baboons; it was found that when β-carotene and alcohol were given together, more hepatic injury was seen than when either of the compounds were given alone (Leo et al., 1992). The dose of β-carotene given to the baboons was between 7.2 and 10.8 mg, which is a similar amount that would be found in an individual taking supplements. The amount of alcohol (50% of dietary energy) was equivalent to the daily intake of an alcoholic person (Patek et al., 1975). In rats, however, the alcoholic dose was less than that to produce hepatic injury (36% of total dietary energy), namely liver lesions (Leo et al., 1992). This raises the possibility of β-carotene interacting with alcohol in humans. As β-carotene has also been shown to increase the incidence of pulmonary cancer in smokers, it has been suggested that because heavy smokers are often heavy drinkers, too, alcohol abuse could be contributing to this effect (Leo and Lieber,

1994). This could be due to pulmonary cells being exposed to high oxygen concentration and β-carotene losing its antioxidant activity and showing autocatalytic pro-oxidant effect at these pressures (Burton and Ingold, 1984), although the exact mechanism is not known. In a study that looked at the combination of β-carotene acetaldehyde, which is the toxic metabolite of ethanol on liver cells in culture, and Hep G2 cells, the results showed that when β-carotene inhibits the activity of mitochondria in human liver tumor cells and when it is given with acetaldehyde, the effect is enhanced. This finding may well help to explain the toxicity that was observed when β-carotene and alcohol were combined in baboons (Leo et al., 1992), rats (Leo and Lieber, 1994), and in humans (Albanes et al., 1997).

A number of interactions have been reported between β-carotene and other antioxidants (see the earlier section titled "Drug Interactions"). Supplementation of β-carotene has been shown to decrease the plasma and hepatic levels of α-tocopherol (vitamin E) in rats (Blakely et al., 1990). It has been suggested that this could be due to β-carotene altering the binding or absorption of α-tocopherol by competing for the same binding sites on the lipoprotein molecules (Pellet et al., 1994). These findings have also been demonstrated in humans. In this study the subjects were given 800 mg of α-tocopherol, and it was demonstrated that their plasma carotenoids decreased (Willett et al., 1983).

GLUCOSAMINE

Glucosamine is an amino monosaccharide formed from glucose and glutamic acid, and is also found occurring naturally within the human body, concentrations being the largest in cartilage, tendons, and ligament. This is due to the fact that it is a precursor of articular cartilage glucosaminoglycan. To gain extra glucosamine, a supplement must be taken as, unlike other nutraceuticals, it is not found within the diet (Sutton et al., 2002).

Glucosamine supplementation of 1500 mg in divided doses daily is believed to be useful in osteoarthritis. This is a result of depletion of the articular cartilage, which may lead to secondary changes in the underlying bone. Patients develop restrictive movement due to stiffness and pain. This condition is largely treated with analgesics such as paracetamol and nonsteroidal antiinflammatory drugs, but with side effects including nausea, diarrhea, and occasionally bleeding and ulceration (Rapport and Lockwood, 2002). Consequently, the prospect of an alternative treatment such as glucosamine is very appealing. Glucosamine is largely regarded to be a safe supplement, although it has been found that gastrointestinal side effects occur in up to 12% of those who take it. These effects included upset stomach, nausea, heartburn, and diarrhea (Murray, 1994).

ADVERSE EFFECTS

In a study of ten healthy human subjects, the results showed that glucosamine affected insulin secretion and its action on humans. This is due to the fact that glucosamine accelerates the hexosamine pathway flux independently of glucose. Glucose toxicity may be mediated by an increased flux through this pathway, and it may induce a reduction in insulin secretion. The study tested glucose-tolerance levels in subjects given either saline, low glucosamine, or high glucosamine concentration infusion. The study found that glucosamine increased both the glucose threshold of glucose-stimulated insulin secretion and plasma fasting glucose levels, thus concluding that glucosamine infusion recapitulates some metabolic features of human diabetes (Monauni et al., 2000).

A similar study was performed on Sprague–Dawley rats. The rats were infused with glucosamine, and then insulin secretion *in vivo* was assessed using the euglycemic hyperinsulinemic clamp technique. This study concluded that glucosamine could indeed cause glucose-induced insulin resistance (Baron et al., 1995).

However, in the most recently reported human study it was demonstrated that oral glucosamine supplementation did not result in clinically significant alterations in glucose metabolism in patients with type 2 diabetes (Scroggie et al., 2003). The problems with formulated products mentioned in the preceding text demonstrate the lengths to which manufacturers need to go to improve standards.

CONCLUSIONS

Carnitine, soy isoflavones, proanthocyanidins, and β-carotene are safely ingested on a daily basis by most people; however, if they are to be promoted as nutritional supplements or developed as drugs, further investigations must be made concerning the toxicity of these plant derivatives. Melatonin and glucosamine are normal components of human metabolism and also may appear safe in initial considerations but at unusual dosages and in chronic administration, they may be shown to exhibit further side effects.

Compared to both conventional pharmaceuticals and complementary medicines, nutraceuticals have lower incidences of adverse effects, drug interactions, and lower toxicity. Overall, however, the risk–benefit use of these products is not nearly as well documented as for conventional pharmaceuticals, and the absence of a wide range of documented adverse effects and drug interactions does not mean that these products are devoid of these properties. Even less information is available relating to the effects of degradation products formed during processing, such as the products formed during lycopene extraction caused by isomerisation and oxidation (Shi and Le Maguer, 2002). From details reported on drug interactions it would obviously be in the interests of patients' safety, if all use of dietary supplementation was made available to healthcare prescribers.

References

Adebowale, A., Liang, Z., and Eddington, N.D. (1999), Nutraceuticals, a call for quality control of delivery systems: A case study with chondroitin and glucosamine, *Journal of Nutraceuticals, Functional & Medical Foods*, 2: 15–30.

Adlercreutz, H., Honjo, H., Higashi, A., Fostis, T., Hamalainen, E., Hasegawa, T., and Okada, H. (1991), Urinary excretion of lignans and isoflavonoid phytoestrogens in Japanese men and women consuming a traditional Japanese diet, *American Journal of Clinical Nutrition*, 54: 1093–1100.

Adlercreutz, C.H., Goldin, B.R., Gorbach, S.L., Hockerstedt, K.A., Watanabe, S., Hamaleinen, E.K., Markkanen, M.H., Makela, T.H., Wahala, K.T., and Adlercreutz, T. (1995), Soybean phytoestrogen intake and cancer risk, *Journal of Nutrition*, 125: 757S–770S.

Ahmed, S., Leo, M.A., and Lieber, C.S. (1994), Interactions between alcohol and beta-carotene in patients with alcoholic liver disease, *American Journal of Clinical Nutrition*, 60: 430–436.

Albanes, H.R., Virtamo, J., Taylor, P.R., Rautalahti, M., Pietinen, P., and Heinonen, O.P. (1997), Effects of supplemental β-carotene, cigarette smoking, and alcohol consumption on serum carotenoids in the alpha-tocopherol, β-carotene cancer prevention study, *American Journal of Clinical Nutrition*, 66: 366–372.

Aleman, C., Rodeiro, I., Noa, M., Menendez, R., Gamez, R., Hernandez, C., and Mas, R. (2001), One-year dog toxicity study of D-002, a mixture of aliphatic alcohols, *Journal of Applied Toxicology*, 21: 179–184.

Anderson, M.A., Slater, M.R., and Hammad, T.A. (1999), Results of a survey of small-animal practitioners on the perceived clinical efficacy and safety of an oral nutraceutical, *Preventive Veterinary Medicine*, 38: 65–73.

Arendt, J. (1997), Safety of melatonin in long-term use (?), *Journal of Biological Rhythms*, 12: 673–681.

Arsenian, M.A., New, P.S., and Cafasso, C.M. (1996), Safety, tolerability, and efficacy of glusose-insulin-potassium-magnesium-carnitine solution in acute myocardial infarction, *American Journal of Cardiology*, 78: 476–479.

Ashar, B. and Vargo, E. (1996), Shark cartilage-induced hepatitis, *Annals of Internal Medicine*, 125: 780–781.

Bagchi, D., Garg, A., Krohn, R.L., Bagchi, M., Bagchi, D.J., Balmoori, J., and Stohs, S.J. (1998), Protective effects of grape seed proanthocyanidins and selected anitoxidants against TPA-induced hepatic and brain lipid peroxidation and DNA fragmentation, and peritoneal macrophage activation in mice, *General Pharmacology*, 30: 771–776.

Baron, A. D., Zhu, J.S., Zhu, J.H., Weldon, H., and Maianu, G. (1995), Glucosamine induces insulin resistance *in vivo* by affecting GLUT 4 translocation in skeletal muscle, *Journal of Clinical Investigation*, 96: 2792–2801.

Bast, A. and Haenen, G.R.M.M. (2002), The toxicity of antioxidants and their metabolites, *Environmental Toxicology and Pharmacology*, 11: 251–258.

Beltz, S.D. and Doering, P.L. (1993), Efficacy of nutritional supplements used by athletes, *Clinical Pharmacy*, 12: 900–908.

Biagiotti, G. and Cavallini, G. (2001), Acetyl-L-carnitine vs. tamoxifen in the oral therapy of Peyronie's disease: A preliminary report, *BJU International*, 88: 63–67.

Bieber, L.L. (1998), Carnitine, *Annual Reviews in Biochemistry*, 57: 261–283.

Blakely, S.R., Grundel, E., Jenkins, M.Y., and Mitchell, G.V. (1990), Alterations in β-carotene and vitamin E status in rats fed β-carotene and excess vitamin A, *Nutrition Research*, 10: 1035–1044.

Bloedon, L.T., Jeffcoat, A.R., Lopaczynski, W., Schell, M.J., Black, T.M., Dix, K.J., Thomas, B.F., Albright, C., Busby, M.G., Crowell, J.A., and Zeisel, S.H. (2002), Safety and pharmacokinetics of purified soy isoflavones: single-dose administration to post menopausal women, *American Journal of Clinical Nutrition*, 76: 1126–1137.

Burton, G.W. and Ingold, K.U. (1984), β-carotene: an unusual type of lipid antioxidant, *Science*, 224: 569–573.

Busby, M.G., Jeffcoat, R.A., Bloedon, L.T., Koch, M., Black, T., Dix, K.J., Heizer, W.D., Thomas, B.F., Hill, J.M., Crowell, J.A., and Zeisel, S.H. (2002), Clinical characteristics and pharmacokinetics of purified soy isoflavones: single-dose administration to healthy men, *American Journal of Clinical Nutrition*, 75: 126–136.

Carta, A., Calvani, M., Bravi, D., and Bhuachalla, S.N. (1993), Acetyl-L carnitine and Alzheimer's disease: pharmacological considerations beyond the cholinergic sphere, *Annals of New York Academy of Science*, 195: 324–326.

Cassidy, A. (1996), Physiological effects of phyto-estrogens in relation to cancer and other human health risks, *Proceedings of the Nutrition Society*, 55: 399–417.

Chang C.W., Chu G., Hinz B.J., and Greve M.D.J. (2003), Current use of dietary supplementation in patients with age-related macular degeneration, *Canadian Journal of Opthalmology*, 38: 27–32.

Chin, R.L., Sporer, K.A., Cullison, B., Dyer, J.E., and Wu, T.D. (1998), Clinical course of gamma-hydroxybutyrate overdose, *Annals of Emergency Medicine*, 31: 716–722.

Chung, K-T., Wei, C.I., and Johnson, M.G. (1998a), Are tannins a double-edged sword in biology and health? *Trends in Food Science and Technology*, 9: 168–175.

Chung, K-T., Wong, T.Y., Wei, C.I., Huang, Y.W., and Lin, Y. (1998b), Tannins and human health: A review, *Critical Reviews in Food Science and Nutrition*, 38: 421–464.

Chiang, H.S., Wang, S.P., and Hsu, F.L. (1992), *In vitro* screening of antimotility effect on human sperm with polyphenolic compounds purified from Chinese herbal medicines, *Advances in Conraceptive Delivery Systems,* 8: 239–246.

Codde, J.P., Beilin, L.J., Croft, K.D., and Vandongen, R. (1985), Study of diet and drug interactions on prostanoid metabolism, *Prostaglandins*, 29: 895–910.

Cohen, M., Josimovich, J., and Brzezinski, A. (1995), *Melatonin: From contraception to breast cancer prevention*, Potomac, M.D: Sheba Press.

Das, N.P., Koay, E., and Ratty, A.K. (1988), *In vivo* effects of (+)-catechin on iron-overloaded rats, *Progress in Clinical and Biological Research*, 280: 215–218.

De Vivo, D.C., Bohan, T.P., Coulter, D.L., Dreifuss, F.E., Greenwood, R.S., Nordli, D.R., Shields, W.D., Stafstrom, C.E., and Tein, I. (1998), L-carnitine supplementation in childhood epilepsy: Current perspectives, *Epilepsia*, 39: 1216–1225.

Dieppe, P. and Chard, J. (2001), Glucosamine for osteoarthritis: magic, hype or confusion?, *British Medical Journal*, 322: 1439–1440.

Dillard, C.J. and German, J.B. (2000), Phytochemicals: nutraceuticals and human health, *Journal of the Science of Food and Agriculture*, 80: 1744–1756.

Dollins, A.B., Zhdanova, I.V., Wurtman, R.J., Lynch, H.J., and Deng, M.H. (1993), Effects of inducing nocturnal serum melatonin concentrations in daytime on sleep, mood, body temperature and performance, *Proceedings of the National Academy of Science*, 91: 1824–1828.

Dyer, J.E. (1991), gamma-Hydroxybutyrate: a health-food product producing coma and seizurelike activity, *American Journal of Emergency Medicine*, 9: 321–324.

Ellaway, C., Williams, K., Leonard, H., Higgins, G., Wilcken, B., and Christadoulou, J. (1999), Rett syndrome: randomized controlled trial of L-carnitine, *Journal of Child Neurology*, 14: 162–167.

Fariello, R.G., Zeeman, E., Golden, G.T., Reyes, P.T., and Ramacci, T. (1984), Transient seizure activity induced by acetylcarnitine, *Neuropharmacology*, 23: 585–587.

Fetrow, C.W. and Avila, J.R. (2001), Efficacy of the dietary supplement *S*-adenosyl-L-methionine, *Annals of Pharmacotherapy*, 35: 1414–1425.

Fine, A.M. (2000), Oligomeric proanthocyanidin complexes: history, structure, and phytopharmaceutical applications, *Alternative Medicine Review*, 5: 144–151.

Fuhrmann, M. (2000), Stereospecific purity of L-carnitine and analytical control of deleterious D-carnitine, *Annals of Nutrition and Metabolism*, 44: 75–76.

Furlong, J.H. (1996), Acetyl-L-carnitine: metabolism and applications in clinical practice, *Alternative Medicine Review*, 1: 85–93.

Garcia-Rubio, L., Garcia- Abad, A.M., Saavedra, R.S., Soler, F., and Miguez, M.P. (1996), Effect of carnitine levels on acute paraquat toxicity, *Toxicology Letters*, 88: 31–32.

German, J.B. and Walzem, R.L. (2000), The health benefits of wine, *Annual Reviews in Nutrition*, 20: 561–593.

Georgala, S., Schulpis, K.H., Georghala, C., and Michas, T. (1999), L-Carnitine supplementation in patients with cystic acne on isotretinoin therapy, *Journal of the European Academy of Dermatology and Venereology*, 13: 205–209.

Gibb, J.W., Bush, L., and Hanson, G.R. (1997), Exacerbation of methamphetamine-induced neurochemical deficits by melatonin, *Journal of Pharmacology and Experimental Therapeutics*, 283: 630–635.

Gortmaker, S.L., Kagan, J., Caspi, A., and Silva, P.A. (1997), Day-length during pregnancy and shyness in children: Results from Northern and Southern hemispheres, *Developmental Psychobiology*, 31: 107–114.

Guardiola-Lemaitre, B. (1997), Toxicology of Melatonin, *Journal of Biological Rhythms*, 12: 697–706.

Harkness, R. and Bratman, S. (2003), *Handbook of Drug-Herb and Drug-Supplement Interactions*, St Louis, MO: Mosby.

Hartter, S., Grozinger, M., and Weigmann, H. (2000), Increased bioavailability of oral melatonin after fluvoxamine coadministration, *Clinical Pharmacology and Therapeutics*, 67: 1–6.

Heyes, M.P., Saito, K., Devinsky, O., and Nadi, N.S. (1994), Kynurenine pathway metabolites in cerebrospinal fluid and serum in complex partial seizures, *Epilepsia*, 2: 251–257.

Hinds, T.S., West, W.L., and Knight, E.M. (1997), Carotenoids and retinoids: a review of research, clinical, and public health applications, *Journal of Clinical Pharmacology*, 37: 551–558.

Hong, Y.G. and Riegler, J.L. (1997), Is melatonin associated with the development of auto immune hepatitis?, *Journal of Clinical Gastroenterology*, 25: 376–378.

Horvath, K., Noker, P.E., Somfai-Relle, S., Glavits, R., Financsek, I., and Schauss, A.G. (2002), Toxicity of methylsulfonylmethane in rats, *Food and Chemical Toxicology*, 40: 1459–1462.

Huether, G., Hajak, G., Reimer, A., Poeggeler, B., Blômer, M., Rodenbeck, A., and Rûther, E. (1992), The metaboloc fate of infused L-tryptophan in men: Possible clinical implications of the accumulation of circulation trypyophan and tryptophan metabolites, *Psychopharmacology*, 109: 422–432.

Ishizaki, A., Sugama, M., and Takeuchi, N. (1999), Usefulness of melatonin for developmental sleep and emotional behavior disorders- studies of melatonin trial on 50 patients with developmental disorders, *No To Hattatsu [Brain and Development]*, 31: 428–437.

Iruela, L.M., Minguez, L., Merino, J., and Monedero, G. (1993), Toxic interaction of *S*-adenosylmethionine and clomipramine, *American Journal of Psychiatry*, 150: 522.

Juniewicz, P.E., Pallante Morell, S., Moser, A., and Ewing, L.L. (1988), Identification of phytoestrogens in the urine of male dogs, *Journal of Steroid Biochemistry*, 31: 987–994.

Kawachi, I., Colditz, G.A., Stampfer, M.J., Willett, W.C., Manson, J.E., Speizer, F.E., and Hennekens, C.H. (1995), Prospective study of shift work and risk of coronary heart disease in women, *Circulation*, 92: 3178–3182.

Kava, R., Meister, K.A., Whelan, E.M., Lukachko, A.M., and Mirabile C. (2002), Dietary supplement safety information in magazines popular among older readers, *Journal of Health Communication*, 7: 13–23.

Kelly, G.S. (1998), L-Carnitine: therapeutic applications of a conditionally-essential amino acid, *Alternative Medicine Review*, 3: 345–360.

Kendler, B.S. (1997), Melatonin: media hype or therapeutic breakthrough?, *Nurse Practitioner*, 22: 66–67, 71–72, 77.

Kitajima, T., Kanbayashi, T., Saitoh, Y., Ogawa, Y., Sugiyama, T., Kaneko, Y., Sasaki, Y., Aizawa, R., and Shimisu, T. (2001), The effects of oral melatonin on the autonomic function in healthy subjects, *Psychiatry and Clinical Neurosciences*, 55: 299–300.

Kraemer, W.J. and Volek, J.S. (2000), L-Carnitine supplementation for the athlete, a new perspective, *Annals of Nutrition and Metabolism*, 44: 75–96.

Kruger, C.L., Murphy, M., DeFreitas, Z., Pfannkuch, F., and Heimbach, J. (2002), An innovative approach to the determination of safety for a dietary ingredient derived from a new source: case study using a crystalline lutein product, *Food and Chemical Toxicology*, 40: 1535–1549.

Kulling, S.E., Rosenberg, B., Jacobs, E., and Metzler, M. (1999), The phytoestrogens coumestrol and genistein induce structural chromosomal aberrations in cultured human peripheral blood lymphocytes, *Archives of Toxicology*, 73: 50–54.

Lamartiniere, C.A., Zhang, J-X., and Cotroneo, M.S. (1998), Genistein studies in rats: potential for breast cancer prevention and reproductive and development toxicity, *American Journal of Clinical Nutrition*, 68: 1400S–1405S.

Landbo, C. and Almdal, T.P. (1998), Interaction between warfarin and coenzyme Q10, *Ugeskrift for Laeger*, 160: 3226–3227.

Larsen, T.M., Toubro, S., Astrup, A. (2003), Efficacy and safety of dietary supplements containing CLA for the treatment of obesity: Evidence from animal and human studies, *Journal of Lipid Research*, 44: 2234–2241.

Lemaître, B. (1997), Toxicology of melatonin, *Journal of Biological Rhythms*, 12: 697–706.

Lenga, R.E. (1988), *The Sigma-Aldrich Library of Chemical Safety Data*, 2nd ed. Milwaukee, WI: Sigma-Aldrich.

Leo, M.A., Kim, C.I., Lowe, N., and Lieber, C.S. (1992), Interaction of ethanol with β-carotene: delayed blood clearance and enhanced hepatotoxicity, *Hepatology*, 15: 883–891.

Leo, M.A. and Lieber, C.S. (1994), Beta carotene, vitamin E and lung cancer, *New England Journal of Medicine*, 331: 612.

Linford, N.J., Yang, Y., Cook, D.G., and Dorsa, D.M. (2001), Neuronal apoptosis resulting from high doses of the isoflavone genistein: Role of calcium and P42/44 mitogen-activated protein kinase, *Journal of Pharmacology and Experimental Therapeutics*, 299: 67–75.

Liu, L.F. (1989), DNA Topoisomerase poisons as anti-tumor drugs, *Annual Reviews in Biochemistry*, 58: 351–375.

Ly, J., Percy, L., and Dhanani, S. (2002), Use of dietary supplements and their interactions with prescription drugs in the elderly, *American Journal of Health-System Pharmacy*, 59: 1759–1762.

McMichael-Philips, D.F., Harding, C., Morton, M., Roberts, S.A., Howell, A., Potten, C.S, and Bundred, N.J. (1998), Effects of soy-protein supplementation on epithelial proliferation in the histologically normal human breast, *American Journal of Clinical Nutrition*, 68: 1431S–1436S.

Mahle, C.D., Goggins, G.D., Agarwal, P., Ryan, E., and Watson, A.J. (1997), Melatonin modulates vascular smooth muscle tone, *Journal of Biological Rhythms*, 12: 690–696.

Maritim, A.C., Moore, B.H., Sanders, R.A., and Watkins, J.B. (2000), Effect of melatonin on phase I and II biotransformation enzymes in streptozotocin-induced diabetic rats, *International Journal of Toxicology*, 19: 277–283.

Martin-Cordero, C., Lopez-Lazaro, M., Pinero, J., Ortiz, T., Cortes, F., and Ayuso, M.J. (2000), Glucosylated isoflavones as DNA topoisomerase poisons, *Journal of Enzyme Inhibition*, 15: 455–460.

Martinez, E., Domingo, P., and Roca-Cusachs, A. (1993), Potentiation of acenocoumarol action by L-carnitine, *Journal of Internal Medicine*, 233: 94.

Mason, P. (2001a), *Dietary Supplements,* 2nd ed., London: Pharmaceutical Press.

Mason, P. (2001b), Isoflavones, *Pharmaceutical Journal*, 266, 16–19.

Mason, P. (2003), What's safe?, *Pharmacy Magazine*, 36.

Medina-Navarro, R., Duran-Reyes, G., and Hicks, J.J. (1999), Pro-oxidating properties of melatonin in the *in vitro* interaction with the singlet oxygen, *Endocrine Research*, 25: 263–280.

Mellert, W., Deckardt, K., Gembardt, C., Schulte, S., Van Ravenzwaay, B., and Slesinski, R.S. (2002), Thirteen-week oral toxicity study of synthetic lycopene products in rats, *Food and Chemical Toxicology*, 40: 1581–1588.

Messina, M.J., Persky, V., Setchell, K.D.R., and Barnes, S. (1994), Soy intake and cancer risk: A review of the *in vitro* and *in vivo* data, *Nutrition and Cancer*, 21: 113–131.

Misra, R.R., Hursting, S.D., Perkins, S.N., Sathyamoorthy, N., Mirsalis, J.C., Riccio, E.S., and Crowell, J.A. (2002), Genotoxicity and carcinogenicity studies of soy isoflavones, *International Journal of Toxicology*, 21: 277–285.

Monauni, T., Zenti, G.M., Cretti, A., Daniels, M.C., Targher, G., Caruso, B., Caouto, M., McClain, D., Prato, S.D., Giaccari, A., Muggeo, M., Bonora, E., and Bonadonna, R.C. (2000), Effects of glucosamine infusion on insulin secretion and insulin action in humans, *Diabetes*, 49: 926–935.

Moore, J.G. and Goo, R.H. (1987), Day and night aspirin-induced gastric mucosal damage and protection of ranitidine in man, *Chronobiology International*, 4: 111–116.

Murray, M.T. (1994), Glucosamine sulphate: effective osteoarthritis treatment, *American Journal of Natural Medicine*, 1: 10–14.

Nagtegaal, J.E., Smits, M.G., Van Der Meer, Y.G., and Fischer-Steenvoorden, M.G.J. (1997), Melatonin: A survey of suspected adverse drug reactions, *Sleep-Wake Research* The Netherlands, 7: 115–118.

Omenn, G.S., Goodman, G.E., Thornquist, M.D., Balmes, J., Cullen, M.R., and Glass, A. (1996), Risk factors for lung cancer and for intervention effects in CARET, the beta-carotene and retinol efficacy trial, *Journal of the National Cancer Institute*, 88: 1550–1559.

Packer, L., Witt, E.H., and Tritschler, H.J. (1995), alpha-Lipoic acid as a biological antioxidant, *Free Radical Biology and Medicine*, 19: 227–250.

Palmer, M.E., Haller, C., McKinney, P.E., Klein-Schwartz, W., Tschirgi, A., Smolinske, S.C., Woolf, A., Sprague, B.M., Ko, R., Everson, G., Nelson, L.S., Dodd-Butera, T., Bartlett, W.D., and Landzberg, B.R. (2003), Adverse events associated with dietary supplements: an observational study, *Lancet*, 361: 101–106.

Paolini, M., Cantelli-Forti, G., Perocco, P., Pedulli, G.F., Abdel-Rahman, S.Z., and Legator, M.S. (1999), Co-carcinogenic effect of β-carotene, *Nature*, 398: 760–761.

Patek, A.J., Toth, I.G., Saunders, M.G., Castro, G.A.M., and Engel, J.J. (1975), Alcohol and dietary factors in cirrhosis, An epidemiological study of 304 alcoholic patients, *Archives of Internal Medicine*, 135: 1053–1057.

Pellet, J.L., Anderson, H.J., Chen, H., and Tappel, A.L. (1994), β-carotene alters vitamin E protection against haem protein oxidation and lipid peroxidation in chicken liver slices, *Journal Nutritional Biochemistry*, 5: 479–484.

Penn, N.W. (1995), Potentiation of morphine analgesic action in mice by β-carotene, *European Journal of Pharmacology*, 284: 191–193.

Polkowsi, K. and Mazurek, A.P. (2000), Biological properties of genistein. A review of the *in vitro* and *in vivo* data, *Acta Poloniae Pharmaceutica*, 57: 135–155.

Rapport, L. and Lockwood, B. (2002), *Nutraceuticals*, London: Pharmaceutical Press.

Rebouche, C.J. (1999), Carnitine, in *Modern Nutrition in Health and Disease,* Shils, M.E, Olson, J.A., Shike, M., and Ross, A.C. (Eds.) 9th ed., Baltimore: Williams and Wilkins, pp. 505–512.

Record, I.R., Jannes, M., Dreosti, I.E., and King, R.A. (1995), Induction of micronucleus formation in mouse splenocytes by the soy isoflavone genistein *in vitro* but not *in vivo*, *Food and Chemical Toxicology*, 33: 919–922.

Robinson, W.A., Dreiling, L., Gonzalez, R., and Balmer, C. (1995), Treatment of human metastatic melanoma with high dose melatonin, *Life Sciences*, 227: 219–225.

Rogers, N.L., Phan, O., Kennaway, D.J., and Dawson, D. (1998), Effect of daytime melatonin administration on neurobehavioral performance in humans, *Journal of Pineal Research*, 25: 47–53.

Rohdewald, P. (2002), A review of the French maritime pine bark extract (Pycnogenol), a herbal medication with a diverse clinical pharmacology, *International Journal of Clinical Pharmacology and Therapeutics*, 40: 158–168.

Ross, J.A., Potter, J.D., Reaman, G.H., Pendergrass, T.W., and Robinson. L.L. (1996), Maternal exposure to potential DNA topoisomerase II inhibitors and infant leukemia: A report from the children's cancer group, *Cancer Causes and Control*, 7: 581–590.

Russell, A.S., Aghazadeh-habashi, A., and Jamali, F. (2002), Active ingredient consistency of commercially available glucosamine sulfate products, *Journal of Rheumatology*, 29: 2407–2409.

Samel, A. (1999), Melatonin and jet lag, *European Journal of Medical Research*, 4: 385–388.

Sato, F., Hagiwara, Y., and Kawase, Y. (1995), Subchronic toxicity of 3-phenylamino alanine, an impurity in L-tryptophan reported to be associated with eosinophilia-myalgia syndrome, *Archives of Toxicology*, 69: 444–449.

Scroggie, D.A., Albright, A., and Harris, M.D. (2003), The effect of glucosamine-chondroitin supplementation on glycosylated hemoglobin levels in patients with type 2 diabetes mellitus: A placebo-controlled, double-blinded, randomized clinical trial, *Archives of Internal Medicine*, 163: 1587–1590.

Semeniuk, J., Shalansky, K.F., Taylor, N., Jastrzebski, J., and Cameron, E.C. (2000), Evaluation of the effect of intravenous l-carnitine on the quality of life in chronic haemodialysis patients (2000), *Clinical Nephrology*, 54: 470–477.

Shi, J. and Le Maguer, M. (2002), Lycopene in tomatoes: chemical and physical properties affected by food processing, *Critical Reviews in Biotechnology*, 20: 293–334.

Simone, F., Pappalardo, G., Maiani, G., Guadalaxara, A., Bugianesi, R., Conte, A.M., Azzini, E., and Mobarhan, S. (2002), Accumulation and interactions of beta-carotene and alpha-tocopherol in patients with adenomatous polyps, *European Journal of Clinical Nutrition*, 56: 546–550.

Singh, R.B. and Aslam, M. (1998), L-carnitine administration in coronary artery disease and cardiomyopathy, *Journal of the Association of Physicians of India*, 46: 801–805.

Sirtori, C.R. (2000), Dubious benefits and potential risk of soy phyto-oestrogens, *Lancet*, 355: 849.

Strick, R., Strissel, P.L., Borgers, S., Smith, S.L., and Rowley, J.D. (2000), Dietary bioflavonoids Induce cleavage in the MLL gene and may contribute to infant leukemia, *Proceedings of the National Academy of Sciences*, 97: 4790–4795.

Sugden, D. (1983), Psychopharmacological effects of melatonin in mouse and rat, *Journal of Pharmacology and Experimental Therapeutics*, 227: 587–591.

Sutton, L., Rapport, L., and Lockwood, B. (2002), Glucosamine: Con or cure?, *Nutrition*, 18: 534–536.

Takeuchi, M. and Edanaga, M. (1972), Antiatherosclerotic agents. 8. Toxicological studies of sodium chondroitin polysulfate. 1. *Oyo Yakuri*, 6: 573–587.

Tallia, A.F. and Cardone, D.A. (2002), Asthma exacerbation associated with glucosamine-chondroitin supplement, *Journal of the American Board of Family Practice*, 15: 481–484.

The Massachusetts Society (1994), The effect of vitamin E and beta carotene on the incidence of lung cancer and other cancers in male smokers, *The New England Journal of Medicine*, 330: 1029–1035.

Torrioli, M., Vernacotola, S., Mariotti, P., Bianchi, E., Calvani, M., De Gaetano, A., Chiurazzi, P., and Neri, G. (1999), Double bind placebo controlled study of L-acetylcarnitine for the treatment of hyperactive behaviour in fragile X syndrome, *American Journal of Medical Genetics*, 87: 366–368.

Vacheron-Trystram, M.N., Cheref, S., Gauillard, J., and Plas, J. (2002), A case report of mania precipitated by use of DHEA, *Encephale*, 28: 563–566.

Valdes, S., Molina, V., Carbajal, D., Arruzazabala, L., and Mas, R. (2000), Comparative study of the antiulcer effects of D-002 with sucralfate and omeprazole, *Revista CENIC, Ciencias Biologicas*, 31: 117–120.

Van den Berg, H. (1998), Effect of lutein on beta-carotene absorption and cleavage, *International Journal for Vitamin and Nutrition Research*, 68: 360–365.

Van den Berg, H. (1999), Carotenoid interactions, *Nutrition Reviews*, 57: 1–10.

Ward, P.M.L., Fasitsas, S., and Katz, S.E. (2002), Inhibition, resistance development, and increased antibiotic and antimicrobial resistance caused by nutraceuticals, *Journal of Food Protection*, 65: 528–533.

Weaver, D.A. (1997), Reproductive safety of Melatonin: A "wonder drug" to wonder about, *Journal of Biological rhythms*, 12: 682–689.

Wechsler, J. (2003), Standards for Supplements, *Pharmaceutical Technology Europe*, 18–20.

Willett, W.C., Stampfer, M.J., Undersood, B.A., Tayler, J.O., and Hennekins, C.H. (1983), Vitamin A, E, and carotene: Effects of supplementation on their plasma levels, *American Journal Clinical Nutrition*, 38: 559–556.

Yamakoshi, J., Saito, M., Kataoka, S., and Kikuchi, M. (2002), Safety evaluation of proanthocyanidin-rich extract from grape seeds, *Food and Chemical Toxicology*, 40: 599–607.

Yang, C., Chang, C., Tseng, C., Liao, N., Chao, S., Wang, Y., Lung, S., and Yeh, H. (1978), Preliminary studies on the pharmacological effect and toxicity of chondroitin sulfate A, *Zhonghua Yixue Zazhi*, 58: 739–742.

Yeh, S.L. (2001), Induction of oxidative DNA damage in human foreskin fibroblast Hs68 cells by oxidized β-carotene and lycopene, *Free Radical Research*, 35: 203–13.

Zvosec, D.L., Smith, S.W., McCutcheon, J.R., Spillane, J., Hall, B.J., and Peacock, E.A. (2001), Adverse events, including death, associated with the use of 1,4-butanediol, *New England Journal of Medicine*, 344: 87–94.

7 Safety Concerns of Genetically Modified Foods

Felicity Goodyear-Smith

CONTENTS

INTRODUCTION

Mankind has been engaged in a constant process of modifying what we eat, both the original foodstuffs and the end products of food processing. We have been intervening in the heredity of plants and animals (and more recently, microorganisms)

0-8493-3516-7/05/$0.00+$1.50

for 10,000 yr, since the move from hunter-gatherer progressed to agriculturist (Diamond, 1998). For several millennia the genetic makeup of plants has thus undergone changes through selective breeding, including the use of techniques such as controlled pollination, hybridization, and, lately, cloning (Halford and Shewry, 2000).

Genetic recombination and transposition of genes can occur randomly and spontaneously within plants, but chromosome rearrangements also can be produced and mutations induced through the use of chemical and irradiation treatments. Cross-species hybrids have become the norm in the botanical world (Royal Society of New Zealand, 2000). Techniques such as embryo culture and *in vitro* fertilization have created hybrids between species and even between genera (Scott and Conner, 1999). Over the past few decades, "standard breeding" has included techniques such as cloning, mutagenesis, hybridization and controlled pollination, fused-cell fusion, and others. For example, the Rio Red pink grapefruit variety is a mutant created using thermal neutron bombardment, and the Seafarer bean is a mutant created by x-ray mutagenesis (Saul, 2002).

The structure of DNA was determined by Watson, Crick, and Wilson in the 1950s. From this, the path was laid to determine the genetic code of organisms, and by the 1970s individual genes had been isolated. From these advances, techniques have been developed to enable genetic coding to be altered in a manner that was not achievable through either natural evolution or selective breeding.

The process of genetic modification (GM), also known as genetic engineering (GE), allows an individual gene (or more commonly a set of a few genes) that controls particular characteristics to be taken out of the DNA of one organism and inserted into the DNA of another organism. Through this technique, genes can be transferred within species and also between species (Burke, 1998). The gene is isolated from a strand of DNA from a cell, changed, and then inserted into another cell of a plant, animal, or microorganism.

According to the New Zealand Royal Commission, GM comprises:

- The deletion, change, or moving of genes within an organism
- The transfer of genes from one organism to another
- The modification of existing genes or the construction of new genes and their incorporation into any organism

There are several different ways by which gene transfer can be achieved. These include the use of microorganisms and use of a technique in which genetic material coated onto the surface of microscopic gold particles are shot into new cells (Halford and Shewry, 2000). GM can also be used to create microorganisms carrying novel genes, which are then used as chemical-producing "factories." These microorganisms may manufacture organic molecules such as vitamins (Schulz and Beubler, 1989; Koizumi et al., 2000) or enzymes (Wesley, 1981; Mala et al., 1998; Ross et al., 2000) for use in food processing, but the end product (for example, the artificial sweetener aspartame) does not itself contain any GM-modified organism.

The GM process differs from traditional selective breeding in three main ways (see Table 7.1). However, the distinction between GM and non-GM is not as precise as it is portrayed. GM technology exists as part of a continuum with previously

TABLE 7.1
Ways the GM Process Differs from Non-GM Breeding

	Selective Breeding and Other Non-GM Techniques	GM Technology
1	Which individual genes are transferred in each episode of selective breeding is a random process	Individual genes are extracted, may undergo some alternation, and then are transferred from one plant or animal into the DNA of another
2	Selective breeding techniques may take many generations to achieve the desired result, and the qualities achieved are dependent on the inherent genetic characteristics of the organisms used	Allows plants or animals to gain new inheritable qualities in a way that is much faster and sometimes much more novel
3	Selective breeding generally involves transfer of genes within species, although techniques such as embryo culture and *in vitro* fertilization have created interspecies and intergenera hybrids, particularly in the botanical realm	Genes may be transferred between different species, genera, or even kingdoms (between plants and animals)

Source: Royal Commission on Genetic Modification (2001). *Report of the Royal Commission on Genetic Modification*. Wellington, New Zealand Ministry for the Environment <http://www.gmcommission.govt.nz/RCGM/index.html> (accessed February 2003).

developed techniques used to alter the genetic makeup of our food. GM is a tool used in a variety of different ways. It may use modified microorganisms as "organic chemical factories" to produce large volumes of a desired food additive through a fermentation process. It may be intraspecies, for example, it may involve taking a gene from a known variety of peach and introducing it into another variety of peach, speeding up a process of genetic change already achieved through selective breeding. It may be interspecies, introducing a gene from one species into another. For instance, a growth hormone of rainbow trout has been transferred into carp eggs, creating a GM variety of carp that is a third larger than a non-GM carp. It may be more extreme, combining genes from very different species that belong to different genera or even kingdoms (plants and animals) — for example, it has been suggested that a gene from the North Atlantic flounder fish could be inserted into a strawberry in order to make the fruit frost resistant (Biotechinfo, 2002).

The rapid development of this science has sparked intense public and professional debate. There are considerable potential benefits — GM technology may provide dramatic improvements in food production, enhancing both the quality and quantity of foods produced and helping to alleviate poverty and starvation. However, there are also significant potential risks. We may be letting a dangerous genie out of the bottle. There may be unforeseen dangers and risks to the health and well-being of individuals, populations, and our planet — a means for multinational organizations to exploit Third World countries with no individual gain (Burke, 1998; Editorial, 1999; Holden, 1999; Millstone et al., 1999; Mowat et al., 1999; Trewavas and Leaver, 1999).

This chapter addresses the safety concerns of GM foods and their possible adverse effects on those who consume them and to our environment, balanced against their possible benefits.

POTENTIAL DANGERS OF GM FOODS

GM foods pose potential risks to individual consumers, human populations, and the environment (see Table 7.2).

RISK OF PRODUCING ALLERGENIC FOODS

In general, there is no evidence that GM foods will be more allergenic or less allergenic than their corresponding conventional foods (Wal, 1999). GM involves the transfer of genes that may code for proteins not normally present. Therefore, it is possible to transfer allergenicity from the host food to the new variety in specific situations. Under these circumstances, an allergen is imparted into a food, and individuals who respond to that allergen may now develop an allergic response to this food to which they previously had no adverse response.

There have been a number of cases of transgenic varieties being tested for allergenicity and no difference in allergen content found between wild and transgenic strains (Lehrer and Reese, 1997). For example, a new variety of corn with altered amino acid composition was tested on subjects known to be corn reactive; and similarly a transgenic soybean with an altered fatty acid profile was tested on soy-allergic individuals, and no differences between the GM and non-GM foods were detected.

There is a case recorded in which an allergen from a known allergenic food has been transferred into another food by GM technology. The case involved the transfer of methionine-rich 2S albumin from Brazil nuts into soybeans. This was conducted to improve the nutritional quality of soybeans, which are relatively deficient in methionine (Nordlee et al., 1996). Because Brazil nuts have a known allergenicity, they had to undergo a comprehensive testing regime. Individuals allergic to Brazil nuts were tested for allergy to the transgenic soybean, and they were found to test positive. Although the soybeans were intended for animal rather than human consumption, the company abandoned the project to ensure that the modified soy did not enter the human food chain (Jones, 1999).

Another issue of concern was the development of StarLink™ corn in 1998 (Dorey, 2000). The gene that encodes for an insecticidal Cry9C protein was inserted in this corn, using a GM bacterium. It was anticipated that Cry9C protein may be allergenic, so the corn was approved for animal feed only. However, in 2000, fragments of the Cry9C DNA were found in corn taco shells in the U.S. These products were then recalled. Subsequent research indicated a possible risk that the Cry9C protein is a potential allergen (Anonymous, 2001), although this has not been confirmed. There was no risk to human health in this case as only the encoding gene, not the protein, was found in the taco shells. Research indicates that the Cry9C protein is present in StarLink corn kernels at a level of 0.3% (and hence would be considerably <0.3% in the taco shells), whereas most allergenic proteins are present

TABLE 7.2
Potential Risks of GM Foods

Potential Risks		Example	Known Outcome
Potential risk to individual consumers	Transfer allergenicity	Transfer of methionine-rich 2S albumin from Brazil nuts into soybeans transferred allergenicity to the soybeans	Project abandoned
	Transfer toxicity	1. Production of L-tryptophan using GM bacterium resulted in batch contaminated by a toxic L-tryptophan dimer	37 deaths and >1500 nonfatal cases of eosinophilia myalgia syndrome (EMS) Unknown whether caused by GM or change in purification process
		2. GM potatoes in research study examining introduction of lectin (which is unpalatable to insects) reported to have caused intestinal changes in rats	Toxic effects disputed — remains contentious Potatoes never destined as a food source
Potential risk to human populations	Development of antibiotic resistance	Marker gene that confers resistance to a certain antibiotic feared to threaten effectiveness of antibiotics in people and animals through horizontal transfer to other organisms. However, marker gene is eliminated in food processing	No resistance problems encountered to date Alternative marker systems are being developed to phase out antibiotic marker use
	Reduced nutritional value	Fears expressed that foods bred with qualities improving marketability may have reduced nutritional value	No cases have emerged to date
Potential risk to the environment	Development of herbicide-resistant weeds	1. Monsanto's GM soybeans resistant to herbicidal action of Roundup	No evidence to date of superweed development
		2. GM oilseed rape plants with herbicide tolerance	Potential resistance transferred to neighboring weeds such as wild radish
	Development of insecticide-resistant insects	Widespread use of the Bt gene might result in the selection of Bt-resistant insects	Reported case of transgenic Bt cotton attacked by cotton bollworm suggesting possible bollworm resistance to Bt toxin. Ten studies of potato, maize, oilseed rape, and sugar beet(with GM traits of resistance to herbicides or insect pests) grown in 12 different habitats found no resistance problems developed

at levels of 1 to 40% (Dorey, 2000). It seems probable that the detected Cry9C genes had resulted from physical contamination, although the possibility of cross-pollination in the field has not been eliminated. However, this case does raise issues about the difficulty of restricting a GM food for animal use when there is a similar non-GM equivalent available as a food for humans (Royal Commission on Genetic Modification, 2001).

When the gene source is from a known allergen, GM researchers maintain that it is possible to determine whether the allergen content of the transgenic line is altered relative to the nontransgenic varieties, using traditional forms of immunological testing. If allergens are transferred to a new variety, this can be established before the food is made available to unsuspecting consumers (Lehrer and Reese, 1997).

It is counterargued by opponents to GM technology that GM may either create or unmask new immunoreactive structures. Traditional forms of immunological testing may not be applicable where the gene source is not a known allergen. Currently, no completely reliable and objective method exists (using either animal testing or chemical analysis) to evaluate or predict the possible allergenicity of a new protein.(Wal, 1999) Research is under way to determine whether food allergens share physicochemical properties that distinguish them from nonallergens. If this is so, these properties may be used as a tool to predict the inherent allergenicity of proteins newly introduced into the food supply by GM. One study has found that major allergens of plant-derived foods such as legumes (peanuts and soybean) are stable to digestion, in comparison to nonallergenic food proteins (for example, spinach) (Astwood James et al., 1996). This means that digestible proteins are unlikely to be allergens.

It should also be noted that this problem is not confined to GM foods. Kiwifruit consumption has increased dramatically in the past few decades because of the development of varieties that are easily cultivated and have good taste as well as a high vitamin content. There has been a corresponding rise in cases of kiwifruit allergy including reactions of laryngeal edema and anaphylactic shock (Pastorello et al., 1998). This problem is not sufficient, however, to deter the public from trying novel fruit and vegetables that appear in the market.

Risk of Producing Toxic Food

There are no established cases of food toxicity resulting from GM. However, there are two cases that have raised concerns.

The first involves an epidemic of eosinophilia myalgia syndrome (EMS) in 1989. This case resulted in 37 deaths and over 1500 nonfatal reported cases, and are clearly linked to ingestion of a specific batch of the food supplement tryptophan manufactured by the Japanese company Showa Denko KK (Slutsker et al., 1990; Kilbourne et al., 1996). The contaminant was identified as a tryptophan dimer L-tryptophan, and this is believed to be the causative agent in EMS (Buss et al., 1996).

L-Tryptophan is an amino acid, normally obtained from dietary protein. In the 1980s, tryptophan became popular as a dietary supplement, promoted for use in conditions such as insomnia and depression. Although tryptophan can be purified from plant and animal proteins, it is obtained more economically in a vat fermentation process that involves growing large amounts of bacteria and then extracting and purifying the amino acid. Originally, L-tryptophan was commercially produced using a naturally occurring bacterium.

Showa Denko had made two changes in their manufacturing process prior to the outbreak. They had started to use a GM bacterium that had an enhanced production of tryptophan. They had also altered the extraction and purification process by reducing the amount of activated carbon used in filtration by 50% (Scott and Conner, 1999a). Because Showa Denko reportedly destroyed their stocks of the GM bacterium strain once the EMS cases began to emerge, it has not been possible to definitively establish whether the dimer L-tryptophan contaminant (and hence the EMS outbreak) specifically arose from the use of the GM bacteria or from an inadequate purification process (Belongia et al., 1990). Although there has been heated scientific debate as to whether the EMS epidemic was triggered by an impurity formed when the manufacturing conditions were modified or was GM-related (Henning et al., 1993; Mayeno and Gleich, 1994; Scott and Conner, 1999b), the answer will remain unknown because the GM strain was destroyed.

The second case of potential transferred toxicity involves GM potatoes used in a research project. Arpad Pusztai, a British scientist, prematurely announced through the media in 1998 the unpublished results of a study describing changes in the intestines of rats that had eaten these GM potatoes (Anonymous, 1999). Considerable public concern ensued regarding the dangers of GM foods. A review of Pusztai's available data conducted by the Royal Society concluded that there were methodological flaws in the research and misinterpretation of these results (Lachmann, 1999). This particular study was examining whether lectins that make some plants unpalatable to insects could have a similar effect if introduced into other plants, and hence these potatoes were not being developed as a food source. This remains a controversial issue (Mowat et al., 1999).

Risks GM Foods may Pose to Human Populations

GM foods may pose threats to human populations. Foodstuffs might be produced that are selected for qualities that improve marketability but have reduced nutritional value. Of greater concern is the possibility that GM processes may lead to development of antibiotic resistance.

To transfer genes using a microorganism, DNA strands are extracted from a cell, proteins known as restriction enzymes are added to break the DNA at particular points to obtain an individual gene, and then this gene is placed inside a bacterial cell. The genes are carried by plasmids, small DNA molecules that are not chromosomal in nature and that can move between cells, carrying the gene with them. Bacteria carrying the added gene can then transfer this new gene to target plant or animal cells whose DNA is recombined to incorporate the new gene. All the cells in the target tissue that are not genetically modified are then destroyed.

To identify cells with the desired gene, an accompanying "marker" gene has been used. Typically these selective markers have been antibiotic-resistant genes rendering the transformed cells resistant to an antibiotic that is toxic to cells not genetically modified (Halford and Shewry, 2000). This means that the antibiotic can be used to kill off all cells that have not undergone GM. The marker gene is later removed from the GM food. The use of antibiotic-resistant markers has been hypothesized to pose a threat to the effectiveness of antibiotics in people and animals through its horizontal transfer to other organisms, including microbes that inhabit humans.

In general, the gene and the enzyme it produces that renders the antibiotic inactive are eliminated in subsequent processing (Advisory Committee on Novel Foods and Processes, 1994). Furthermore, generally the antibiotics used are not those used to treat human infections. However, as a precautionary measure, concerns that this resistance could be transferred to human pathogens (British Medical Association, 1999) have led to regulatory controls placed on further development of this product in Britain.

The World Health Organization and several European expert advisory panels have extensively investigated the issue of GM markers rendering human antibiotics ineffective. Their general conclusion is that the impact on human health is effectively zero. The issue still does need to be addressed on a case-by-case basis because it depends on the nature of each particular genetic modification (Australia and New Zealand Food Authority, 2000). Alternative marker systems are being developed to phase out the use of antibiotic markers to meet these concerns even if this step is shown to be needless (Advisory Committee on Novel Foods and Processes, 1994).

Risks GM Foods may Pose to the Environment

There are concerns that the development of GM crops with resistance to herbicides may result in this resistance being passed on to weeds through cross-pollination, leading to the development of herbicide-resistant "superweeds" (Nestle, 1998).

A well-publicized case is that of Monsanto's genetically modified soybeans that are able to resist the herbicidal action of Roundup. This food has been extensively produced in the U.S., it has undergone comprehensive testing, and to date there appears to be no evidence that it is causing superweed development.

However, there are reports that GM oilseed rape plants have passed on their herbicide tolerance to neighboring weeds (Mikkelsen et al., 1996). The cross-pollination would have to occur with either wild rapeseed plants (intraspecific crosses) or with its wild relatives (interspecific crosses). These are quite numerous for rape — for example, wild radish has been shown to form hybrids with rape, but the germination rate of these hybrids is low and their fitness is reduced (Darmency et al., 1998). The potential for the development of resistant weeds could however increase with future generations of this wild plant because the advantageous (herbicide-resistant genes) could assist the wild radish and its progeny to survive.

GM crops are bioengineered to contain the naturally occurring insect toxin *Bacillus thuringiensis* (Bt). Bt has a long history of use, particularly in the organic food industry, to protect agricultural and forestry crops from insect pests. Concern has been expressed that widespread use of the Bt gene might result in the selection of Bt-resistant insects. There is one report of large-scale plantings of transgenic Bt cotton suffering attack by cotton bollworms, one of three pests that the crops were supposed to kill. This has heightened fears that the insects will eventually develop resistance to the Bt toxin (Kaiser, 1996).

One long-term study involving 4 different crops (potato, maize, oilseed rape, and sugar beet) grown in 12 different habitats and monitored over 10 yr found that the GM crops (with GM traits of resistance to herbicides or insect pests) were no more invasive nor persistent than their conventional (non-GM) counterparts (Crawley et al., 2001).

POTENTIAL BENEFITS OF GM FOODS

Given these potential risks to individuals, human populations, and the environment, the advantages of pursuing GM alternatives needs to be explored (see Table 7.3). GM foods may primarily benefit commercial interests. However, there are also possible direct gains to consumers.

Producing Foods with Reduced Allergenicity and Lower Toxin Levels

Both selective breeding and GM techniques can result in random and unexpected genetic events. A difference is that GM techniques introduce only one or a few genes into the crop, whereas conventional cross-hybridization has the potential for multiple introduction of undesirable genes (Feldbaum, 1999; Trewavas and Leaver, 1999). It is known what protein a specific gene makes. The novel protein in transgenic plants, therefore, can be quantified, tested, and evaluated for its possible allergic or toxic qualities to an extent that is not possible with conventional breeding methods.

It is, therefore, possible to produce foods with reduced allergenicity. Hypersensitive people can have life-threatening allergic reactions to foods such as cereals, nuts, milk, and eggs. In the field of allergy immunotherapy, transgenic varieties of grasses, birch, and oilseed rape have been created with reduced anaphylactic activity by susceptible individuals (Singh et al., 1999). Patients are given increasing amounts of the disease-eliciting allergens in order to develop reduced allergen-specific responsiveness. Creation of edible cereals with similar reduced hypersensitivity in reactive people should also be possible.

Some people have a hypersensitivity to cow's milk. Proteases from several different microorganisms have been used to produce hypoallergenic casein, with an antigenicity about 10^4 times lower than that of standard cow's casein, for possible use in many kinds of milk-containing foods and hypoallergenic baby formulae (Nakamura et al., 1991).

There has been a hypoallergenic rice developed with the same palatability and nutritional value as normal rice for people who suffer from rice allergy (Arai, 1993).

Improving the Nutritional Value of Food

The nutritional value of food may be increased through GM technology. In Third World countries, vitamin A and iron deficiency are prevalent. These deficiencies cause considerable morbidity (including blindness) and mortality. Rice is the staple

TABLE 7.3
Potential Advantages of GM Foods

Potential Advantages		Examples	Known Outcome
Potential advantages to individual consumers	Foods with reduced allergenicity	1. Production of GM grasses, birch, and oilseed rape with reduced anaphylactic activity by susceptible individuals	Currently used in allergy immunotherapy to develop reduced allergen-specific responsiveness (not foods)
		2. Production of hypoallergenic varieties of cereals, nuts, milk, and eggs (currently life-threatening in hypersensitive people)	Hypoallergenic rice developed with same palatability and nutritional value as normal rice Hypoallergenic casein developed for possible use in milk-containing foods and hypoallergenic baby formulae
	Foods with improved nutritional value	GM rice (golden rice) with increased β-carotene (precursor for vitamin A) to reduce morbidity including blindness and mortality from vitamin A deficiency	Developed but not yet released into Third World countries
	Foods with medicinal properties	1. Disease prevention	Add plant sterols to margarine for cardiac protection GM potato with increased starch and decreased water content halves oil absorbed during frying
		2. Disease intervention	Reduced lactose milk for people with lactose intolerance GM brewer's yeast to produce reduced calorie beer for diabetics
		3. Medicinal foods	Potential GM enhancement of medicinal components of foods such as green tea, garlic, honey, and ginseng
		4. Nutritional supplements	GM technology used to produce extracts, concentrates, or synthetic versions of food substances (vitamins, minerals, amino acids, herbs, or synthetic nutrients) produced in controlled dose forms such as powders, capsules, or tablets

TABLE 7.3 (continued)
Potential Advantages of GM Foods

Potential Advantages		Examples	Known Outcome
Potential advantages to human populations	Foods with improved taste	GM lipoxygenase-free soybean	Soybeans have less beany flavor and bitter taste, as well as increased storage stability
	Improve crop yield in countries beset with overpopulation and poverty	1. GM crops with increased yield and high nutritional quality 2. GM crops requiring less water to thrive 3. GM crops to grow in saline soils	Currently under research
	Acceptable food alternatives	GM-produced chymosin	Cheese production for vegetarians who do not eat cheese made with rennet
Potential advantages to the environment	Manufacture of organic chemicals used in food production without requiring extraction from plants or animals	GM microorganisms for large-scale production of proteases	Production of proteases for cheese-making, baking, tenderizing of meat, and production of soy products.
		Large-scale production of aspartame	Widely used as noncalorific artificial sweetener
		GM synthesis of vitamins	Production of riboflavin (vitamin B_2)
		GM-produced food flavorings	GM vanilla flavoring counters worldwide decline in vanilla production
	Pest and disease-resistant crops	Less residue in food and reduction of chemical pollution of environment	Glyphosate-tolerant crops allow use of single safe herbicide and reduction in total herbicide use

food for much of the world's population, and standard milled Indica rice is very low in vitamin A. A strain of rice has been developed with increased β-carotene (the precursor for vitamin A) and enhanced iron content. This transgenic rice, known as "golden rice," is hoped to provide a cheap accessible form of supplementation for these deficiencies (Ye et al., 2000).

Foods with Medicinal Properties

Both conventional and GM means are being used to develop foods designed to have health-promoting effects. This covers a gamut of "health promotion" possibilities including health maintenance and disease prevention, performance improvement, disease risk reduction, symptom self-treatment, and disease intervention. Such products may be classified as functional foods, nutraceuticals, medicinal foods, and dietary supplements, although precise definitions of these terms have not been established and one may blend with another.

Specialized nutraceutical foods are being developed with a range of potential benefits for human health. The word nutraceuticals generally refers to products that enhance nutrition (for example, vitamin A-enriched rice).

A functional food is one intended for consumption as part of a normal diet but contains modifications that may contribute to disease prevention (such as adding plant sterols to margarine). Another example is a GM potato that has been developed with an increased starch and decreased water content, which can halve the amount of oil absorbed during frying (Rose, 1993).

Some people suffer from lactose intolerance, caused by a deficiency of the lactase enzyme, which is normally produced in the small intestine. Digestion of lactose in milk requires the presence of lactase. People who cannot digest lactose and, therefore, malabsorb it suffer from symptoms of nausea, diarrhea, and abdominal pain. GM research is currently developing the expression of the lactase enzyme in bovine mammary glands, which converts the lactose to glucose and galactose. The resultant milk has the appearance and qualities of standard milk but greatly reduced lactose levels (Whitelaw, 1999).

GM use of glucoamylase genes from various microorganisms has produced a brewer's yeast with an altered ability to degrade starch. This means that the resultant beer contains less digestible carbohydrate, producing a reduced calorie beer for diabetics (Schulz and Beubler, 1989). GM yeasts are used to change the vitamin content of beer.

There are numerous examples of foods with potential medicinal properties, with the possibilities expanding rapidly. Although they may prove not to be the "panaceas" previously acclaimed, many traditional food remedies do have valid pharmaceutical actions.

One example is the research on green tea (from the plant *Camellia sinensis*), which suggests that it may have antimicrobial, immunostimulatory, anticarcinogenic, antiinflammatory, and cardiovascular-protective capacities (Sato and Miyata, 2000a). Green tea may have some antimicrobial capacity against various bacteria. Garlic (*Allium sativum*) has also been shown to have antimicrobial action against bacteria, as well as antifungal properties (Sato and Miyata, 2000b). The antimicrobial activity of garlic is attributed to allicin, which has also been shown to have cholesterol-lowering effects. Honey may also have antimicrobial properties, although its antimicrobial ability differs depending on its floral source (Sato and Miyata, 2000c). Studies on ginseng indicate the immunostimulatory and anticarcinogenic properties of this plant, which contains a compound (ginsenoside) with apparent immunologically active properties (Sato and Miyata, 2000d).

It is possible that genetic modification could enhance the medicinal properties of these foods. For example, the level of allicin in garlic could be increased to improve its cholesterol-lowering capacity. GM foods are also being developed with increased concentrations of components such as carotenoids, omega fatty acids, and flavonoids. There is growing evidence that these phytonutrients have a positive impact on human health (Kochian and Garvin, 1999).

Products that contain extracts, concentrates, or synthetic versions of food substances, such as vitamins, minerals, amino acids, herbs, or synthetic nutrients, are designated dietary supplements. These are usually sold in controlled dose forms such as powders, capsules, or tablets. Some of these also may be produced using GM technology. For example, a GM bacterium is used to synthesize aspartame, a widely used noncalorific artificial sweetener (Wesley, 1981; Mala et al., 1998).

Foods with Improved Taste

GM foods can be developed with improved flavor. For example, soybean products are frequently used as milk substitutes in products such as soy yoghurt, soy milks, and soy ice cream, as well as in traditional foods such as tofu. However, soybeans have a beany flavor and bitter taste, which is unpalatable to many people. This flavor is caused by lipoxygenase isozymes. A GM soybean has been developed that is lipoxygenase-free. Soybeans lacking lipoxygenases, therefore, have a taste that is more acceptable to many consumers. Furthermore, these soybeans have increased storage stability. (Kitamura, 1995).

Production of Organic Chemicals Used in Food Production

Microorganisms carrying novel genes may serve as factories for large-scale production of desired organic chemicals used in food production. GM of microorganisms allows genes to be mixed from several different organisms to produce hybrid structures. This engineered strain is then used *in vivo* to produce the desired metabolites. Genetically engineered synthesis is a growing field, which involves a microorganism acquiring the genes with their recombination *in vitro* and subsequent production of the desired compounds without requiring a host organism (Roessner and Scott, 1996).

There is an increasing number of possible applications of microbes in the production of ingredients for food through the use of specific enzymes and fermentation processes. Proteases are degradative enzymes that catalyze the hydrolysis of proteins into amino acids and are found in all living organisms. These enzymes have wide application within the food industry and have been routinely used for centuries in processes such as cheese-making, baking, tenderizing of meat, and production of soy products such as soy sauce. Protease extraction from plant and animal sources is costly in time and resources; hence, the genetic manipulation of microorganisms for protease production is an important alternative source (Mala et al., 1998). Proteases now can be commercially produced using bacteria, fungi (particularly useful in cheese-making), and viruses (particularly for medical rather than food applications). This has considerable potential benefit to the environment because it enables large-scale production of these ingredients without the need to grow the large number of plants and animals previously needed from their extraction.

Microorganisms are used to produce the enzyme chymosin. This enzyme cleaves to milk protein (casein) to allow curdling in cheese-making. Its traditional source is from rennet extracted from calves' stomachs. GM chymosin has been produced by isolating the chymosin gene and introducing it into a microorganism (Mala et al., 1998), and then manufacturing large quantities of the enzyme through fermentation. Recombinant chymosin is a commonly used alternative to calf rennet. Some vegetarians who do not eat cheese made with rennet find cheese produced using recombinant chymosin an acceptable alternative (Ross et al., 2000).

Vitamins can also be synthesized by microorganisms carrying specified genes. A microorganism has been used to carry plasmid-containing riboflavin biosynthetic genes, resulting in a manyfold increase in the production of riboflavin (vitamin B_2) (Koizumi et al., 2000).

Food flavorings can also be produced through genetic manipulation of microorganisms. There is a worldwide decline in vanilla production, and increasing volumes of vanilla flavoring are now produced by GM food fermentation organisms (Narbad and Walton, 1998). This frees up agricultural land for other crops.

Reducing the Use of Chemicals in the Growing Process of Foods

If pest and disease-resistant crops are developed, there can be a resultant reduction in use of pesticides. This is hoped to lead to less residue in the food and a reduction of chemical pollution of the environment (Potrykus et al., 1995). Glyphosate-tolerant crops allow the use of a single safe herbicide instead of the need for more toxic chemicals. This led to a 50% reduction of total herbicide use in some cases (Halford and Shewry, 2000).

Improving Processing and Marketing Qualities of Food

Many advantages of GM food primarily benefit food growers and manufacturers. The food industry can profit from GM technology in any number of ways. Productivity can be improved by improving sustainability, efficiency, or cost-effectiveness in food production. The shelf life of a product can be extended through changes made to properties affecting food storage. There has been extensive use of a GM slow-ripening tomato in the manufacture of tomato paste (Tucker, 1993). The softening process has been retarded by introduction of a gene that neutralizes the pectin-degrading enzyme. The resultant tomatoes remain firm and sweet for 7 to 10 d longer than ordinary varieties (Rose, 1992).

Problems of overpopulation and poverty in developing countries face a plateau in the improvement rate of crop production yield with an exponential increase in the rate of demand (McLaren, 2000). GM crops may help provide a stable and sustained production of high-quality food by increasing the yield and improving the nutritional quality (Potrykus et al., 1995). Crops may be developed that require less water to thrive or that can grow in saline or polluted soil (Glenn et al., 1999).

These developments may be driven by the commercial interests of the food industry. However, there may also be gains to consumers from increased yield of GM foods in the form of a greater continuity of supply and possible lower prices (Jones, 1999; Scott and Conner, 1999a,b; Halford and Shewry, 2000.

THE BENEFIT–RISK RATIOS

SAFETY OF GM FOODS COMPARED WITH FOOD FROM NON-GM VARIETIES

It can be argued that most conventional foods have not been tested. Our food crops have been developed using conventional breeding methods for centuries or even millennia and, hence, assumed to be safe (Halford and Shewry, 2000). However, numerous plants contain toxins, evolved to protect them against predators. For this reason, many conventional foods contain toxins naturally and are poisonous to humans if not prepared properly. Examples include kidney beans, cassava, and Jamaican akee, all of which require some form of processing before consumption to remove the toxin. Green potatoes contain poison, and dried beans require soaking and cooking in fresh water to remove the toxic lectins they contain. Despite their potential toxicity, these foods are staples for large sections of the world population.

Traditional breeding may carry risks to health. For example, a new variety of celery bred to be more disease resistant caused skin rashes on a large percentage of grocery workers who handled the vegetable (Berkley et al., 1986). Foods can be contaminated by toxins produced by fungal activity before harvesting or during storage. Mycotoxins such as aflatoxin carry a carcinogenic risk (Campbell and Stoloff, 1974). A fungal contamination of rye by *Claviceps purpurea* causes ergot poisoning, outbreaks of which have caused illness and deaths throughout the centuries, now largely prevented by the use of modern fungicides and vigilant monitoring. Ironically, organic foods that are not treated with fungicides are more at risk of containing mycotoxins.

Food poisoning by microorganisms, especially bacterial contamination, poses by far the greatest risk to health. The U.S. Centers for Disease Control and Prevention (CDC) estimates that each year 76 million people get sick, more than 300,000 are hospitalized, and 5,000 Americans die as a result of food-borne illnesses (Morbidity and Mortality Weekly Report, 2001). The worldwide annual death toll from food poisoning is unknown. Common causative agents include salmonella and staphylococcal contamination, as well as botulism and toxins related to specific food types such as ciguatera fish or mushrooms. It can therefore be argued that food poisoning poses a real, significant and ongoing threat, which is potentially preventable and of vastly greater magnitude than the potential risk from GM foods.

Genetic modification is a very precise process. The products of introduced genes are easily identifiable. The protein produced should be capable of being purified, tested, and monitored in the new food. In foods created by conventional breeding in contrast, the process is much more random and varied, making identification of changes in the composition of the food very difficult or impossible (Halford and Shewry, 2000). The source of most modern hybrids is from mutagenesis through chemical and irradiation treatments. These processes result in the recombination of large numbers of genes in unpredictable ways. Given this, GM foods may be safer than foods derived from these non-GM food techniques and hence require no more testing than that needed for conventional foods.

TABLE 7.4
Regulatory Requirements for Various Consumables

Consumable	Requirements
Pharmaceutical agent (drug)	Statutory regulation regarding efficacy and safety
GM foods	Regulatory requirements regarding laboratory and field testing before release
Dietary supplements	Regulated as food — no testing required
Functional foods	Regulated as food — no testing required
Novel conventional foods	No testing required

Risks Involved with Foods Compared with Pharmaceuticals

It could be contended that testing of GM foods should be as stringent as that required for pharmaceuticals (see Table 7.4). Introduction of a new medical intervention requires considerable and stringent evidence about its efficacy (how effectively it produces the desired health effect) and its safety (the relative frequency and severity of possible adverse effects). There are usually clear-cut, desirable outcomes that can be measured in regard to the development of a new drug. Its potential health risk from adverse side effects is offset by the desired medical effects for a particular health problem. Ideally, the benefit–risk ratio strongly favors the use of the drug in appropriate cases.

Usually, a particular medication is not intended for general human consumption but only for the subpopulation of people who require it for health reasons. Pharmaceutical products may be subjected to a number of restrictions of use: to specific ages (for example, may be deemed not suitable for children or the elderly); to patients suffering from conditions the particular drug is promoted to assist; and with defined upper limits for dosage and duration of use. The exact chemical composition of the drug is known, which may greatly facilitate its testing and monitoring.

Foods, in contrast, often contain hundreds of different compounds. Their exact composition is usually unknown. Upon introduction into the market, the potential exists for a foodstuff to be consumed by the entire population, including the very young and the very old. There may be little or no control on how much or how often a particular food is consumed by some people, in contrast to the tight regulations for pharmaceutical agents.

However, what differentiates a food from a medicine is becoming increasingly indistinct, with the development of a number of products that are consumed for both health and medicinal benefits (Murphy, 1997). In general, the development, sale, and monitoring of pharmaceuticals are much more stringently regulated than for foods. In New Zealand statute, dietary supplements are regulated as food not medicine. Other products such as functional foods do not have statutory regulation. However, there is no clear distinction between what should be considered a food and what should be considered a drug. Many of the specific criteria for a drug (such as a particular dose for a given age and for a given condition) do not apply to functional foods for which there may be neither control on who consumes the product nor on the amount or frequency with which it is consumed. Whereas GM foods are relatively stringently regulated, whole foods from traditional sources or new sources other than GM are not subjected to testing at all.

Benefit–Risk Ratio

Risk assessment involves estimating both the probability and the consequence of an adverse event. Critics of GM food promote the precautionary principle, which although variously defined, effectively states that where potential adverse effects are not fully understood, an activity should not proceed. However, no activity is risk free, and no food is risk free.

There is currently no evidence that GM foods present significant threat to individual safety, although the potential exists for the creation of a GM food containing unexpected allergens or toxins. Any potential risk needs to be compared with known risks of everyday living. Food from any source may pose risks, both short-term (contain infective organisms, toxins, or allergens, or cause choking) or long-term (for example, high-cholesterol diets may contribute to heart disease).

Risk to the environment is a different and more complex issue, and issues such as the potential development of superweeds or pesticide-resistant insects need to be assessed on a case-by-case basis.

The benefit–risk ratio needs to take into consideration who is taking the risk and who is getting the benefit. Currently, it is likely that GM food producers and the multinational companies that buy their produce benefit more than individual consumers. New foods are likely to be more acceptable to consumers if they can identify direct benefits to themselves (food that is cheaper, tastes better, or improves health) rather than gains by the food industry.

Moral and Ethical Issues

A high degree of public concern has been expressed toward GM technology, especially GM food. GM foods are developed primarily for commercial gain, not for the direct benefit of consumers. Large corporations are seen to be motivated by their own financial interests and suspected of underestimating or ignoring risks to human health to protect their own profits. Public mistrust of commercial industries has some validity. The tobacco industry is an example known to all.

There are many ways that the food industry can threaten human morbidity and mortality. Some risks are well known, and regulations exist to prevent contamination and ensure food safety. Unforeseen risks also occur. For example, a new variant form of the degenerative central nervous system disease Creutzfeldt–Jakob Disease (CJD) emerged in Britain that appeared to be linked with consumption of beef infected with bovine spongiform encephalopathy (BSE), the condition popularly known as mad cow disease. It seems that the BSE epidemic resulted from feeding contaminated meat and bone meal to the animals. Government agencies were highly criticized in an official report of minimizing the risk to humans and too tardy in mounting a response (Royal Commission on Genetic Modification, 2001).

There are also religious or spiritual considerations. Objections to GM technology include views that species are sacrosanct. Combining the genetic material of two species that do not breed "in nature" is considered immoral, against God's will and potentially ecologically dangerous. The proliferation and public acceptability of interspecies horticultural hybrids that have been created over the past few centuries counters this argument to some degree, but it may be a more valid concern with the combination of much more dissimilar species.

Anti–GM food proponents often illustrate their case with emotive images such as a potato with frogs legs (Butler, 1999) or a pumpkin with chicken legs (Kibbell, 1999). These are designed to act on people's emotions rather than their intellects. People confronted with these images may not realize that much of GM food technology involves production of an enzyme or other food component, or enhancement of an existing characteristic of a food, and not necessarily the amalgamation of two vastly different food groups. There is poor understanding regarding the role microorganisms play in much GM technology. These emotional images, however, do raise the important ethical issue of where to "draw the line." For some people, it may be acceptable to cross different plants that do not normally interbreed — but not different animals or plants with animals; for others, tampering with any species may represent a violation of their religious beliefs.

On the broader scale, genetic technology such as cloning and xenotransplantation (transplantation of tissues or organs from animals to humans for medical reasons) raises huge moral dilemmas regarding the nature of what it is to be "human" and the sanctity of human life, but beyond the scope of this discussion.

Testing and Monitoring

There is need for rigorous and continuous testing, including the use of animal trials where appropriate, to ensure that all innovative food products are not hazardous. Close monitoring of the chemical purity of biotechnology-derived products should occur, particularly following any significant changes to the manufacturing process (Mayeno and Gleich, 1994). Testing and monitoring need to consider the broader issues relating to the effects on the environment with rigorous safeguards to prevent uncontrolled release of GM organisms that may have detrimental effects on the ecosystem.

The U.S. Food and Drug Administration (FDA) decided in 1999 that GM crops should receive the same consideration for health risks as any other new crop plant (Center for Food Safety and Applied Nutrition, 1992). This recommendation is not accepted by opponents to GM foods who argue that existing systems of checks and regulations are not adequate, and believe that GM foods need additional testing compared to conventionally produced novel foods (Millstone et al., 1999; Trewavas and Leaver, 1999).

The risks involved in consuming foods (both GM and non-GM produced), functional foods, dietary supplements, and medicines are similar and vary only with the concentration of a component within the product and the quantity and frequency of its consumption. Therefore, an argument can be made for a common regulation of products for human consumption, with the same standards of testing, monitoring, and ongoing surveillance applying to all. The assessment should be undertaken on a case-by-case basis, taking into account individual variables and circumstances. This is the position reached by the New Zealand Royal Commission on Genetic Modification (Royal Commission on Genetic Modification, 2001) and is in line with the recommendations of the comprehensive report compiled in Britain (Nuffield Council on Bioethics, 1999) that advocated local solutions for local people.

Clearly, any regulatory process must be transparent. The preferable model would appear to follow that used in pharmaceutical regulation in which the testing and monitoring are done by the company. The role of the government is to set the criteria and protocols and ensure strict adherence to the regulations.

CONCLUSIONS

There are millennia-long processes of human intervention in the heredity of plants, animals, and, more recently, microorganisms. GM technology forms part of this continuum and allows for the transfer of genes between different species, genera, or even kingdoms in a way not previously possible.

There are potential risks of foods produced by this process. These include risks to individual consumers — introduction of an allergen or a toxin to a food. Potential threats to human populations include production of foodstuffs with improved marketability but reduced nutritional value, or the development of antibiotic resistance through the use of marker genes. GM crops may threaten the environment if their resistance to herbicides is passed on to weeds though cross-pollination, resulting in herbicide-resistant superweeds, or if their insect pests become resistant to existing pesticides.

Potential advantages of GM foods for individuals include foods with reduced potential toxicity, hypoallergenicity, increased nutritional value, or enhanced health-promoting effects. Potential gains for human populations include improvement in both the quality and quantity of foods produced to help alleviate poverty and starvation. Increased yield of GM foods may give consumers greater continuity of supply at possible lower prices. Drought- and temperature-resistant or saline-tolerant crops may allow food production on land currently not utilized. The environment might benefit from development of pest and disease-resistant crops that reduce pesticide use.

Most conventional foods have not been tested, and many contain natural toxins requiring processing before consumption, yet these are staples for large sections of the world population. Foods can be contaminated by mycotoxins or other microorganisms before harvesting or during storage. Food from any source may pose both short-term and long-term risks. The precise nature of GM technology means that the products of introduced genes are identifiable and, hence, can be more easily purified, tested, and monitored for in the new food.

Development of pharmaceuticals requires evidence about its efficacy and safety before release, with the benefit–risk ratio strongly favoring the use of the drug in appropriate cases. The boundary between food and medicine is becoming increasingly blurred, with the development of a number of products consumed for health and medicinal benefits. Currently, regulation is much more stringent for the pharmaceutical than the food industry. However, with the development of functional foods and dietary supplements, there is no clear distinction between what is food and what is drug.

Risk assessment requires the estimation of both the probability and the consequence of an event. Although the precautionary principle might be evoked, no activity is risk free. To date, there is no evidence that GM foods present a threat to consumer safety, although potential risk to the environment is more uncertain. Comparison should be made between perceived potential risks and known risks of everyday living, taking into account who is taking the risk and who is getting the benefit. New foods are likely to be more acceptable to consumers if they can identify direct benefits for themselves rather than gains by the food industry.

Apart from scientific considerations, people may choose to avoid GM foods for ethical, cultural, or religious reasons. Labeling appropriate foods "GM-free" therefore may be a desirable marketing strategy. All innovative food products need rigorous and continuous testing, including the use of animal trials, to ensure that they are not hazardous. There are similar risks involved in consuming foods (both GM and non-GM produced), functional foods, dietary supplements, and medicines; these commodities vary only with the concentration of a component within the product and the quantity and frequency of consumption. A case can be made for common regulation of products for human consumption, with the same standards of testing, monitoring, and ongoing surveillance applying to all. To single out GM from non-GM is an unhelpful, irrational, and potentially dangerous policy. This assessment should be undertaken on a case-by-case basis, taking into account individual variables and circumstances. This regulatory process must be transparent.

The goals to improve the quantity, quality, and safety of the food supply are laudable, but it is also recognized that the primary motivation of the biofood industry is financial gain. The consumption of GM foods may not be inherently less safe than foodstuffs developed through traditional breeding techniques. However, it must also be recognized that public distrust runs high.

Genetic modification of food has both potential benefits and potential dangers; this is no different from many other scientific developments. It is desirable that the issues undergo comprehensive public and professional debate. However, it is important that this discussion is informed by science and that claims of both benefits and risks are evidence based. Objectivity is a prerequisite. The process must neither be driven by the vested interest of the biotechnical multinational companies on the one hand nor ill-informed public fears on the other.

ACKNOWLEDGMENTS

Some of the material for this chapter has been adapted from Goodyear-Smith, F.A. (2001). Health and safety issues pertaining to genetically modified foods, *Australia and New Zealand Journal of Public Health,* 25, (4): 371–375; (2003, in press). *Risks and Benefits of Genetically Modified Foods: An Overview, in VIII. Culture collections, legal aspects and biosafety, Handbook of Fungal Biotechnology,* 2nd ed., Dilip K. Arora, Paul D. Bridge, and Deepak Bhatnagar Eds., Marcel Dekker, New York, ch. 91.

References

Advisory Committee on Novel Foods and Processes (1994). *Report on the use of antibiotic resistance markers in genetically modified food organisms.* London, ACNFP (accessed).

Anonymous (1999). Health risks of genetically modified foods, *Lancet,* 353: 1811.

Anonymous (2001). Advisory panel finds medium likelihood that StarLink corn could be allergenic, *World Food Regulation Review,* 10: 23–24.

Arai, S. (1993). Hypoallergenic rice, *Denpun Kagaku,* 40: 177–181.

Astwood James, D., N. Leach John et al. (1996). Stability of food allergens to digestion *in vitro*, *Nature Biotechnology,* 14: 1269–1273.

Australia and New Zealand Food Authority (2000). *GM foods and the consumer: ANZFA's safety assessment process for genetically modified foods.* Canberra/Wellington, ANZFA <http://www.anzfa.gov.au/> (accessed June 2002).

Belongia, E.A., C.W. Hedberg et al. (1990). An investigation of the cause of the eosinophilia-myalgia syndrome associated with tryptophan use, *New England Journal of Medicine,* 323: 357–365.

Berkley, S.F., A.W. Hightower et al. (1986). Dermatitis in grocery workers associated with high natural concentrations of furanocoumarins in celery, *Annals of Internal Medicine,* 105: 351–355.

Biotechinfo (2002). *Genetically modified food*, Ireland's biotechnology information source <http://www.biotechinfo.ie/content/content.asp?section_id=29&language_id=1> (accessed).

British Medical Association (1999). *The impact of genetic modification on agriculture, food and health.* London, BMA (accessed).

Burke, D. (1998). Why all the fuss about genetically modified food? *British Medical Journal,* 16: 1845–1846.

Buss, W.C., J. Stepanek et al. (1996). EBT, a tryptophan contaminant associated with eosinophilia myalgia syndrome, is incorporated into proteins during translation as an amino acid analog, *Autoimmunity,* 25: 33–45.

Butler, P. (1999). The myths of genetically modified foods, *Soil and Health,* July: 12–13.

Campbell, T.C. and L. Stoloff (1974). Implication of mycotoxins for human health, *Journal of Agricultural and Food Chemistry,* 22: 1006–1015.

Center for Food Safety and Applied Nutrition (1992). 57 FR 22984, *Federal Register.* Washington, Center for Food Safety and Applied Nutrition. 57 <http://vm.cfsan.fda.gov/~lrd/fr92529b.html> (accessed October 2002).

Crawley, M., S. Brown et al. (2001). Transgenic crops in natural habitats, *Nature,* 409: 682–683.

Darmency, H., E. Lefol et al. (1998). Spontaneous hybridizations between oilseed rape and wild radish, *SO — Molecular Ecology,* 7: 1467–1473.

Diamond, J. (1998). *Guns, germs and steel: A short history of everybody for the last 13,000 years*. London: Vintage, 480 p.

Dorey, E. (2000). Taco dispute underscores need for standardized tests, *Nature Biotechnology,* 18: 1136–1137.

Editorial (1999). Health risks of genetically modified foods, *Lancet,* 353: 1811.

Feldbaum (1999). Health risks of genetically modified foods, *Lancet,* 354: 69.

Glenn, E.P., J.J. Brown et al. (1999). Salt tolerance and crop potential of halophytes, *Critical Reviews in Plant Sciences,* 18: 227–255.

Halford, N.G. and P.R. Shewry (2000). Genetically modified crops: methodology, benefits, regulation and public concerns, *British Medical Bulletin,* 56: 62–73.

Henning, K.J., E. Jean-Baptiste et al. (1993). Eosinophilia-myalgia syndrome in patients ingesting a single source of L-tryptophan, *Journal of Rheumatology,* 20: 273–278.

Holden, P. (1999). Safety of genetically modified foods is still dubious, *British Medical Journal,* 318: 332.

Jones, L. (1999). Science, medicine, and the future: genetically modified foods, *British Medical Journal,* 318: 581–584.

Kaiser, J. (1996). Pests overwhelm Bt cotton crop, *Science,* 273: 423.

Kibbell, J. (1999). GE and organics, *Soil and Health,* July: 46.

Kilbourne, E.M., R.M. Philen et al. (1996). Tryptophan produced by Showa Denko and epidemic eosinophilia-myalgia syndrome, *Journal of Rheumatology — Supplement,* 46: 81–88; discussion 89–91.

Kitamura, K. (1995). Genetic improvement of nutritional and food processing quality in soybean, *JARQ — Japan Agricultural Research Quarterly,* 29: 1–8.

Kochian, L.V. and D.F. Garvin (1999). Agricultural approaches to improving phytonutrient content in plants: an overview, *Nutrition Reviews,* 57: S13–S18.

Koizumi, S., Y. Yonetani et al. (2000). Production of riboflavin by metabolically engineered *Corynebacterium ammoniagenes*, *Applied Microbiology and Biotechnology,* 53: 674–679.

Lachmann, P. (1999). Health risks of genetically modified foods, *Lancet,* 354: 69.

Lehrer, S.B. and G. Reese (1997). Recombinant proteins in newly developed foods: identification of allergenic activity, *International Archives of Allergy and Immunology,* 113: 122–124.

Mala, B.R., M.T. Aparna et al. (1998). Molecular and biotechnological aspects of microbial proteases, *Microbiology and Molecular Biology Reviews,* 62: 597–635.

Mayeno, A.N. and G.J. Gleich (1994). Eosinophilia-myalgia syndrome and tryptophan production: a cautionary tale, *Trends in Biotechnology,* 12: 346–352.

McLaren, J.S. (2000). The importance of genomics to the future of crop production, *Pest Management Science,* 56: 573–579.

Mikkelsen, T.R., B. Andersen et al. (1996). The risk of crop transgene spread, *Nature,* 380: 31.

Millstone, E., E. Brunner et al. (1999). Beyond "substantial equivalence," *Nature,* 410: 525–526.

Morbidity and Mortality Weekly Report (2001). *Diagnosis and Management of Foodborne Illnesses*. Washington, D.C., Centers for Disease Control and Prevention: 1–69 <http://www.cdc.gov/mmwr/preview/mmwrhtml/rr5002a1.htm> (accessed February 2003).

Mowat, A., A.J. Fitzgerald et al. (1999). GM debate, *Lancet,* 354: 1725–1733.

Murphy, M. (1997). Nutraceuticals, functional foods and medical foods: commentary and caveats, *Journal of Nutraceuticals, Functional and Medical Foods,* 1: 73–99.

Nakamura, T., Y. Syukunobe et al. (1991). Development of a hypoallergenic peptide from bovine casein by enzymic hydrolysis, *Japanese Journal of Zootechnical Science,* 62: 683–691.

Narbad, A. and N. Walton (1998). Vanilla flavour biotechnology, *Food Ingredients and Analysis International,* 18: June16–18.

Nestle, M. (1998). Food biotechnology: labelling will benefit industry as well as consumers, *Nutrition Today,* 33: 6.

Nordlee, J.A., S.L. Taylor et al. (1996). Identification of a Brazil-nut allergen in transgenic soybeans, *New England Journal of Medicine,* 334: 688–692.

Nuffield Council on Bioethics (1999). *Genetically modified crops: the ethical and social issues*. London, Nuffield Foundation <http://www.nuffieldbioethics.org/filelibrary/pdf/gmcrop.pdf> (accessed October 2002).

Pastorello, E.A., A. Conti et al. (1998). Identification of actinidin as the major allergen of kiwi fruit, *Journal of Allergy and Clinical Immunology,* 101: 531–537.

Potrykus, I., P.K. Burkhardt et al. (1995). Genetic engineering of Indica rice in support of sustained production of affordable and high quality food in developing countries, *Euphytica,* 85: 441–449.

Roessner, C.A. and A.I. Scott (1996). Genetically engineered synthesis of natural products: from alkaloids to corrins, *Annual Review of Microbiology,* 50: 467–490.

Rose, P. (1992). Good breeding shows, *Food Manufacture International,* 9: 12–13.

Rose, P. (1993). The search for the super spud, *Food Manufacture International,* 10: 39.

Ross, R.P., C. Stanton et al. (2000). Novel cultures for cheese improvement, *Trends in Food Science and Technology,* 11: 96–104.

Royal Commission on Genetic Modification (2001). *Report of the Royal Commission on Genetic Modification.* Wellington, New Zealand Ministry for the Environment <http://www.gmcommission.govt.nz/RCGM/index.html> (accessed February 2003).

Royal Society of New Zealand (2000). *Submission to the Royal Commission on Genetic Modification Part A: biological sciences in NZ.* Wellington, Royal Society of New Zealand <http://www.rsnz.govt.nz/news/gene/submissionA.php> (accessed January 2002).

Sato, T. and G. Miyata (2000a). The nutraceutical benefit, part i: green tea, *Nutrition,* 16: 315–317.

Sato, T. and G. Miyata (2000b). The nutraceutical benefit, part iv: garlic, *Nutrition,* 16: 787–788.

Sato, T. and G. Miyata (2000c). The nutraceutical benefit, part iii: honey, *Nutrition,* 16: 468–469.

Sato, T. and G. Miyata (2000d). The nutraceutical benefit, part ii: ginseng, *Nutrition,* 16: 391–392.

Saul, D. (2002). Genetically modified food: what is it, and what are the implications for health? *New Ethicals Journal,* 12: 11–13.

Schulz, R. and A. Beubler (1989). International status of the application of genetically manipulated yeasts in the brewing industry, *Lebensmittel-Industrie,* 36: 62–65.

Scott, P. and A. Conner (1999a). GE — the need for scientific honesty in debate part 1, *New Zealand Medical Association Newsletter,* 3–4.

Scott, P. and A. Conner (1999b). GE — the need for scientific honesty in debate part 2, *New Zealand Medical Association Newsletter,* 3,6.

Singh, M.B., N. de Weerd et al. (1999). Genetically engineered plant allergens with reduced anaphylactic activity, *International Archives of Allergy and Immunology,* 119: 75–85.

Slutsker, L., F.C. Hoesly et al. (1990). Eosinophilia-myalgia syndrome associated with exposure to tryptophan from a single manufacturer, *Journal of the American Medical Association,* 264: 213–217.

Trewavas, A. and C. Leaver (1999). Conventional crops are the test of GM prejudice, *Nature,* 401: 640.

Tucker, G.A. (1993). Improvement of tomato fruit quality and processing characteristics by genetic engineering, *Food Science and Technology Today,* 7: 103–108.

Wal, J.M. (1999). Assessment of allergic potential of (novel) foods, *Nahrung,* 43: 168–174.

Wesley, P. (1981). Applied genetics and the food industry, *Food Development,* 15: 30–33.

Whitelaw, B. (1999). Toward designer milk, *Nature Biotechnology,* 17: 135–136.

Ye, X., S. Al-Babili et al. (2000). Engineering the provitamin A (beta-carotene) biosynthetic pathway into (carotenoid-free) rice endosperm, *Science,* 287: 303–305.

8 Food Contaminants and Children: Cause for Concern

Lynn R. Goldman

CONTENTS

ABSTRACT

Globally, many children lack sufficient quantities of food, and assuring that food is safe for children is a challenge everywhere. Children may be more susceptible than adults to hazards in food, yet safety standards have been set to protect adults. Consideration of children's risks begins with an understanding that there are critical windows of exposure during development; that early exposures can have lifelong consequences; and that exposures may be proportionately greater. Immature immune systems confer greater risks to children when they are exposed to certain pathogens. Risks of early childhood exposure to substances that cause cancer, birth defects, endocrine disruption, developmental neurotoxicity, and allergy are discussed. A number of substances in food may be particularly hazardous to children whether present in food as natural products, as pathogens that have been introduced, or as

0-8493-3516-7/05/$0.00+$1.50

chemicals (food additives, chemicals migrating from packaging, residues of veterinary drugs, and residues of environmental pollutants or pesticides that were applied to the food). There are a number of emerging issues, including the presence in our food supply of recently discovered contaminants (brominated flame retardants, perfluorinated compounds, and acrylamide) and the increased use of genetically engineered crops for food production.

INTRODUCTION

Assuring an adequate supply of safe food is a concern of all countries, especially when it comes to meeting the needs of children. According to the Food and Agriculture Organization of the United Nations (FAO), in 2001 approximately 791 million people in the developing world, 18%, were food insecure. The FAO projects that 680 million people in developing countries may be food insecure by 2010. More importantly, the International Food Policy Research Institute (IFPRI) forecasts that one out of every four children in developing countries will be malnourished in 2020 (Wiebe et al., 2001). The need to feed increasing populations of children places increasing demands on the productivity of the food sector, as well as on efforts to reduce losses due to contamination of food. The most vital factor having an effect on food security, especially in developing countries, is domestic food production. Financial constraints limit import of safe food in many developing countries (Shapouri and Rosen, 2001). Whereas efforts are under way to increase food production, it also is important for the health of children to increase the safety of food.

This chapter reviews the full array of food safety hazards to children. In order to ensure the safety of food for children, fitting food safety measures must be implemented during production, harvesting, processing, and distribution, and up to consumption. Food safety risks differ greatly among countries. There is great variability in systems to assure food safety, but most countries employ an element of risk assessment in decisions about management of food safety risks. From the scientific standpoint, it is thus important to understand the dimensions of risks to children in order to assure that such efforts adequately protect them. Another commonality is the incorporation of precautionary principles or approaches in determining food safety standards. In this regard, our understanding of risks to children and other vulnerable populations is critical.

SUSCEPTIBILITY OF CHILDREN

It is well known that children are not just little adults when it comes to their interaction with hazardous agents in food. Initially, the allowable levels of contaminants and other substances in food were established based on risks to healthy adult males and not accounting for risks to other populations. Figure 8.1 shows how the interplay between critical windows of exposure, differential exposure rates and metabolism, and timing can lead to increased susceptibility of children.

Critical Windows of Exposure to Hazards: The fetus and infant have immature, developing organ systems, especially the nervous, immune, and endocrine systems,

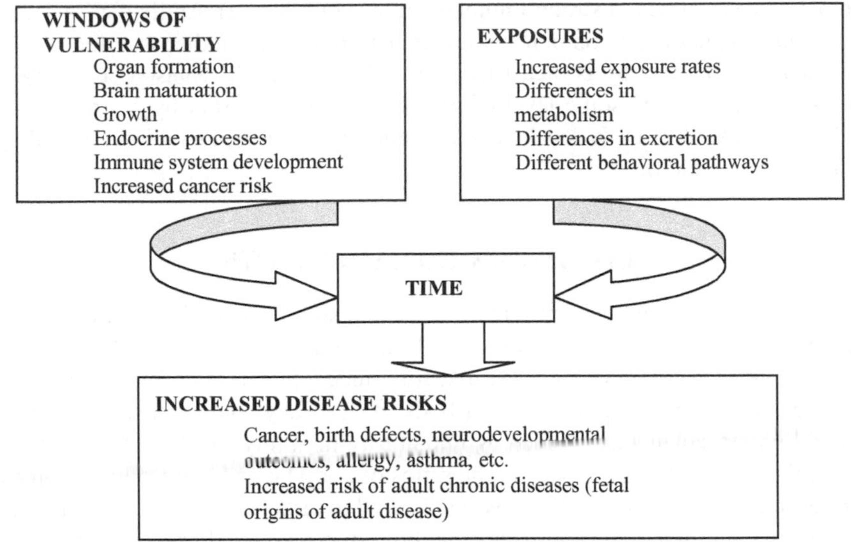

FIGURE 8.1 Susceptibility of children to early life exposures. Exposures in infancy and early childhood are critical in the development of childhood disease, as well as in providing the fetal origins of adult disease. Timing of exposure is critical in this regard. Additionally, exposures may be greater for body mass because of greater intake rates, differences in metabolism and excretion, and differences in behavior that cause increased exposures (e.g., hand-to-mouth behavior in young children).

so that there are "windows of exposure" during which development can be disrupted in ways that may result in serious and irreversible long-term effects (Selevan et al., 2000). During gestation, children can be exposed when contaminants are passed from the mother to the baby via the placenta. A number of adverse outcomes have been documented, including miscarriage, birth defects, low birth weight and preterm birth, and infections, as well as causation of chronic diseases later in life such as childhood cancer and birth defects. Of particular concern is that organ systems are rapidly developing so that toxic substances and pathogens can adversely affect brain development. Such exposure windows exist with various outcomes through adolescence.

Fetal Origins of Adult Disease: It is known that children who are born with low birth weight are more likely to have chronic health problems later in life such as obesity, diabetes, and cardiovascular disease; however, it is not known whether environmental exposures are involved in this. The recognition of the fetal origins of adult disease has increased awareness of the need to reduce harmful exposures during that time of development (Barker et al., 2002). The likelihood that effects with longer latency periods (such as cancer), will be manifested is greater for children than for adults because children have more years of potential life (Bearer, 1995).

Immature Immune Systems: Certainly, one of the reasons for increased susceptibility of newborns and young infants to pathogens in food is their immature immune systems. Listeria and toxoplasma can cause serious systemic infections, fetal demise, and severe neonatal infections when the mother is infected during pregnancy (and the baby cannot mount a strong immune defense on its own). Antibodies and other important immunologic components are transferred from mother to infant via the placenta and, if the baby is breast-fed, through breast milk (Hanson et al., 2003). Breast feeding confers some level of protection from infection by gastrointestinal pathogens as well as from development of allergic diseases, including food allergy (Hanson et al., 2003).

CHILDREN'S EXPOSURE RATES

Infants and young children consume a greater quantity of foods per unit of body weight. Most significant contributions are from certain foods that infants and young children prefer to consume, such as milk, fruit juices, cereals, and fruits. This means that they will receive a larger dose per body mass of any contaminants in the food. Altered pharmacokinetics — such as immaturity in ability to detoxify and excrete contaminants and the differences in the ratios of body fat and extracellular water — can result in increase in effective doses delivered to target organs. Breast-fed infants may be exposed to food contaminants via those that are passed from mother to baby through breast milk. (However, it is considered that the benefits of breast feeding greatly outweigh the risks.) As infants grow and mature they switch to a diet of solid foods, milk and milk formulas, and juices and thus have direct exposure to contaminants in food. Exposures in infancy, such as to allergens and pathogens in food, probably are related to the development of asthma as well as to food allergy (Gern et al., 1999). Adolescence is another critical time when growth is rapid and maturational steps such as puberty and brain development may be adversely affected.

CHILDREN'S HEALTH HAZARDS

Cancer: The hormone diethylstilbestrol (DES) caused vaginal cancer in females born to women who were prescribed DES during pregnancy to prevent miscarriage. Follow-up studies of male children are under way (Giusti et al., 1995; Herbst, 1987; Palanza et al., 2001). Toxicology studies have provided limited information about transplacental and childhood carcinogenesis. Not surprisingly, carcinogens for adults also cause cancer *in utero* (Olshan et al., 2000); generally these are of the same types as are caused in adult animals but, in many cases, occurring at a greater frequency than with dosing later in life. It is not understood whether this is due to alterations in susceptibility, effective dose to target organs, or time taken for tumor development (Anderson et al., 1985; Tomatis, 1989).

Of particular interest is whether exposure of germ cells to carcinogens and mutagens can lead to cancer and other adverse health effects in children. In experimental studies, there are models for tumors in offspring caused by mutations in germ cells (Daher et al., 1998; Tomatis, 1989; Tomatis et al., 1992). Maternal germ cell exposures have not been studied in human populations. This would be quite difficult because the ova are developing *in utero*. In view of the lack of epidemiological evidence on germ-cell exposures and cancers, based on biological and toxicological studies, this area needs further research.

Probably the most important exposure time for carcinogenesis is at the fetal and neonatal stages. A number of factors that may be involved have been identified: (1) numbers of target cells at risk, (2) sensitivity to cell killing, (3) effects of rate of cell division on fixation of mutation before repair can occur, (4) ability to repair DNA damage, (5) expansion of clones of mutated cells as part of normal ontogeny, (6) presence of undifferentiated stem cells, (7) development of differentiated characteristics, including the ability to carry out metabolic activation of chemicals, (8) metabolic detoxification by placenta or maternal tissues, (9) metabolic detoxification by the perinate itself, and (10) immaturity of the endocrine and immunological systems. Since many of these factors are not independent of each other, one should interpret the evidence with caution and it is likely that some factors are species, strain, tissue, and agent specific (Anderson et al., 2000). Of these, a few factors clearly stand out as having more evidence supporting them. Certainly, there is evidence that the greater number of target cells, as well as more rapid cell division, and an embryonic environment that is more conducive to expansion of clones of cancerous cells are important factors (Anderson et al., 2000). A decreased rate of metabolic detoxification by the newborn is clearly an issue as well (Ginsberg et al., 2002).

Despite the theoretical concern about early life exposure to carcinogens, regulatory cancer studies conducted for the U.S. Environmental Protection Agency (EPA) and the U.S. Food and Drug Administration (FDA) rely on testing of mature animals. Appropriate animal studies for early life exposure are difficult to design because of the vulnerability of the fetus to other toxic effects and the difficulty in measuring exposure to the fetus. Newer molecular techniques to monitor genetic and protein changes in response to environmental exposures may provide more powerful measurements of exposure as well as precancer changes, which in turn might shed more light on this issue. Additionally, the EPA in its recent cancer guidelines has proposed to apply an uncertainty factor of ten when extrapolating cancer risk for mutagenic carcinogens from adults to the fetus and child through age 2 and a factor of three for children ages 2 through 15 (U.S. Environmental Protection Agency, Risk Assessment Forum, 2003).

Birth Defects: Birth defects are the primary cause of infant mortality (<12 months) in the U.S. High level contamination of food (e.g., polychlorinated biphenyls (PCBs) as described in the following text) has been linked to birth defects; however, food nutritional composition is also an important factor — for example, the discovery of the role of folic acid in prevention of neural tube defects (Werler et al., 1993).

Endocrine Disruption: More recently, there are questions whether or not endocrine-active chemicals may have adverse effects on child development (Goldman and Koduru, 2000) and whether certain malformations involving the reproductive system might be due to such effects (Paulozzi et al., 1997). However, it is important to keep in perspective that endocrine disruption is a mode of action and not an adverse effect *per se*; what is of concern is whether such endocrine disruption leads to developmental or reproductive abnormalities or cancer.

There are three major endocrine systems that have been shown to be affected by endocrine disruptors in laboratory experiments. *Environmental estrogens* (Table 8.1) in animal studies are associated with feminization of male offspring after

TABLE 8.1
Endocrine Disruptors Found in Food: Hormonal Systems and Examples

Endocrine Effect	Substances (Examples)
Estrogenic	DES (very potent), pesticide methoxychlor, bisphenol A (plasticizer), genistein (natural compound in soy genistein), and certain PCBs and dioxins
Antiandrogenic	Pesticides DDE (DDT metabolite) and vinclozolin, phthalate plasticizers
Thyroid	Certain PCBs and dioxins

Note: A number of hazardous substances are capable of mimicking or blocking the effects of endogenous hormones. Such modes of action for toxicity are particularly of concern for the developing fetus and infant.

in utero exposure (e.g., decreased anogenital distance, decreased sperm count) as well as later-life increases in cancer risks in reproductive organ systems. There are few human studies that have definitely shown effects, except in the case of DES at higher doses than those that occur in food. There are numerous theoretical or hypothesized risks, including increases in hypospadias, undescended testicles, and lowered sperm counts due to prenatal exposure to males; increases in premature thelarche (breast development) or puberty among exposed girls; and increased risk of estrogen-responsive cancers (e.g., cancers of breast and ovary). *Environmental antiandrogens* in animal studies cause feminization in male offspring exposed *in utero* and have not been as clearly associated with hazards to females. There are no human studies that clearly demonstrate these impacts; however, there is a strong theoretical basis to conclude that, at appropriate levels of exposure, men could be affected by *in utero* exposures. *Environmental thyroid chemicals* can interfere with normal thyroid hormone function in laboratory animals. In the case of humans, it has been suggested that the known neurodevelopmental effects of PCBs are correlated with such a thyroid hormonal mechanism. However, human studies have not consistently found associations between PCB exposure levels and thyroid hormone status in newborns.

Neurotoxicity: Successively, lead (Needleman et al., 1979), PCBs (Rogan and Gladen, 1991) and methylmercury (Grandjean et al., 1997) have been shown to cause subtle impacts on IQ (intelligence quotient) and cognition in children, at levels much lower than levels that cause effects in adults. Several critical components of brain development have been described, as shown in Table 8.2. Environmental agents may perturb development in any of these domains. These processes are tightly controlled via DNA expression and neurotransmitter signals. Neuronal components are developing at various rates at different times during gestation and early life (Rice and Barone, 2000).

Allergy: Food allergy and other adverse reactions to food also occur more commonly among children than adults. Environmental factors, broadly including indoor and outdoor allergens and pollutants, infectious disease, breast feeding, and diet are also believed to play an important role in asthma prevalence and are likely to be involved with the increases in the disease over time (Weiss, 1998). Early viral

TABLE 8.2
Critical Components of Brain Development That May Be Disrupted through Exposure to Neurotoxic Agents

Component	Time Interval
Growth of neurons	*In utero* through years 1–3
Proliferation of neurons	*In utero* through adolescence
Migration of recently proliferated neurons from central to cortical regions	Following proliferation
Differentiation of neuroblasts to mature forms	Following migration
Synaptogenesis or formation of connections between neurons	*In utero* through adolescence
Gliogenesis or formation of support cells	*In utero* through adolescence
Myelogenesis or formation of myelin sheaths for neurons	*In utero* through adolescence
Apoptosis or programmed cell death	*In utero* through adolescence

Note: The brain of the developing fetus and young child is susceptible to the effects of a number of neurotoxic agents, largely because of the very precisely timed series of complex developmental processes that are occurring during this time.

Reference: Rice and Barone, 2000.

infections seem to play a role in development of asthma in childhood and currently there is much research under way to show how infections, immunizations, *in utero* exposures, and breast feeding may influence development of the immune system and ultimately the prevalence of asthma (Bjorksten, 1999; Martinez and Holt, 1999).

FOOD SAFETY HAZARDS FOR CHILDREN

The food safety hazards that are of particular concern to children include natural products, pathogens, chemicals, pesticides, radiation products, and food alterations via biotechnology. Table 8.3 lists the number of types of hazards that may be of particular concern for children.

Natural Products

Hazardous substances can enter food as a result of natural processes. Many substances produced by nature are very toxic to people, and children are often the most vulnerable. One of the most toxic substances known is botulinum toxin, which is produced by the bacterium *Clostridium botulinum*. Botulism spores can reside dormant in the environment for years and reproduce in anaerobic environments. In the U.S., there are infant deaths every year from botulinum spores that can be naturally present in honey and other products. Infants apparently are susceptible because *C. botulinum* spores are able to reproduce in their gastrointestinal tracts, thereby producing the toxin that causes paralysis and, if not treated, death. The average annual rate of infant botulism was about 69/million/yr compared to 0.8 to 13/million/yr in older age groups during 1992 to 1998 (Chang et al., 2003). The incidence of infant botulism in developing countries is unknown.

TABLE 8.3
Food Safety Hazards of Particular Concern for Children

Type of Hazard	Examples
Natural products	*Clostridium botulinum* toxin
	Natural carcinogens
	Phytoestrogens
	Mycotoxins
Pathogens	Viruses (Norwalk virus, rotavirus)
	Bacteria (*Salmonella enteriditis, Listeria monocytogenes, E. coli* O157:H7, *Campylobacter pylori*, shigella)
	Protozoa (cyclospora, cryptosporidia, *Toxoplasma gondii*)
Chemicals	Direct food additives (tartrazine, sulfites)
	Indirect food additives (phthalates, bisphenol A)
	Veterinary drugs and food additives (antibiotics, hormones)
	Inadvertent chemicals and metals (PCBs, dioxins, lead, methylmercury, polybrominated flame retardants, perfluorinated compounds), acrylamide
Pesticides	Insecticides (DDT, methoxychlor, lindane, chlordane, organophosphates, carbamates)
	Herbicides (atrazine, alachlor)
	Fungicides
Biotechnology	Enhancement of hazardous proteins already present in food
	Transfer of hazardous substances between species
	Inadvertent creation of new proteins that are possibly hazardous

Note: Food safety hazards for children span a broad spectrum of classes of pathogens and chemicals.

Natural Carcinogens: There are numerous natural carcinogens in food, and in 1996 the U.S. National Academy of Sciences published a report suggesting that such natural carcinogens may confer a risk equivalent to chemical and pesticide contaminants in food (National Research Council, 1996). However, as is the case for chemical carcinogens, there are many unanswered questions about the actual cancer risk conferred by these carcinogens and particularly whether the risk might be greater for children.

Phytoestrogens: Recently, it has been noted that many foods contain high levels of so-called plant estrogens or phytoestrogens. In animal testing, these plant estrogens can show significant physiologic and even structural effects, especially on the developing organism. Although there has been much conjecture about possible risks and benefits from these substances to humans, sufficient evidence has not accumulated to make firm conclusions. Concerns have been voiced about exposures to infants (via soy formula) and about the potential for long-term sequelae (Tuohy, 2003).

Mycotoxins: A number of toxins are produced by certain molds found in foods such as peanuts and corn (Morgan and Fenwick, 1990). Aflatoxin, a well-known mycotoxin, is produced by the *Aspergillus* fungus. People are primarily exposed to aflatoxin via the diet. The International Agency for Research on Cancer (IARC) has concluded that aflatoxin is an important risk factor for hepatocellular carcinoma, based on epidemiological studies conducted in Asia, where the rate of hepatocellular carcinoma is quite high. Aflatoxins are important for children's health because they occur in foods that are frequently eaten at relatively high levels by children.

Two other fungal toxins are commonly found and are regulated by the EPA: patulin and fumonisin. Whether there is increased susceptibility of children to these toxins is unknown. However, disproportionate exposure is a concern, given their presence in foods frequently consumed by children. Patulin occurs in apple juice products when rotten or partially rotten apples have been used (Jackson et al., 2003). The FDA has set an action level for patulin in juice of 50 μg/kg; at higher levels it is possible for infants and small children to receive excessive doses. Toxicology studies find hazards of chronic toxicity, reproductive toxicity, and possible immunotoxicity (U.S. FDA, Center for Food Safety and Applied Nutrition, 2000b), although no human data are available. Fumonisin is a mycotoxin found worldwide in corn and corn products (U.S. FDA, Center for Food Safety and Applied Nutrition, 2000a). In the U.S., they have been found to be at very low levels in children's breakfast cereals as well as corn-based snack foods. Fumonisins are toxic to a number of organ systems in animals, including the brain, lungs, liver, kidney, and heart; they cause liver cancers. No comparable data are available for humans.

PATHOGENS

Pathogens in food include virus, bacteria, toxin producers, and parasites. Food-borne pathogens are often particularly risky for children. In the U.S., there are an estimated 76 million illnesses, 325,000 hospitalizations, and 5,000 deaths per year due to food-borne pathogens; most of these are among the elderly and the very young (Mead et al., 1999). In the U.S., the Norwalk virus is the most commonly found cause of food-borne illness, and other viruses (rotavirus and astrovirus), and certain parasites (campylobacter and giardia) also play a major role. Three pathogens, salmonella, *Listeria monocytogenes*, and *Toxoplasma gondii*, are responsible for 1,500 deaths each year, more than 75% of those caused by known pathogens; other pathogens that also play an important role include Norwalk virus, *campylobacter pylori,* and *E. coli* O157:H7.

As is shown in Table 8.4, of the food-borne pathogens that commonly cause death in the U.S., most affect children disproportionately (American Academy of Pediatrics, 2003). Many bacteria and certain parasites are particularly virulent for children. *Salmonella enteriditis, Listeria monocytogenes, E. coli* O157:H7, shigella, and *Campylobacter pylori* are among the many food-borne pathogens that pose risks to young children. Very young children may be more at risk if they lack immunity to the pathogen. If they do develop severe food-borne illness, they are more likely to become dehydrated from vomiting and diarrhea and to die from the sequelae of dehydration. Recently in the U.S., declines have been reported in the incidence of bacterial food-borne illnesses, including yersinia (49%), *Listeria monocytogenes* (35%), *Campylobacter pylori* (27%), and *Salmonella enteriditis* (15%) (Pinner et al., 2003). These declines may be due to new rules regarding control of hazards in meat production.

TABLE 8.4
Pathogens of Special Concern for Children: Incubation Time, Symptoms, and Age Group

Pathogen	Incubation	Symptoms	Age
Salmonella enteriditis	6–72 h	Bacteremia, fecal infections	<5, >70
Listeria monocytogenes	21 d	Stillbirth, preterm delivery, neonatal sepsis, meningitis	Neonate/fetus
Toxoplasma gondii	7 d	Systemic infection, meningitis, oculitis, still birth	Fetus
Norwalk virus	12–96 h	Diarrhea, vomiting, severe dehydration	<4
Campylobacter pylori	1–7 d	Septicemia, meningitis, severe dehydration	Neonate
E. Coli O157:H7	3–4 d	Bloody diarrhea, hemolytic uremic syndrome (5–10% with complications)	Children

Note: A number of pathogens are of particular risk for life-threatening disease for children, including a number of bacteria (*Salmonella enteriditis, Listeria monocytogenes, Campylobacter pylori,* and *E. Coli* O157:H7; protozoans (*Toxoplasma gondii*); and viruses (Norwalk virus). In addition, any organism causing acute gastroenteritis is life threatening to infants under circumstances where there is poor nutrition, inadequate immunity (e.g., HIV infection), and lack of availability of prompt treatment (such as oral rehydration therapy).

Reference: American Academy of Pediatrics, 2003.

Chemicals

A number of nonfood substances added to food may pose risks to children:

- Colorings, flavorings, and other deliberate chemical additives
- Indirect food additives from packaging
- Veterinary drugs and feed additives
- Inadvertent chemical additives, such as polychlorinated biphenyls, dioxins, metals
- Residues of pesticides such as dichlorodiphenyltrichloroethane (DDT), which persist in the environment years after use

Food Additives: Children may have adverse reactions to certain food additives. For example, tartrazine (also known as FD&C [food dye and coloring] Yellow No. 5) is a food and beverage dye, mostly used in highly processed foods. Some people are sensitive to it and can develop hives, urticaria, or asthma exacerbations postingestion. The flavor enhancer monosodium glutamate (MSG) is associated with a number of acute symptoms among those who are sensitive to it: headache, nausea, diarrhea, sweating, chest tightness, and a burning sensation along the back of the neck. Sulfites are used to preserve foods and are found in many foods. Because sulfites can cause severe allergy and asthma symptoms among those who are sensitized (Wuthrich, 1993), they are regulated in the U.S. and in Europe.

Indirect Food Additives: These enter the food supply through processes such as manufacturing, packing, packaging, transporting, or storing food; they are not present in food by intent. The FDA recognizes more than 3,000 of such substances. Many of these are "food contact substances," which are mostly derived from packaging materials. Recently, certain plasticizer substances have drawn increased attention as food contact items. One is a class of plasticizers called phthalates, which are used in soft plastics (Table 8.5). Certain phthalates were recently removed from pacifiers, nipples, and teething toys in Europe and the U.S. (voluntarily), but they are allowed as indirect additives to food. All toxicology data cited in Table 8.5 are from animal studies because there are very few studies on humans, and there is uncertainty about levels of exposure that are risky to children. Most phthalates are antiandrogen (androgen-blocking) chemicals toxic to cells that foster developing sperm in laboratory animals (Sertoli cells). In 2000, higher phthalate levels were reported associated with premature breast development (thelarche) in Puerto Rico, but this finding is yet to be confirmed [Colon et al., 2000]. Also in 2000, the U.S. Centers for Disease Control and Prevention (CDC) reported higher exposures than had been suspected in women of childbearing age; however, cosmetics and not food was suspected as the source of this exposure. In the U.S., two phthalate compounds, DEHP (diethylhexylphthalate) and DINP (diisononylphthalate), were used in pacifiers, baby bottle nipples, and teething toy in the past. These were removed from such uses by the U.S. Consumer Products Safety Commission (CPSC) (voluntarily in the case of DINP). The FDA has allowed the use of phthalates in food contact items and has reported that exposures are very low.

The second plasticizer commonly used in food containers is bisphenol A, which is used in certain hard plastics. Bisphenol A is weakly estrogenic. It has been used in hard plastic food containers, water bottles, and even baby bottles. There are no data on current human exposure levels in the general population. Small amounts of Bisphenol A do migrate from plastic containers to food and water; however, the FDA has determined that these amounts are safe. Estrogenic effects have clearly been demonstrated in tissue culture cells and in animal studies but not in human studies. Low-dose effects (reduced prostate weight in male offspring) have been demonstrated in some but not all studies (Melnick et al., 2002).

Veterinary Drugs and Feed Additives: There are also deliberate additives in animal feed. Of concern today is the use of antibiotics in agriculture. In the U.S., 50% of antibiotics are used in veterinary practice or as additives to animal feeds. Such use may be generating antibiotic-resistant organisms in food and in the environment, organisms that are themselves pathogenic or may be capable of transferring resistance to pathogens. Such uses range from veterinary treatments of single or groups of sick animals, treatment of entire groups of animals to prevent spread of illness from sick animals, and use as growth enhancers. Worldwide, the increased use of antibiotics in animal feeds and antimicrobial products in food products, in general, is expanding the natural selection pressure for strains that are resistant against antibiotics. Today bacterial strains with resistance and multiresistance to antibiotics have become an increasing problem, and multiresistant salmonella strains have been found in connection with chicken production. Recent studies suggest that

TABLE 8.5
Phthalates in Use in the U.S.: Major Uses and Known Toxicities

Phthalate Compound	Major Uses	Toxicity
Butyl benzyl phthalate (BBP)	Vinyl tiles, food conveyor belts, artificial leather, automotive trim, and traffic cones	Antiandrogen, male reproductive malformations (inadequate data to assess carcinogenicity)
Di-(2-ethylhexyl) phthalate (DEHP)	Building products (flooring, roof coverings, wallpaper, coatings, and wire insulation), car products (upholstery, car seats, underbody coating, and trim), clothing (footwear, raincoats), food packaging, children's products (toys, crib bumpers), and medical devices and tubing. In the U.S., not used in nipples, teethers, pacifiers, rattles, but used in toys for older children	Antiandrogen, male reproductive malformations. Reasonably anticipated to be a human carcinogen (NTP)
Di-isodecyl phthalate (DIDP)	Coverings on wires and cables, artificial leather, toys, carpet backing, and pool liners	Malformations (but not of reproductive tract) (inadequate data to assess cancer risks)
Di-isononyl phthalate (DINP)	Garden hoses, pool liners, flooring tiles, tarps, and toys including baby toys and teething rings	Antiandrogen, male reproductive malformations; causes liver tumors in animals
Di-*n*-butyl phthalate (DBP)	Latex adhesives, cosmetics, and other personal care products, cellulose plastics (used in food packaging), and dyes	Antiandrogen, male reproductive and other malformations, reduced fertility (inadequate information for cancer risks)
Di-*n*-di-isodecyl phthalate (DnHP)	Tool handles, dishwasher baskets, flooring, gloves, flea collars, and food processing conveyer belts	Antiandrogen, reduced fertility in males and females (inadequate information on cancer risks)
Di-*n*-octyl phthalate (DnOP)	Flooring, carpet tiles, tarps, pool liners, and garden hoses. FDA-approved as a food additive and in food containers and conveyor belts	(Inadequate data to assess toxicity)

Note: Phthalates are a family of chemicals with numerous uses in the plastics industry (as plastics softeners for example). Their toxicity recently was evaluated by the National Toxicology Program.

Source: From Kavlock, R., Boekelheide, K., Chapin, R., Cunningham, M., Faustman, E., Foster, P., Golub, M., Henderson, R., Hinberg, I., Little, R., Seed, J., Shea, K., Tabacova, S., Tyl, R., Williams, P., and Zacharewski, T. (2002a), NTP Center for the Evaluation of Risks to Human Reproduction: phthalates expert panel report on the reproductive and developmental toxicity of butyl benzyl phthalate, *Reproductive Toxicology*, 16: 453–87, Kavlock, R., Boekelheide, K., Chapin, R., Cunningham, M., Faustman, E., Foster, P., Golub, M., Henderson, R., Hinberg, I., Little, R., Seed, J., Shea, K., Tabacova, S., Tyl, R., Williams, P., and Zacharewski, T. (2002b), NTP Center for the Evaluation of Risks to Human Reproduction: phthalates expert panel report on the reproductive and developmental toxicity of di(2-ethylhexyl) phthalate, *Reproductive Toxicology*, 16: 529–653, Kavlock, R., Boekelheide, K., Chapin, R., Cunningham, M., Faustman, E., Foster, P., Golub, M., Henderson, R., Hinberg, I., Little, R., Seed, J., Shea, K.,

TABLE 8.5 (continued)
Phthalates in Use in the U.S.: Major Uses and Known Toxicities

Tabacova, S., Tyl, R., Williams, P., and Zacharewski, T. (2002c), NTP Center for the Evaluation of Risks to Human Reproduction: phthalates expert panel report on the reproductive and developmental toxicity of di-isodecyl phthalate, *Reproductive Toxicology*, 16: 655–78; Kavlock, R., Boekelheide, K., Chapin, R., Cunningham, M., Faustman, E., Foster, P., Golub, M., Henderson, R., Hinberg, I., Little, R., Seed, J., Shea, K., Tabacova, S., Tyl, R., Williams, P., and Zacharewski, T. (2002d), NTP Center for the Evaluation of Risks to Human Reproduction: phthalates expert panel report on the reproductive and developmental toxicity of di-isononyl phthalate, *Reproductive Toxicology*, 16: 679–708; Kavlock, R., Boekelheide, K., Chapin, R., Cunningham, M., Faustman, E., Foster, P., Golub, M., Henderson, R., Hinberg, I., Little, R., Seed, J., Shea, K., Tabacova, S., Tyl, R., Williams, P., and Zacharewski, T. (2002e), NTP Center for the Evaluation of Risks to Human Reproduction: phthalates expert panel report on the reproductive and developmental toxicity of di-n-butyl phthalate, *Reproductive Toxicology*, 16: 489–527; Kavlock, R., Boekelheide, K., Chapin, R., Cunningham, M., Faustman, E., Foster, P., Golub, M., Henderson, R., Hinberg, I., Little, R., Seed, J., Shea, K., Tabacova, S., Tyl, R., Williams, P., and Zacharewski, T. (2002f), NTP Center for the Evaluation of Risks to Human Reproduction: phthalates expert panel report on the reproductive and developmental toxicity of di-n-hexyl phthalate, *Reproductive Toxicology*, 16: 709–19; Kavlock, R., Boekelheide, K., Chapin, R., Cunningham, M., Faustman, E., Foster, P., Golub, M., Henderson, R., Hinberg, I., Little, R., Seed, J., Shea, K., Tabacova, S., Tyl, R., Williams, P., and Zacharewski, T. (2002g), NTP Center for the Evaluation of Risks to Human Reproduction: phthalates expert panel report on the reproductive and developmental toxicity of di-n-octyl phthalate, *Reproductive Toxicology*, 16: 721–34.

these effects may be irreversible. Efforts are under way in Europe and the U.S. to control this risk; however, development of resistant strains of pathogens anywhere in the world is of concern everywhere because pathogens know no boundaries (Shea, 2003). The addition of hormones to the food supply has also been of concern, particularly in Europe where the use of some hormones is not allowed in meat production. Although there is no agreement about the nature and magnitude of the risks from such hormones currently in use, certainly the experience with DES as a pharmaceutical agent would indicate that the life stage at most risk, theoretically, would be the developing fetus.

Inadvertent Chemical Constituents

PCBs and Dioxins: Polychlorinated dibenzo-*p*-dioxins (dioxins) and polychlorinated dibenzofurans (furans) are inadvertently produced in the manufacture of certain chemicals and by incineration. PCBs were manufactured for use as fire retardants and in electrical transformers and capacitors. Dioxin, furans, and PCBs are persistent and bioaccumulative chemicals and are found at the highest levels in fish in contaminated areas. Several are recognized as endocrine disruptors. They are fat soluble and are not readily metabolized or excreted by humans. They can be transferred to the fetus from maternal fat stores. They are also transferred to human milk; however, breast milk is nonetheless the preferred food for infants because of its many benefits.

Potential chronic effects are particularly of concern to the developing fetus. The carcinogenic potential of dioxins has been identified by the EPA and one dioxin, tetrachlorodibenzo-*p*-dioxin (TCDD), has been upgraded by IARC and the U.S. National Toxicology Program (NTP) to "human carcinogen." Certain PCB and dioxin compounds in food also have been associated with developmental neurotoxicity. Infants whose mothers ate contaminated fish from Lake Michigan during pregnancy were somewhat smaller for gestational age, had shortened gestation, smaller head circumference, and were of lower birth weight (Jacobson et al., 1990). Studies of newborns whose mothers' body burden of PCBs and dioxins more nearly reflects that of the general population than fish eaters, suggest that although the neurological effects are not severe, higher exposure is associated with hypotonia and increased incidence of abnormally weak reflexes (Jacobson et al., 1990). Follow-up studies of these and other populations in the U.S. indicate that there are subtle impacts on IQ of school children due to higher exposure to PCBs prenatally (Jacobson and Jacobson, 1997). These findings have been confirmed in Holland as well, in two different cohorts (Huisman et al., 1995; Lanting et al., 1998; Patandin et al., 1998; Patandin et al., 1999).

In the U.S., fish are the major source of PCBs in the diet. States in which PCB-contaminated fish have been found issue fish advisories to inform sports fishermen about the locations of contaminated fish and the limits for consumption. Specific guidelines are issued to advise pregnant women, lactating mothers, and young children regarding avoiding or limiting consumption of such fish.

Lead: Exposure to lead from food sources comes about in a number of ways, including use of lead-soldered cans (banned in many countries); leaching of lead into foods from low-fired ceramics and other food containers with lead; and deposition of lead in the air onto crops. Children are most susceptible to the toxic effects of lead. Lead toxicity affects almost every organ system — most importantly, the central nervous system, peripheral nervous system, kidneys, and blood. At high blood lead levels (>70 μg/dl), lead may cause encephalopathy and death in children. Low-level lead exposure inhibits both prenatal and postnatal growth, impairs hearing acuity, and impairs cognition. A number of studies have found that for every increase of 10 to 15 μg/dl blood lead, within the range of 5 to 35 μg/dl, there is a lowering of the mean IQ of children by 2–4 points (National Research Council, 1993a). Recent evidence suggests that the effects of early lead exposure can persist, and that there may be no threshold for lead impact on IQ. The current public health target for blood lead reduction in children in the U.S. is 10 μg/dl but the CDC has discussed lowering this level to 5 μg/dl. Efforts in the U.S. have resulted in a sharp decline in average blood lead levels among children during the last 20 yr. However, lead levels among poor children are on an average four times higher than levels among more affluent children (Brody et al., 1994).

Methylmercury: Methylmercury is formed from elemental mercury by microorganisms in the environment (Goldman and Shannon, 2001). Methylmercury is found in fish and is especially hazardous to the fetus. Highly exposed populations in Japan and Iraq had shown severe congenital neurotoxic effects (Amin-Zaki et al., 1974; Goscinska, 1965). Recent studies have identified significant neurotoxicity of mercury to the fetus even with moderate levels of maternal exposure (Grandjean et al., 1997; National Research Council, 2000), although one major study has been interpreted as being negative (Davidson et al., 1998). In 1999, the CDC reported that the geometric mean total blood mercury level in the U.S. population was four times higher for women of childbearing age than for children, indicating that exposure *in utero* would be more significant. Most of this mercury is considered to consist of methylmercury, which is the most toxic to the fetus (Centers for Disease Control and Prevention, 2001).

Most states in the U.S. have issued advisories to limit the consumption of certain fish species in areas with methylmercury contamination and to reduce exposures to pregnant and lactating women as well as infants and toddlers due to sports fishing in such areas. Such advisories usually include women of child-bearing age whether or not they currently are pregnant, in order to protect the first few weeks of pregnancy. In terms of commercially marketed fish, the fresh water fish with the highest average levels of methylmercury are pike and walleyes. Marine fish with higher levels of methylmercury include shark, swordfish, king mackerel, tilefish, and large tuna. The FDA now recommends that pregnant women, nursing mothers, women who may become pregnant, and young children should not consume shark, swordfish, king mackerel, or tilefish. Two states in the U.S. (Washington and Connecticut) now additionally recommend limits for the consumption of tuna.

Pesticides

Pesticides include insecticides, herbicides, fungicides, rodenticides, and fumigants. Although these products have benefits such as increasing crop yields, they may be toxic, especially to people who mix and apply the chemicals; to farmworkers; and (to a much lesser extent) to people consuming foods with pesticide residues. Additionally, persistent pesticides may contaminate food for years after they have been taken out of commerce. Around 600 pesticides are approved by the EPA for use in food; only around 200 are routinely monitored in the U.S. food supply. Generally, residue levels are within regulatory limits.

In 1993, a National Research Council report, "Pesticides in the Diets of Infants and Children"(National Research Council, 1993b), concluded that the EPA did not adequately assure that children's health would be protected in standards setting. First, the EPA did not factor in the dietary patterns of young children. Second, it did not assure that sensitive endpoints, such as developmental neurotoxicity, developmental immunotoxicity, and endocrine effects, were assessed. Third, the report concluded that the EPA erred by failing to account for cumulative exposures to pesticides sharing the same mode of action, as well as aggregate exposures from nonfood as well as food pesticide uses. The NRC committee suggested that the EPA should apply an additional factor of ten to account for the susceptibility of children. In the U.S., this report resulted in efforts to improve dietary surveys for children, pesticide-residue monitoring for foods that children commonly eat, and newer dietary intake models that give a more accurate picture of children's diets. EPA incorporated several child health endpoints to the list of studies required for approval of a food-use pesticide. In 1996, the U.S. Congress enacted the Food Quality Protection Act (FQPA) (Food Quality Protection Act, 1996). This required an additional tenfold "FQPA factor" to protect children, as well as assessments of cumulative and aggregate risk of pesticides. In the U.S., all food pesticide standards are being reassessed on a schedule to be completed by 2006. Such reassessments have resulted in cancellations of most household organophosphate pesticide products (for example, chlorpyrifos and diazinon). These were determined to have greater potential for exposure to children, as well as cumulative impacts, given that there were dozens of organophosphate pesticides on the market. At the same time, a number of tolerances for foods that children preferably eat (e.g., apples) were eliminated or reduced (for example, chlorpyrifos and methyl parathion). Cumulative assessments under FQPA are not completed at press time and the total impact of FQPA is uncertain at present (to be completed by 2006, as required in the statute.) FQPA also requires screening and testing of pesticides for potential to disrupt the endocrine system; such a program is still under development at the EPA at press time (Goldman and Koduru, 2000).

EMERGING ISSUES

Brominated Flame Retardants: These prevent foams used in mattresses, cars and building materials, and television and computer casings from catching fire. When heated or in contact with water, these additives can leach out of the plastic and ultimately accumulate within the food chain. In Sweden, brominated flame retardants have been found to be accumulating rapidly in breast milk over time.

Perfluorooctanyl Sulfonate: A chemical used to produce a range of products, including the stain repellant Scotchguard®, this was found to persist in the environment and to accumulate in human and animal tissues. The manufacturer conducted tests that identified toxic effects to developing animals at high doses. In consequence, in the U.S. this chemical is to be phased out on a voluntary basis (Wood et al., 2000).

Biotechnology: Hazardous substances can be created or enhanced as a result of the application of plant-breeding technologies, especially genetic engineering. Genetic engineering has added to the tools of conventional plant breeding to make it possible to combine genes from completely unrelated species. Genes encode proteins, which in turn create the structures and the functional components of plants. Genetic engineering has been used to modify numerous plant traits, including flavor, ripening, color, nutritional aspects, and pesticidal properties. Some of the theoretical health risks of biotechnology include: amplifying the expression of proteins that are toxic to humans; creating new toxic proteins; amplifying or creating proteins that interfere with the uptake of nutrients, movement of a food allergen from a source (where it is recognized and can be avoided) to another food (where its presence would be unknown); and creation of new food allergens. These concerns are particularly acute for children, given that children are more susceptible to developing food allergy and that there is a potential difficulty in detecting novel allergens in food (Bernstein et al., 2003). New genetic engineering efforts involve ever more creative use of the technology to produce vaccines, pharmaceuticals, and chemicals. The technology shows the promise of producing these materials in a more sustainable manner, including reducing allergens and enhancing nutrients. Yet, there are significant concerns over what would happen if such nonfood traits were transferred to food crops.

Acrylamide: Food preparation can create toxic products. In April 2002 investigators in Sweden reported that acrylamide is produced in preparation of starchy foods such as potatoes. An industrial chemical, acrylamide is known to be a potent carcinogen and, in uncontrolled occupational exposures, to cause neurotoxicity. The risks to children are unknown; however, in the U.S. the FDA has reported finding acrylamide in a number of processed food products (snacks, bread) frequently eaten by children.

CONCLUSIONS

Children need an adequate supply of safe and healthful food. To assure the safety of food for children, it is necessary to take into account their unique exposures and susceptibilities. Whereas we are clearly just beginning to understand the nature of risks to children from various exposures, it is equally clear that children are more vulnerable at certain life stages. Only by factoring in the exposures and hazards to children can we assure that food is safe to eat throughout the lifespan.

ABBREVIATIONS

CDC — U.S. Centers for Disease Control and Prevention
CPSC — U.S. Consumer Products Safety Commission
DES — Diethylstilbestrol
EPA — U.S. Environmental Protection Agency
FAO — UN Food and Agriculture Organization
FDA — U.S. Food and Drug Administration
FQPA — Food Quality Protection Act
IARC — International Agency for Research on Cancer
IQ — Intelligence quotient
μg/dl — Micrograms/deciliter
PCBs — Polychlorinated biphenyls

References

American Academy of Pediatrics (2003), *Red Book: Report of the Committee on Infectious Diseases*, Elk Grove Village, IL: AAP.

Amin-Zaki, L., Elhassani, S., Majeed, M.A., Clarkson, T.W., Doherty, R.A., and Greenwood, M. (1974), Intra-uterine methylmercury poisoning in Iraq, *Pediatrics*, 54: 587–595.

Anderson, L., Donovan, P., and Rice, J. (1985), Risk assessment for transplacental carcinogenesis, in A. Li (Ed.) *New Approaches in Toxicity Testing and Their Application in Human Risk Assessment*, New York: Raven Press.

Anderson, L.M., Diwan, B.A., Fear, N.T., and Roman, E. (2000), Critical windows of exposure for children's health: cancer in human epidemiological studies and neoplasms in experimental animal models, *Environmental Health Perspectives*, 108 Suppl. 3: 573–594.

Barker, D.J., Eriksson, J.G., Forsen, T., and Osmond, C. (2002), Fetal origins of adult disease: strength of effects and biological basis, *International Journal of Epidemiology*, 31: 1235–1239.

Bearer, C.F. (1995), How are children different from adults?, *Environmental Health Perspectives*, 103 Suppl. 6: 7–12.

Bernstein, J.A., Bernstein, I.L., Bucchini, L., Goldman, L.R., Hamilton, R.G., Lehrer, S., Rubin, C., and Sampson, H.A. (2003), Clinical and laboratory investigation of allergy to genetically modified foods, *Environmental Health Perspectives*, 111: 1114–1121.

Bjorksten, B. (1999), The environmental influence on childhood asthma, *Allergy*, 54: 17–23.

Brody, D.J., Pirkle, J.L., Kramer, R.A., Flegal, K.M., Matte, T.D., Gunter, E.W., and Paschal, D.C. (1994), Blood lead levels in the U.S. population. Phase 1 of the Third National Health and Nutrition Examination Survey (NHANES III, 1988 to 1991) [see comments] [published erratum appears in JAMA 1995 July 12, 274(2):130], *Journal of the American Medical Association*, 272: 277–283.

Buzby, J. (2001), Effects of food-safety perceptions on food demand and global trade, *Changing Structure of Global Food Consumption and Trade*, Washington, D.C.: USDA Economic Research Service.

Centers for Disease Control and Prevention (2001), Blood and Hair Mercury Levels in Young Children and Women of Childbearing Age — United States, 1999, *Morbidity and Mortality Weekly Report*, 50: 140–143.

Chang, M.H., Glynn, M.K., and Groseclose, S.L. (2003), Endemic, notifiable bioterrorism-related diseases, United States, 1992–1999, *Emerging Infectious Diseases*, 9: 556–564.

Colon, I., Caro, D., Bourdony, C.J., and Rosario, O. (2000), Identification of phthalate esters in the serum of young Puerto Rican girls with premature breast development, *Environmental Health Perspectives,* 108: 895–900.

Daher, A., Varin, M., Lamontagne, Y., and Oth, D. (1998), Effect of pre-conceptional external or internal irradiation of N5 male mice and the risk of leukemia in their offspring, *Carcinogenesis*, 19: 1553–1558.

Davidson, P.W., Myers, G.J., Cox, C., Axtell, C., Shamlaye, C., Sloane-Reeves, J., Cernichiari, E., Needham, L., Choi, A., Wang, Y., Berlin, M., and Clarkson, T.W. (1998), Effects of prenatal and postnatal methylmercury exposure from fish consumption on neurodevelopment: outcomes at 66 months of age in the Seychelles Child Development Study [see comments], *Journal of the American Medical Association*, 280: 701–707.

Food Quality Protection Act (1996), U.S. Congress.

Gern, J.E., Lemanske, R.F., Jr., and Busse, W.W. (1999), Early life origins of asthma, *Journal of Clinical Investigations*, 104: 837–843.

Ginsberg, G., Hattis, D., Sonawane, B., Russ, A., Banati, P., Kozlak, M., Smolenski, S., and Goble, R. (2002), Evaluation of child/adult pharmacokinetic differences from a database derived from the therapeutic drug literature, *Toxicological Science*, 66: 185–200.

Giusti, R.M., Iwamoto, K., and Hatch, E.E. (1995), Diethylstilbestrol revisited: a review of the long-term health effects, *Annals of Internal Medicine*, 122: 778–788.

Goldman, L.R. and Koduru, S. (2000), Chemicals in the environment and developmental toxicity to children: A public health and policy perspective, *Environmental Health Perspectives*, 108 Suppl. 3: 443–448.

Goldman, L.R. and Shannon, M.W. (2001), Technical report: mercury in the environment: implications for pediatricians, *Pediatrics*, 108: 197–205.

Goscinska, Z. (1965), Degenerative brain syndrome (Minimata disease) caused by alkyl mercury compounds, *Helvetica Paediatrica Acta*, 20: 216–221.

Grandjean, P., Weihe, P., White, R.F., Debes, F., Araki, S., Yokoyama, K., Murata, K., Sorensen, N., Dahl, R., and Jorgensen, P.J. (1997), Cognitive deficit in 7-year-old children with prenatal exposure to methylmercury, *Neurotoxicology and Teratology*, 19: 417–428.

Hanson, L.A., Korotkova, M., Lundin, S., Haversen, L., Silfverdal, S.A., Mattsby-Baltzer, I., Strandvik, B., and Telemo, E. (2003), The transfer of immunity from mother to child, *Annals of New York Academy of Sciences*, 987: 199–206.

Herbst, A.L. (1987), The effects in the human of diethylstilbestrol (DES) use during pregnancy, *Princess Takamatsu Symposia*, 18: 67–75.

Huisman, M., Koopman-Esseboom, C., Fidler, V., Hadders-Algra, M., van der Paauw, C.G., Tuinstra, L.G., Weisglas-Kuperus, N., Sauer, P.J., Touwen, B.C., and Boersma, E.R. (1995), Perinatal exposure to polychlorinated biphenyls and dioxins and its effect on neonatal neurological development, *Early Human Development*, 41: 111–127.

Jackson, L.S., Beacham-Bowden, T., Keller, S.E., Adhikari, C., Taylor, K.T., Chirtel, S.J., and Merker, R.I. (2003), Apple quality, storage, and washing treatments affect patulin levels in apple cider, *Journal of Food Protection*, 66: 618–624.

Jacobson, J.L. and Jacobson, S.W. (1997), Evidence for PCBs as neurodevelopmental toxicants in humans, *Neurotoxicology*, 18: 415–424.

Jacobson, J.L., Jacobson, S.W., and Humphrey, H.E. (1990), Effects of exposure to PCBs and related compounds on growth and activity in children, *Neurotoxicology and Teratology*, 12: 319–326.

Kavlock, R., Boekelheide, K., Chapin, R., Cunningham, M., Faustman, E., Foster, P., Golub, M., Henderson, R., Hinberg, I., Little, R., Seed, J., Shea, K., Tabacova, S., Tyl, R., Williams, P., and Zacharewski, T. (2002a), NTP Center for the Evaluation of Risks to Human Reproduction: phthalates expert panel report on the reproductive and developmental toxicity of butyl benzyl phthalate, *Reproductive Toxicology*, 16: 453–487.

Kavlock, R., Boekelheide, K., Chapin, R., Cunningham, M., Faustman, E., Foster, P., Golub, M., Henderson, R., Hinberg, I., Little, R., Seed, J., Shea, K., Tabacova, S., Tyl, R., Williams, P., and Zacharewski, T. (2002b), NTP Center for the Evaluation of Risks to Human Reproduction: phthalates expert panel report on the reproductive and developmental toxicity of di(2-ethylhexyl) phthalate, *Reproductive Toxicology*, 16: 529–653.

Kavlock, R., Boekelheide, K., Chapin, R., Cunningham, M., Faustman, E., Foster, P., Golub, M., Henderson, R., Hinberg, I., Little, R., Seed, J., Shea, K., Tabacova, S., Tyl, R., Williams, P., and Zacharewski, T. (2002c), NTP Center for the Evaluation of Risks to Human Reproduction: phthalates expert panel report on the reproductive and developmental toxicity of di-isodecyl phthalate, *Reproductive Toxicology*, 16: 655–678.

Kavlock, R., Boekelheide, K., Chapin, R., Cunningham, M., Faustman, E., Foster, P., Golub, M., Henderson, R., Hinberg, I., Little, R., Seed, J., Shea, K., Tabacova, S., Tyl, R., Williams, P., and Zacharewski, T. (2002d), NTP Center for the Evaluation of Risks to Human Reproduction: phthalates expert panel report on the reproductive and developmental toxicity of di-isononyl phthalate, *Reproductive Toxicology*, 16: 679–708.

Kavlock, R., Boekelheide, K., Chapin, R., Cunningham, M., Faustman, E., Foster, P., Golub, M., Henderson, R., Hinberg, I., Little, R., Seed, J., Shea, K., Tabacova, S., Tyl, R., Williams, P., and Zacharewski, T. (2002e), NTP Center for the Evaluation of Risks to Human Reproduction: phthalates expert panel report on the reproductive and developmental toxicity of di-n-butyl phthalate, *Reproductive Toxicology*, 16: 489–527.

Kavlock, R., Boekelheide, K., Chapin, R., Cunningham, M., Faustman, E., Foster, P., Golub, M., Henderson, R., Hinberg, I., Little, R., Seed, J., Shea, K., Tabacova, S., Tyl, R., Williams, P., and Zacharewski, T. (2002f), NTP Center for the Evaluation of Risks to Human Reproduction: phthalates expert panel report on the reproductive and developmental toxicity of di-n-hexyl phthalate, *Reproductive Toxicology*, 16: 709–719.

Kavlock, R., Boekelheide, K., Chapin, R., Cunningham, M., Faustman, E., Foster, P., Golub, M., Henderson, R., Hinberg, I., Little, R., Seed, J., Shea, K., Tabacova, S., Tyl, R., Williams, P., and Zacharewski, T. (2002g), NTP Center for the Evaluation of Risks to Human Reproduction: phthalates expert panel report on the reproductive and developmental toxicity of di-n-octyl phthalate, *Reproductive Toxicology*, 16: 721–734.

Lanting, C.I., Patandin, S., Fidler, V., Weisglas-Kuperus, N., Sauer, P.J., Boersma, E.R., and Touwen, B.C. (1998), Neurological condition in 42-month-old children in relation to pre- and postnatal exposure to polychlorinated biphenyls and dioxins, *Early Human Development*, 50: 283–292.

Martinez, F. and Holt, P. (1999), Role of microbial burden in aetiology of allergy and asthma, *Lancet*, 354: 12–15.

Mead, P.S., Slutsker, L., Dietz, V., McCaig, L.F., Bresee, J.S., Shapiro, C., Griffin, P.M., and Tauxe, R.V. (1999), Food-related illness and death in the United States, *Emerging Infectious Diseases*, 5: 607–625.

Melnick, R., Lucier, G., Wolfe, M., Hall, R., Stancel, G., Prins, G., Gallo, M., Reuhl, K., Ho, S.M., Brown, T., Moore, J., Leakey, J., Haseman, J., and Kohn, M. (2002), Summary of the National Toxicology Program's report of the endocrine disruptors low-dose peer review, *Environmental Health Perspectives*, 110: 427–431.

Morgan, M.R. and Fenwick, G.R. (1990), Natural foodborne toxicants, *Lancet*, 336: 1492–1495.

National Research Council (1993a), *Measuring Lead Exposure in Infants, Children and Other Sensitive Populations*, Washington, D.C.: National Academy Press.

National Research Council (1993b), *Pesticides in the Diets of Infants and Children*, Washington, D.C.: National Academy Press.

National Research Council (1996), *Carcinogens and Anticarcinogens in the Human Diet: A Comparison of Naturally Occurring and Synthetic Substances*, Washington, D.C.: National Academy Press.

National Research Council (2000), *Toxicological Effects of Methylmercury*, Washington, D.C.: National Academy Press.

Needleman, H.L., Gunnoe, C., Leviton, A., Reed, R., Peresie, H., Maher, C., and Barrett, P. (1979), Deficits in psychologic and classroom performance of children with elevated dentine lead levels, *New England Journal of Medicine*, 300: 689–695.

Olshan, A.F., Anderson, L., Roman, E., Fear, N., Wolff, M., Whyatt, R., Vu, V., Diwan, B.A., and Potischman, N. (2000), Workshop to identify critical windows of exposure for children's health: cancer work group summary, *Environmental Health Perspectives*, 108 Suppl 3: 595–597.

Palanza, P., Parmigiani, S., and vom Saal, F.S. (2001), Effects of prenatal exposure to low doses of diethylstilbestrol, o,p′DDT, and methoxychlor on postnatal growth and neurobehavioral development in male and female mice, *Hormonal Behavior*, 40: 252–265.

Patandin, S., Koopman-Esseboom, C., de Ridder, M.A., Weisglas-Kuperus, N., and Sauer, P.J. (1998), Effects of environmental exposure to polychlorinated biphenyls and dioxins on birth size and growth in Dutch children, *Pediatrics Research*, 44: 538–545.

Patandin, S., Lanting, C.I., Mulder, P.G., Boersma, E.R., Sauer, P.J., and Weisglas-Kuperus, N. (1999), Effects of environmental exposure to polychlorinated biphenyls and dioxins on cognitive abilities in Dutch children at 42 months of age [see comments], *Journal of Pediatrics*, 134: 33–41.

Paulozzi, L.J., Erickson, J.D., and Jackson, R.J. (1997), Hypospadias trends in two U.S. surveillance systems, *Pediatrics*, 100: 831–834.

Pinner, R.W., Rebmann, C.A., Schuchat, A., and Hughes, J.M. (2003), Disease surveillance and the academic, clinical, and public health communities, *Emerging Infectious Diseases*, 9: 781–787.

Rice, D. and Barone, S., Jr. (2000), Critical periods of vulnerability for the developing nervous system: evidence from humans and animal models, *Environmental Health Perspectives*, 108 Suppl. 3: 511–533.

Rogan, W.J. and Gladen, B.C. (1991), PCBs, DDE, and child development at 18 and 24 months, *Annals of Epidemiology*, 1: 407–413.

Selevan, S.G., Kimmel, C.A., and Mendola, P. (2000), Identifying critical windows of exposure for children's health, *Environmental Health Perspectives*, 108 Suppl. 3: 451–455.

Shapouri, S. and Rosen, S. (2001), Review of 67 low-income developing countries, *Food Security Assessment: Regional Overview*, Washington, D.C.: USDA Economic Research Service.

Shea, K.M. (2003), Antibiotic resistance: what is the impact of agricultural uses of antibiotics on children's health? *Pediatrics*, 112: 253–258.

Tomatis, L. (1989), Overview of perinatal and multigeneration carcinogenesis, *IARC Science Publication*, 96: 1–15.

Tomatis, L., Narod, S., and Yamasaki, H. (1992), Transgeneration transmission of carcinogenic risk, *Carcinogenesis*, 13: 145–151.

Tuohy, P.G. (2003), Soy infant formula and phytoestrogens, *Journal of Paediatric Child Health*, 39: 401–405.

U.S. Environmental Protection Agency, Risk Assessment Forum (2003), Supplemental Guidance for Assessing Cancer Susceptibility from Early-Life Exposure to Carcinogens (External Review Draft), Washington, D.C.: US EPA.

U.S. Food and Drug Administration, Center for Food Safety and Applied Nutrition (2000a), Background Paper in Support of Fumonisin Levels in Corn and Corn Products Intended for Human Consumption, Washington, D.C.: FDA.

U.S. Food and Drug Administration, Center for Food Safety and Applied Nutrition (2000b), Patulin in Apple Juice, Apple Juice Concentrates and Apple Juice Products, Washington, D.C.: FDA.

Weiss, S.T. (1998), Environmental risk factors in childhood asthma, *Clinical Experiments in Allergy*, 28 Suppl. 5: 29–34; discussion 50–51.

Werler, M.M., Shapiro, S., and Mitchell, A.A. (1993), Periconceptional folic acid exposure and risk of occurrent neural tube defects [see comments], *Journal of the American Medical Association*, 269: 1257–1261.

Wiebe, K., Ballenger, N., and Pinstrup-Anderson, P. (2001), Meeting Food Needs in the 21st Century: How Many and Who Will Be at Risk? (Chapter 1) *Who Will Be Fed In The 21st Century? Challenges for Science and Policy*, Baltimore: Johns Hopkins University Press.

Wood, A. and Clarin, W. (2000), 3M to phase out PFOS, *Chemical Week*, May: 9.

Wuthrich, B. (1993), Adverse reactions to food additives, *Annals of Allergy*, 71: 379–384.

9 Celiac Disease: A New Paradigm of an Immune-Mediated Disorder Due to Dietary Gluten

Michelle Maria Pietzak and Alessio Fasano

CONTENTS

0-8493-3516-7/05/$0.00+$1.50

ABSTRACT

Celiac disease is an immune-mediated disorder that occurs in genetically predisposed individuals who ingest gluten. Gluten is a dietary protein found in the grains wheat, barley, and rye. In a susceptible person, this gluten-containing diet can lead to the development of an autoimmune enteropathy, causing malabsorption of carbohydrates, proteins, fats, and critical vitamins and minerals. Classically, celiac disease was thought to occur in childhood, after the introduction of gluten in the diet. These children often exhibit gastrointestinal symptoms such as diarrhea, gaseousness, weight loss, and chronic abdominal pain. However, recent research indicates that this disease may present in adulthood with symptoms outside the gastrointestinal tract. Patients with celiac disease may present with extraintestinal symptoms, association with other autoimmune diseases (such as type I diabetes or autoimmune thyroiditis), or may simply have a positive family history for the disease. Testing serum antibodies to gluten and the tissue transglutaminase can screen for the presence of celiac disease. However, the "gold standard" for confirmation of the diagnosis remains a small intestinal biopsy combined with the patient's clinical response to a gluten-free diet. Clinicians must maintain a high index of suspicion for this disease, as it may present with a myriad of symptoms that closely mimic other diseases. If diagnosed early, the nutritional and malignant complications of long-standing celiac disease can be completely avoided by strict adherence to the gluten-free diet.

INTRODUCTION

Celiac disease has been formally known by many names in the medical literature, including gluten-sensitive enteropathy, gliadin-sensitive enteropathy, and celiac sprue (to differentiate it from tropical sprue). This disease can be defined as a permanent intolerance to the gliadin fraction of wheat protein and related alcohol-soluble proteins, called prolamins, found in rye and barley. The disorder occurs in genetically susceptible individuals who ingest these proteins, leading to an autoimmune enteropathy that self-perpetuates as long as these food products remain in the diet.

Making the diagnosis of celiac disease can be difficult because its clinical features can be highly variable. The disease may present at any age, but peaks appear to occur during the second year of life, adolescence, and middle age. Many adult patients exhibit symptoms for years without being diagnosed correctly. Likewise, the severity of the disease can be highly variable, with some patients experiencing severe diarrhea and weight loss, and others having no gastrointestinal symptoms whatsoever. Given its immune-mediated nature, celiac disease may affect districts other than the gastrointestinal tract, including the neurologic, endocrine, orthopedic, reproductive and hematologic systems. Thus, a patient may present with signs and symptoms that are completely unrelated to the gastrointestinal tract. Because of this, multiple physicians may see patients with celiac disease over many years before the disorder is correctly identified.

The presentations of celiac disease can be divided into six categories: classical gastrointestinal form, late-onset gastrointestinal form, extraintestinal form, asymptomatic form, presenting with associated conditions, and a latent form. It has been proposed that the varying clinical manifestations could best be described using an iceberg analogy (Logan, 1992). The patients who have the classic gastrointestinal form represent the "tip of the iceberg," in that they are the obvious, visible patients with the disease. What is concerning about this model is that the majority of patients are "submerged" and may experience chronic ill health and multiple complications of the disease without ever being correctly diagnosed.

Classic Gastrointestinal Form

The classic gastrointestinal form occurs in infants between the ages of 6 and 18 months, after the introduction of gluten-containing foods into the diet. The onset is usually insidious, with diarrhea and fat malabsorption, which can lead to anorexia and weight loss. Physical signs of protein-calorie malnutrition may be present, such as poor muscle bulk, peripheral edema, decreased subcutaneous fat stores, and abdominal distension and gaseousness due to lactose malabsorption. The child may also be anemic due to iron or folate deficiency, and radiographs may reveal rickets or osteopenia due to vitamin D and calcium malabsorption. The diagnosis of celiac disease is often prompt in this population due to this obvious gastrointestinal presentation.

Late-Onset Gastrointestinal Form

These patients exhibit gastrointestinal symptoms outside the classic age range of infancy, with intermittent diarrhea, weight loss, short stature, and often abdominal distension. However, patients may complain of constipation rather than diarrhea, and often have upper gastrointestinal tract symptoms such as nausea, dyspepsia, and gastroesophageal reflux. These patients are often misdiagnosed with irritable bowel syndrome, lactose intolerance, recurrent abdominal pain of childhood, or Crohn's disease.

Extraintestinal Form

Once thought to be primarily a disease of the gastrointestinal tract, we now know that celiac disease may present with signs and symptoms in many other organ systems. These include the musculoskeletal system, the skin and mucous membranes, the reproductive system, the hematologic system, the hepatobiliary system, and the central nervous system.

The Musculoskeletal System

One of the most common presenting features of celiac disease in childhood is idiopathic short stature. It is thought that 10% of children with idiopathic short stature, whose height is far below their genetic predisposition for unclear reasons, have celiac disease (Cacciari et al., 1983). Short stature may be the only symptom of the disease. Radiographs can reveal delayed bone age, rickets, osteomalacia, or even osteoporosis in advanced disease. These individuals may experience delayed puberty, and stimulation testing may reveal growth hormone deficiency (Verkasalo et al., 1987). The potential to achieve normal stature and avoid permanent bone loss are good if the patient is diagnosed with celiac disease before puberty and placed on a gluten-free diet (Prader et al., 1969). After the growth plates have closed, however, individuals will have short stature for life. The osteoporosis is also irreversible after a critical mass of bone has been lost.

Celiac disease can also manifest itself with dental enamel defects. People with untreated celiac disease in childhood may experience multiple areas of caries at an early age and in atypical locations. Up to 30% of adults with celiac disease have been reported to have dental enamel hypoplasia (Smith and Miller, 1979). These enamel defects are linear, and occur symmetrically in all four quadrants (Aine, 1986). The precise cause of the enamel hypoplasia is uncertain. Interestingly, it affects only the permanent dentition, which forms before the age of 7 yr while the child is ingesting gluten (as opposed to the primary dentition, which forms *in utero*).

Joint pain is a common presenting complaint of celiac disease. The disease is also seen in higher incidence in patients with rheumatologic disorders such as systemic lupus erythematosus, and rheumatoid arthritis (Lepore et al., 1996). In the adolescent patient, arthritis may be the only presenting symptom (Mäki et al., 1988). Clubbing of the fingers and toes (broad digits with abnormally curved nails) has also been reported.

One of the leading causes of morbidity in adults diagnosed with celiac disease is the development of osteoporosis. The potential health implications are severe, and include vertebral fractures, kyphosis, hip fractures, and Colles' fracture of the lower radius. Although the osteopenia begins in childhood, early therapy with a gluten-free diet will prevent progression and may even reverse bone loss in adults and children (Valdimarsson et al., 1996b; Mora et al., 1999). However, severe osteoporosis from celiac disease diagnosed late in life will not improve on a gluten-free diet because a critical amount of bone has been lost (Meyer et al., 2001).

The Skin and Mucous Membranes

The skin and mucous membranes are obvious sites for the expression of celiac disease; however, they are often overlooked. The classic skin manifestation is dermatitis herpetiformis, which occurs in about 5% of patients between the ages of 15 to 40 yr and is rarely seen in children. Its hallmark is severe pruritus, which does not respond well to topical or systemic medications. The lesions are symmetrical in distribution, appearing as erythematous blisters on the face, elbows, back, buttocks, and knees. A biopsy of the normal appearing skin next to the affected area demonstrates the characteristic histology of granular IgA deposits (Fry, 1995). Dermatitis herpetiformis is now known to be pathognomonic for celiac disease, and once a skin biopsy has proven the diagnosis, an intestinal biopsy is unnecessary. However, if patients with dermatitis herpetiformis do have a small-bowel biopsy, there is usually histologic correlation with flattened or blunted villi. Dermatitis herpetiformis resolves with a gluten-free diet. Flares of dermatitis herpetiformis and pruritus may occur when gluten is inadvertently ingested, alerting the patient to scrutinize the diet for contamination.

Other manifestations of celiac disease seen in the skin and mucous membranes include urticaria (hives), psoriasis, and aphthous stomatitis (Scala et al., 1999; Michaelsson et al., 2000; Catassi et al., 1996). In the patient with untreated celiac disease, aphthous stomatitis always correlates with histologic changes in the small intestine. However, since aphthous ulcers are commonly seen in other types of autoimmune and inflammatory diseases of the gastrointestinal tract, as well as with many common viral infections, they are not pathognomonic for celiac disease.

The Reproductive System

Women with undiagnosed celiac disease may first present to an obstetrician or gynecologist, as they have difficulty becoming pregnant and an 8.9-times higher risk for spontaneous abortions (Auricchio et al., 1988; Gasbarrini et al., 2000; Ciacci et al., 1996). Males with untreated celiac disease may also experience infertility due to gonadal dysfunction (Farthing et al., 1983). These adults may undergo painful, expensive, and exhaustive infertility studies without discovery of an etiology. Whether these individuals experience reproductive challenges due to nutritional factors alone is unclear. Many of these women will also have a history of delayed menarche, similar to the patients described earlier with short stature and delayed puberty.

The Hematologic System

A routine complete blood-cell count may reveal many hematologic abnormalities in an untreated patient with celiac disease. Anemia, leukopenia, and thrombocytopenia have all been reported. The anemia is usually microcytic and hypochromic due to iron deficiency (Carroccio et al., 1998). A macrocytic anemia should warrant an investigation into gut malabsorption of B12 or folic acid. Because fat-soluble vitamins are also malabsorbed in this disorder, vitamin K deficiency may occur, resulting in an increased risk for spontaneous bruising and bleeding.

The Hepatobiliary System

Diseases of the liver, such as autoimmune hepatitis and chronic transaminasemia, have been reported with celiac disease in both children and adults (Maggiore et al., 1986; Leonardi et al., 1990; Davison, 2002). Children with celiac disease who adhere to a strict gluten-free diet have had resolution of these biochemical abnormalities associated with hepatocyte damage (Vajro et al., 1993).

The Central Nervous System

Recent studies have shed some light regarding how celiac disease may affect the central nervous system. Patients with this disorder are 20 times more likely than the general population to suffer from epilepsy, and often have associated cerebral and cerebellar calcifications that can be imaged by computed tomography and magnetic resonance imaging (Gobbi et al., 1992). Cerebellar degeneration with resulting ataxia ("gluten associated ataxia") is a well-described entity (Pellecchia et al., 1999). Focal white-matter lesions in the brain have been reported to occur in children with celiac disease and are thought to be either ischemic in origin as a result of vasculitis or caused by inflammatory demyelination (Kieslich et al., 2001).

Parents of children with celiac disease have reported improvements in irritability, separation anxiety, emotional withdrawal, and autistic-like behaviors on the gluten-free diet (Fabiani et al., 1996). There are also some early data to suggest the use of a gluten-free, casein-free diet in children with autism and dyslexia (Knivsberg, 1997; Knivsberg et al., 2001; Knivsberg et al., 2002). Whether or not children with autism have a higher incidence of celiac disease, or whether children with undiagnosed celiac disease have a higher incidence of autism, remains to be proven. However, people with Down syndrome, who often have autistic-like behaviors, are at higher risk for celiac disease (Book et al., 2001). Depression, dementia, and schizophrenia are also more common in untreated celiacs. It has been hypothesized that gluten may be broken down into small peptides that may cross the blood–brain barrier and interact with morphine receptors, leading to alterations in conduct and perceptions of reality (Hadjivassiliou et al., 1996).

Asymptomatic Form

In the asymptomatic form of this disorder, patients lack the classic signs and symptoms as described earlier, which can occur in either the gastrointestinal tract or extraintestinal systems. Upon closer inspection, these asymptomatics may reveal evidence of trace vitamin or mineral deficiencies, such as anemia, short stature, low bone density, or low serum fat–soluble vitamin levels. These patients are usually first- or second-degree relatives of a biopsy-proven celiac, although many give no family history of the disease. Often a family history of other autoimmune disorders is obtained. These patients are usually identified through mass serologic screenings (Troncone et al., 1996; Ferguson et al., 1993).

Associated Conditions

Patients with other autoimmune disorders as well as some syndromes are at significantly higher risk of developing celiac disease (Table 9.1). Similarly, patients diagnosed with celiac disease early in life have a significantly lower risk of developing other autoimmune disorders than patients diagnosed later in life if they adhere to a strict gluten-free diet (Ventura et al., 1999). People with Down or Turner syndromes are often short and are at increased risk for many autoimmune diseases (such as diabetes mellitus, Crohn's disease, and thyroid disease), making the diagnosis of celiac disease in this population exceptionally challenging (Aman, 1999). A recent multicenter study from Italy strongly suggests screening all children with Down syndrome for celiac disease, regardless of the presence or absence of symptomatology (Bonamico et al., 2001).

TABLE 9.1
Medical Conditions Associated with Celiac Disease

Condition	Percent with Celiac Disease	References
Dermatitis herpetiformis	100%	Fry, 1995; Ruenala and Collin, 1997
Dental enamel defects	19–30%	Smith and Miller, 1979; Aine, 1986
First- or second-degree relatives	5–20%	Mylotte et al., 1974; Stevens et al., 1975; Pietzak et al., 2002b
Down syndrome	4–14%	Book et al., 2001; Bonamico et al., 2001
Diabetes type I	3.5–13%	Pietzak et al., 2002a, 2002b; Thain et al., 1974; Collin et al., 1989
Short stature	9–10%	Cacciari et al., 1983; Verkasalo et al., 1987; [illegible] et al., 1989
Turner syndrome	4–8%	Rujner et al., 2001; Bonamico et al., 1998; Ivarsson et al., 1999; Aman, 1999
Arthritis	1.5–7.5%	Lepore et al., 1996; Maki et al., 1988; Lubrano et al., 1996
Lactose intolerance	7%	Pietzak et al., 2002b
Selective IgA deficiency	7%	Collin et al., 1992b
Primary biliary cirrhosis	6%	Kingham and Parker, 1998
Idiopathic dilated cardiomyopathy	5.7%	Curione et al., 1999
Irritable bowel syndrome	4%	Pietzak et al., 2002b
Autoimmune thyroid disease	4%	Collin et al., 1994
Sjogren's syndrome	3%	Collin et al., 1992a
Epilepsy	2%	Gobbi et al., 1992; Cronin et al., 1998

Latent Form

People who are latent for celiac disease have positive serology (see following discussion on screening tests) with normal small-bowel biopsies. It is believed that over time, with further ingestion of gluten, these individuals will develop the disease with subsequent classic abnormal histology (Collin et al., 1999).

COMPLICATIONS OF CELIAC DISEASE

Undiagnosed, and therefore untreated, patients with celiac disease have an increased morbidity and mortality due to its associated conditions. Due to its multisystemic effects, these patients are at risk for chronic ill health, permanently stunted growth, infertility, osteoporosis, and malignancy. Patients may also incur increased health care costs because of multiple subspecialist visits and tests performed until the correct diagnosis is obtained. The mortality rate at every age is twofold greater in untreated celiacs (Corrao et al., 2001). However, there is no increase in mortality with strict adherence to the gluten-free diet. The increase in mortality is due to the association of untreated celiac disease with gastrointestinal malignancies, specifically non-Hodgkin's lymphoma of the small bowel. The prevalence of this malignancy has been reported to be as high as 10 to 15% of adult celiac patients noncompliant with the diet (Swinson et al., 1983). The Italian Working Group on Coeliac Disease and Non-Hodgkin's Lymphoma recently reported that the odds ratio overall for non-Hodgkin's lymphoma associated with celiac disease was 3.1, with odds ratios of 16.9 for gut lymphoma and 19.2 for T-cell lymphoma, respectively (Catassi et al., 2002). The risk for lymphoma decreases to that of the general population on the gluten-free diet (Holmes, 1989).

SCREENING FOR CELIAC DISEASE WITH SERUM ANTIBODIES

The previously utilized screening tests for malabsorption disorders, such as examining the stool for fat and performing intestinal permeability tests using lactulose or mannitol, are aspecific markers for celiac disease. Serum immunologic tests with increased sensitivity and specificity are now the screens of choice. The clinician must be aware that no screening test is perfect and that the gold standard for confirming the diagnosis of celiac disease remains a small-intestinal biopsy combined with the patient's clinical response to a gluten-free diet. Therefore, a patient with classic signs and symptoms should have a small-bowel biopsy even if the serology is not highly suggestive.

The current serologic tests consist of antigliadin, antiendomysial, and antitissue transglutaminase antibodies. These antibodies vary in sensitivity, specificity, and positive and negative predictive values (Table 9.2). False negative results may occur with an inexperienced lab, or if a patient is very young, IgA deficient, or on a gluten-free diet. Thus, these antibodies need to be correlated with each other and with the patient's clinical status.

ANTIGLIADIN ANTIBODIES

The antigliadin (AGA) IgG and IgA recognize gliadin, a small antigenic portion of the gluten protein (Case, 2001). AGA IgG has very good sensitivity, whereas AGA IgA has very good specificity, and therefore their combined use provided the first reliable screening test when these antibodies became available during the late 1970s. A strength of the AGA antibodies is that they are automated ELISA (enzyme-linked immunosorbent assay) tests, and the results are independent of observer variability. Unfortunately, many normal individuals without the disease will have an elevated AGA IgG, causing much confusion among practitioners and perhaps unnecessary procedures. The AGA IgG is useful in screening IgA-deficient individuals, as the other antibodies are usually of the IgA class. It is estimated that 0.2 to 0.4% of the general population has selective IgA deficiency, whereas 2 to 3% of celiac patients are IgA deficient (Cataldo et al., 1998). If a patient's celiac panel is positive *only* for AGA IgG, this is *not* suggestive for celiac disease if the patient is IgA sufficient (corrected for age). Markedly elevated AGA IgG (greater than three to four times the upper limit of normal) suggests increased gut permeability to gluten, which can be seen in gastrointestinal infections, Crohn's disease, allergic gastroenteropathy, and autoimmune enteropathy.

TABLE 9.2
Sensitivity, Specificity, and Positive and Negative Predictive Values of the Serologic Markers for Celiac Disease

Test	Sensitivity	Specificity	PPV	NPD
Antigliadin IgA	53–100	65–100	28–100	65–100
Antigliadin IgG	57–100	42–98	20–95	41–88
Antiendomysial IgA	75–100	96–100	80–100	80–100
Guinea pig tissue transglutaminase IgA	85–100	21–100	67–100	97–100
Guinea pig tissue transglutaminase IgG	90–97	95–98	83	—
Human tissue transglutaminase IgA	93–98.5	98–99	—	—
Human tissue transglutaminase IgG	47	98	—	—
Human tissue transglutaminase IgG and IgA combined[a]	100	98	—	—

[a]via dot blot assay

Source: From Gandolfi, L., Bocca, A.L., and Pratesi, R. (2000), Screening of celiac disease in children attending the outpatient clinic of a university hospital, *Journal of Pediatric Gastroenterology and Nutrition,* 31:S212; Lindquist, B.L., Rogozinski, T., Moi, H., Danielsson, D., and Olcen, P. (1994), Endomysium and gliadin IgA antibodies in children with coeliac disease, *Scandinavian Journal of Gastroenterology,* 29:452–456; Grodzinsky, E., Jansson, G., Skogh, T., Stenhammar, L., and Falth-Magnusson. K. (1995), Anti-endomysium and anti-gliadin antibodies as serological markers for coeliac disease in childhood: a clinical study to develop a practical routine, *Acta Paediatrica,* 84:294–298; Cataldo, F., Ventura, A., Lazzari, R., Balli, F., Nassimbeni, G., and Marino, V. (1995), Antiendomysium antibodies and coeliac disease: solved and unsolved questions. An Italian multicentre study, *Acta Paediatrica,* 84:1125–1131; Russo, P.A., Chartrand, L.J., and Seidman, E. (1999), Comparative analysis of serologic screening tests for the initial diagnosis of celiac disease, *Pediatrics,* 104:75–78; Valdimarsson, T., Franzen, L., Grodzinsky, E., Skogh, T., and Strom, M. (1996a), Is small bowel biopsy necessary in adults with suspected celiac disease and IgA anti-endomysium antibodies?, *Digestive Diseases and Sciences,* 41:83–87; Troncone, R. and Ferguson, A. (1991), Anti-gliadin antibodies, *Journal of Pediatric Gastroenterology and Nutrition,* 12:150–158; Baldas, V., Tommasini, A., Trevisiol, C., Berti, I., Fasano, A., Sblattero, D., Bradbury, A., Marzari, R., Barillari, G., Ventura, A., and Not, T. (2000), Development of a novel rapid non-invasive screening test for celiac disease, *Gut,* 47:628–631; Troncone,R., Maurano, F., Rossi, M., Micillo, M., Greco, L., Auricchio, R., Salerno, G., Salvatore, F., and Sacchetti, L. (1999), IgA antibodies to tissue transglutaminase: an effective diagnostic test for celiac disease, *The Journal of Pediatrics,* 134:166–171; Lock,R., Pitcher, M., and Unsworth, D. (1999), IgA anti-tissue transglutaminase as a diagnostic marker of gluten sensitive enteropathy, *Journal of Clinical Pathology,* 52:274–277; Vitoria, J., Arrieta, A., Arranz, C., Ayesta, A., Sojo, A., Maruri, N., and Garcia-Masdevall, M. (1999), Antibodies to gliadin, endomysium, and tissue transglutaminase for the diagnosis of celiac disease, *Journal of Pediatric Gastroenterology and Nutrition,* 29:571–574; Gillett, H. and Freeman, H. (2000), Comparison of IgA endomysium antibody and IgA tissue transglutaminase antibody in celiac disease, *Canadian Journal of Gastroenterology,* 14:668–671.

Antiendomysial Antibodies

The antiendomysial (AEA) IgA immunofluorescent antibody is an excellent screening test with high sensitivity and specificity. The tests was developed in the early 1980s and rapidly gained use as part of a “celiac panel” by commercial labs in combination with AGA IgG and IgA. Its major drawbacks are false negatives in young children (under 2 years of age) and patients with IgA deficiency. Also, the substrate for this antibody was initially monkey esophagus, making it expensive and unsuitable for screening large numbers of individuals. Recently, human umbilical cord has been used as an alternative substrate (Not et al., 1998). The subjective nature of the AEA assay may lead to false negative values in inexperienced hands and unacceptable variability between laboratories (Murray et al., 1997).

Tissue Transglutaminase Antibodies

In 1997, tissue transglutaminase (tTG) was discovered to be the autoantigen for celiac disease (and therefore responsible for AEA positivity) (Dieterich et al., 1997). The human recombinant tTG ELISA is highly sensitive and reasonably specific, and both correlate well with AEA IgA and biopsy (Troncone et al., 1999; Sulkanen et al., 1998). tTG represents a major improvement over AEA because it is inexpensive, rapid, not subjective, and can be performed using a dot blot technique on a single drop of blood (Sblattero et al., 2000). These characteristics make this test ideally suited for mass screenings. In the future, the tTG dot blot could be performed in the general practitioner's office, much like the now-routine finger-stick hematocrit.

Current Guidelines on the Use of Serology

The IgA class human anti-tTG antibody, coupled with a determination of total serum IgA if the anti-tTG level is very low, currently seems to be the most cost-effective way to screen for celiac disease. Due to the wide clinical spectrum of this autoimmune disorder, the tTG assay should be able to find a place in the repertoire of the primary care physician. AEA should be used as a confirmatory, prebiopsy test. AGA determinations should be restricted to the diagnostic work-up of younger children and patients with IgA deficiency.

DEFINITIVE DIAGNOSIS WITH SMALL-INTESTINAL BIOPSY

Despite recent advances in serologic testing, confirmation of clinical suspicion or a positive serology screen still requires a small-intestinal biopsy. Before the advent of fiberoptic and chip technology for endoscopes, biopsies of the jejunum were obtained using a Crosby spring-loaded capsule passed orally under fluoroscopic guidance. Most biopsies are performed today using a flexible endoscope, which has the advantages of being able to directly visualize the mucosa to look for changes consistent with celiac disease (such as "notching" or "scalloping" of the small bowel folds [Vanderhoff, 1997; Jabbari et al., 1995], or lymphonodular hyperplasia) and other lesions (such as ulcers, esophagitis, gastritis, or duodenitis), which may explain the patient's symptomatology. The disadvantages of the flexible endoscope include the requirements for sedation (and in small children, general anesthesia) and the fact that the tissue obtained is smaller, more superficial, and more proximal. Because the capsule biopsies obtain a larger piece of tissue from more distal small bowel,

TABLE 9.3
Conditions Other Than Celiac Disease That May Cause Small-Bowel Villous Atrophy

- Autoimmune enteropathy
- Bowel ischemia
- Chronic gastroenteritis
- Collagenous sprue
- Congenital enteropathies
- Crohn's disease
- Drugs
- Eosinophilic gastroenteritis
- Graft vs. host disease
- Immunodeficiency syndromes
- Intestinal bacterial overgrowth
- Intractable diarrhea of infancy
- Mediterranean lymphoma
- Milk or soy protein intolerance
- Parasites
- Peptic duodenitis
- Postenteritis enteropathy
- Protein-calorie malnutrition
- Radiation
- Tropical sprue
- Viral gastroenteritis (rotavirus)

Source: Vanderhoof, J.A. (1997), Diarrheal disease in infants and children, in P.E. Hyman (Volume Ed.) and M. Feldman (Series Ed.) *Gastroenterology and Hepatology: The Comprehensive Visual Reference. Volume 4: Pediatric GI Problems*, Philadelphia: Churchill Livingstone, pp. 6.1–6.17; Holmes, G. and Catassi C (Eds.) (2000), *Coeliac Disease*, Oxford: Health Press; Dahms, B.B. (1993), Gastrointestinal mucosal biopsy, in R. Wyllie and J.S. Hyams (Eds.), *Pediatric Gastrointestinal Disease*, Philadelphia: W.B. Saunders, pp. 1057–1077; Patey, N., Scoazec, J.Y., Cuenod-Jabri, B., Canioni, D., Kedinger, M., Goulet, O., and Brousse, N. (1997), Distribution of cell adhesion molecules in infants with intestinal epithelial dysplasia (tufting enteropathy), *Gastroenterology,* 113:833–843.

some authors advocate its use preferentially for the diagnosis of celiac disease. It is recommended that the gastroenterologist obtain a minimum of 4 to 5 biopsies from distinct areas because of the patchy nature of the disease. The diagnosis may be missed due to sampling error.

The characteristic small-bowel histology is subtotal or total villous atrophy. Marsh et al. described a spectrum of changes in the intestinal mucosa, which has become the accepted standard (Marsh, 1988). The Marsh criteria describe four patterns of mucosal immunopathology, all of which can be consistent with the disease: (1) Type 0 (preinfiltrative), without detectable inflammation or changes in the crypt or villous architecture; (2) Type 1 (infiltrative), with an increase in the intraepithelial lymphocytes, but without detectable changes in the crypt or villous architecture; (3) Type 2 (hyperplastic), with inflammation, villous blunting, and an increased crypt or villous height ratio; and (4) Type 3 (destructive), with severe inflammation, flat villi, and hyperplastic crypts.

The clinician needs to remember that villous atrophy can be caused by a wide variety of gastrointestinal conditions (Table 9.3). Correlation between serology and the patient's response to a gluten-free diet is imperative to confirm the diagnosis.

An individual's HLA determination can be useful in ruling out celiac disease in questionable cases if the patient is found to be neither DQ2 nor DQ8 positive (Rujner et al., 2001).

REFRACTORY SPRUE

A minority of adults with biopsy-proven celiac disease will continue to have symptoms despite vigorous adherence to a gluten-free diet. The diagnosis of refractory sprue, however, should be one of exclusion. A thorough dietary history must first be taken to rule out the inadvertent ingestion of gluten as the etiology for the persistence of symptoms. Compliance with the gluten-free diet can also be assessed by measurement of the immunologic markers for the disease. For example, persistently elevated antigliadin IgG and IgA may indicate either noncompliance or that trace amounts of gluten from contaminated products are being consumed. One should consider a repeat biopsy in the patient who continues to have gastrointestinal complaints for more than one year after strict adherence to the diet. If there is histologic evidence of villous atrophy, the patient should be further evaluated for both the infectious and noninfectious conditions listed in Table 9.3. Other causes of nonresponsiveness in adults include the long-standing complications of the disease, such as pancreatic insufficiency (Carroccio et al., 1997) and T-cell lymphoma (Swinson et al., 1983; Catassi et al., 2002). As many as 75% of adults with refractory sprue may have an aberrant clonal intraepithelial T-cell population, which is associated with a malignancy currently classified as "cryptic enteropathy-associated T-cell lymphoma" (Cellier et al., 2000). In addition to a gluten-free diet, patients with refractory sprue frequently require immunosuppression with steroids, azathioprine, or cyclosporin (Vaidya et al., 1999; Rolny et al., 1999).

GLUTEN AS A "TOXIN"

Celiac disease is the only autoimmune disease for which we know the trigger — gluten. Therefore, removal of the offending agent from the diet results in complete symptomatic and histologic resolution (except in the case of refractory sprue, as discussed above). Although intestinal toxins have been classically thought of as those elaborated by enteric bacterial pathogens (such as *Vibrio cholera, Escherichia coli, Salmonella, Shigella, Clostridium perfringens*, and many others [Fasano, 2002]), gluten could also be considered a "toxin" based upon its mechanism of action in the small bowel and perhaps even in the central nervous system (Hadjivassiliou et al., 1996). The agents responsible for intestinal damage, leading to an autoimmune response, are called prolamins. Prolamins are the storage proteins located in the seeds of different grains. Gluten is the general name for the prolamins found in wheat (gliadin), rye (secalin), barley (horedin), and oats (avenin). Gluten is important in baked goods because it plays a critical role in leavening, forming the structure of the dough, and in holding the baked product together. Prolamins are also found in corn and rice, but they do not elicit an immune reaction in the intestines of patients with celiac disease (Case, 2001).

It has been shown that gluten deamidation by tissue transglutaminase enhances the recognition of gliadin peptides by HLA DQ2 or DQ8 T cells in genetically predisposed subjects. This reaction seems to initiate the cascade of autoimmune responses that are responsible for the intestinal mucosal destruction through the production of cytokines and matrix metalloproteinases (Molberg et al., 1998; Pender et al., 1997). These reactions imply that gliadin or its breakdown peptides or both cross, in someway, the intestinal epithelial barrier and reach the lamina propria of the intestinal mucosa, where they are recognized by antigen-presenting cells. However, under normal physiologic conditions, the intestinal epithelial barrier functions as the major organ of defense against foreign antigens, toxins, and macromolecules entering the host via the enteric route. The key elements that dictate the intestinal permeability are the intercellular tight junctions. Our group has recently reported that zonulin, a molecule that induces tight junction disassembly, is upregulated during the acute phase of celiac disease (Fasano et al., 2000), suggesting that the reported opening of tight junction at the early stage of the disease (Madara and Trier, 1980; Schulzke et al., 1998) could be mediated by zonulin.

We have also demonstrated that gliadin activates the zonulin-signaling pathway (Figure 9.1) in normal intestinal epithelial cells *in vitro*. The cellular response observed only a few minutes after gliadin incubation was characterized by significant cytoskeleton reorganization with a redistribution of actin filaments mainly in the intracellular subcortical compartment. Spectrofluorometry experiments revealed that such cytoskeleton reorganization was associated with an increment of F-actin amount secondary to an increased rate of intracellular actin polymerization. Experiments performed in Ussing chambers linked these cytoskeletal changes to a zonulin-dependent tight junction disassembly and then increased intestinal permeability (Clemente et al., 2003).

Based on these results and preliminary data generated by using intestinal tissues from both celiac patients in remission and healthy controls (Drago et al., 2002), one could hypothesize a possible gliadin mechanism of action leading to a zonulin-mediated increase in actin polymerization and intestinal permeability. Enterocytes, exposed to gliadin, physiologically react by secreting zonulin in the intestinal lumen. While in normal intestinal tissues this secretion is self-limited (Clemente et al., 2003), in celiac gut tissues the zonulin system is chronically upregulated (Fasano et al., 2000). This leads to a sustained increase in intestinal permeability to macromolecules, including gliadin, from the lumen to the lamina propria. Thus, gliadin is deamidated by the tissue transglutaminase and then recognized by HLA-DQ2 or DQ8 bearing antigen presenting cells, triggering the autoimmune reaction in genetically susceptible subjects.

TREATMENT OF CELIAC DISEASE WITH THE GLUTEN-FREE DIET

As stated above, celiac disease is the only autoimmune disease for which we know of a dietary environmental trigger. Thus, treatment for the disease is completely nutritional. Table 9.4 lists the foods that are to be eliminated from the gluten-free

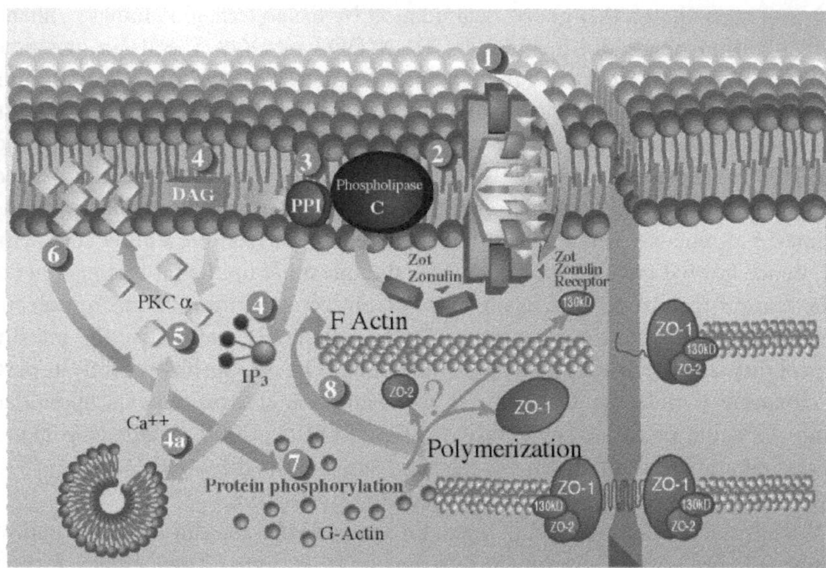

FIGURE 9.1 Proposed zonulin intracellular signaling leading to the opening of intestinal tight junctions. Zonulin interacts with a specific surface receptor (1) whose distribution within the intestine varies. The protein is then internalized and activates phospholipase C (2) that hydrolyzes phosphatidyl inositol (3) to release inositol 1,4,5-tris phosphate (PPI-3) and diacylglycerol (DAG) (4). PKCα is then activated (5) either directly (via DAG) (4) or through the release of intracellular Ca^{++} (via PPI 3) (4a). PKCα catalyzes the phosphorylation of target proteins, with subsequent polymerization of soluble G-actin in F-actin (7). This polymerization causes the rearrangement of the filaments of actin and the subsequent displacement of proteins (including ZO1) from the junctional complex (8). As a result, intestinal tight junctions become looser.

diet. Triticale (a combination of wheat and rye), kamut, and spelt (Forssell and Wieser, 1995) are considered toxic. Other forms of wheat, such as bulgur, couscous, einkorn, farina, and semolina (durum), are also not allowed. Any food product that contains rye, barley, or malt (a partial hydrolysate of barley) (Ellis et al., 1994) has prolamins, which are considered harmful. In general, a food product that includes the word "wheat" (cracked wheat, wheat bran, wheat grass, wheat germ, whole wheat) or "malt" (barley malt, malt extract, malt flavoring, malt syrup) in its name is considered to contain gluten. Buckwheat, however, is not directly related to the Triticum family and is safe for patients with celiac disease to ingest. Distilled ingredients (such as vinegar and alcohol) are rendered gluten free by the distillation process. However, beverages made with barley (such as beer, ale, lager, and some rice and soy drinks) are not allowed (Case, 2001).

Whether or not oats should be eliminated from the gluten-free diet remains controversial. The prolamin of oats, avenin, only accounts for 5 to 15% of the total seed protein. This is in marked contrast to gliadin, which comprises about 50% of

TABLE 9.4
Foods that Need to be Eliminated from the Gluten-Free Diet

Barley (Hordeum vulgare)	Kamut (*Triticum polonicum*)
Bran	Malt (derived from Barley)
Bulgur	Oats (controversial, please refer to text)
Couscous (endosperm of *Durum* wheat)	Rye (*Secale cereale*)
Einkorn wheat (*Triticum monococcum*)	Semolina (*Durum* wheat)
Emmer wheat (*Triticum dicoccon*)	Spelt (Dinkel)
Farro	Triticale (wheat-rye hybrid)
Farina	Wheat
Graham flour	

Source: From Case, S. (2001) *Gluten-Free Diet: A Comprehensive Resource Guide*, Saskatchewan, Canada: Centax Books; Forssell, F. and Wieser, H. (1995), Spelt wheat and coeliac disease, *Zeitschrift fur Lebensmittel Untersuchung und — Forschung,* 201: 39–49; Ellis, H.J., Doyle, A.P., Day, P., Wiesser, H., and Ciclitira, P.J. (1994), Demonstration of the presence of coeliac-activating gliadin-like epitopes in malted barley, *International Archives of Allergy and Immunology,* 104: 308–310.

the wheat protein (Holmes and Catassi, 2000). Avenin does not elicit the same immune response as gliadin and is thought by some to be safe for patients with celiac disease to ingest. A recent study in the U.S. on children newly diagnosed with celiac disease, who were allowed oats, demonstrated symptomatic and histologic resolution of the disease comparable to those who were denied oats (Hoffenberg et al., 2000). However, because oats are often crop rotated, harvested, and milled with wheat, the risk for contamination with wheat gluten is always a consideration.

The good news for patients with celiac disease is that many food products are naturally gluten free. A healthy diet that includes vegetables, fruits, meats, eggs, and dairy products is perfectly safe. Meats need to be cooked without breading or gluten-containing seasonings. Many patients (both with and without celiac disease) are lactose intolerant, and should preferentially choose "partially digested" diary items such as yogurt and cheeses instead of ice cream and fresh whole milk. However, the coating of some cheeses may contain gluten. Because the lactase enzyme is located on the tip of the villus, newly diagnosed celiacs may have an acquired lactase deficiency, which will improve as the small-bowel mucosa heals. Alternatives to the gluten-containing grains that can safely be used in cereals and in baking flours include amaranth, bean flour, buckwheat (kasha), corn (maize), millet, nuts, potato, quinoa, rice, soybean, sorghum, teff, tapioca, and wild rice.

Because of food-labeling practices in certain countries, especially in North America, it may not be obvious that a food item contains gluten. Products to question include those labeled "wheat-free" (which may contain other harmful grains) and ingredients that do not state their source (such as flavorings, spices, starch, or hydrolyzed vegetable protein). Gluten is often used as a flavoring in candy, sauces, seasonings, soups, and salad dressings, and as a filler in vitamins and medications

TABLE 9.5
Ingredients that Need to be Questioned for Gluten Content

Ingredient	Source (Based on North American Food Manufacturers)
Dextrin	Usually derived from corn or tapioca
Flavorings	Gluten-containing grains are almost never used, with the exception of barley malt, which is usually indicated on the label
Hydrolyzed plant protein (HPP)	May be derived from gluten-containing grains
Hydrolyzed vegetable protein (HVP)	May be derived from gluten-containing grains
Modified food starch	Corn, potato, tapioca, and rice are the usual sources; however, it can be made from wheat
Seasonings	A blend of flavoring agents, often using a carrier such as cereal flour or starch
Spices	Pure spices do not contain gluten, but imitation spices may have gluten-containing fillers

Source: From Case, S. (2001) *Gluten-Free Diet: A Comprehensive Resource Guide*, Saskatchewan, Canada: Centax Books; Forssell, F. and Wieser, H. (1995), Spelt wheat and celiac disease, *Zeitschrift fur Lebensmittel Untersuchung und — Forschung,* 201: 39–49; Ellis, H.J., Doyle, A.P., Day, P., Wiesser, H., and Ciclitira, P.J. (1994), Demonstration of the presence of coeliac-activating gliadin-like epitopes in malted barley, *International Archives of Allergy and Immunology,* 104: 308–310.

(Case, 2001; Crowe and Falini, 2001). Table 9.5 lists the ingredients to be questioned on food labels, with their usual sources in North America. Often, the only way for patients to be certain that a specific product is gluten free is to call the manufacturer directly. Toiletries, such as shampoos, conditioners, and skin care products, are not harmful as long as they are not ingested.

Patients with advanced celiac disease may require other forms of nutritional support. Malabsorption of fat-soluble vitamins (A, D, E, and K), folate, B12, and iron may require supplementation until the intestines have healed. Patients, especially women, with celiac disease also usually require calcium supplementation. Checking bone density early and treating with appropriate doses of vitamin D and calcium in addition to the gluten-free diet may prevent further bone loss leading to osteoporosis and fractures.

CONCLUSION

Celiac disease is a relatively common disorder that is underdiagnosed. This is most likely due to its protean manifestations that can occur outside of the gastrointestinal tract. Although measurement of serum antibodies may provide a useful screening tool for this disease, the gold standard of diagnosis remains the intestinal biopsy. Removal of the dietary trigger for this autoimmune condition provides a complete symptomatic and histologic response in most patients. On the cellular level, gluten appears to mediate its damage via the upregulation of zonulin, which in turn opens the tight junctions between cells by causing cytoskeletal rearrangements. Thus, the toxic fragments of gluten can gain access to the lamina propria of the gut and trigger the immune response in genetically susceptible individuals. The observation that these interactions with gluten occur in individuals who do not have celiac disease raises the issue as to whether gluten is toxic to everyone, and reinforces the notion that humans have not evolved to eat these grains.

References

Aine, L. (1986), Dental enamel defects and dental maturity in children and adolescents with celiac disease, *Proceedings of the Finnish Dental Society,* 82:227–229.

Aman, J. (1999), Prevalence of coeliac disease in Turner syndrome, *Acta Paediatrica,* 88:933–936.

Auricchio, S., Greco, L., and Troncone, R. (1988), Gluten-sensitive enteropathy in childhood, *Pediatric Clinics of North America,* 35:157–187.

Baldas, V., Tommasini, A., Trevisiol, C., Berti, I., Fasano, A., Sblattero, D., Bradbury, A., Marzari, R., Barillari, G., Ventura, A., and Not, T. (2000), Development of a novel rapid non-invasive screening test for celiac disease, *Gut,* 47:628–631.

Bonamico, M., Bottaro, G., Pasquino, A.M., Caruso-Nicoletti, M., Mariani, P., Gemme, G., Paradiso, E., Ragusa, M.C., and Spina, M. (1998), Celiac disease and Turner syndrome, *Journal of Pediatric Gastroenterology and Nutrition,* 26:496–499.

Bonamico, M., Mariani, P., and Danesi, H.M. (2001), Prevalence and clinical picture of celiac disease in Italian Down's syndrome patients: a multicenter study, *Journal of Pediatric Gastroenterology and Nutrition,* 33:139–143.

Book, L., Hart, A., Black, J., Feolo, M., Zone, J.J., and Neuhausen, L.S. (2001), Prevalence and clinical characteristics of celiac disease in Downs syndrome in a US study, *American Journal of Medical Genetics,* 98:70–74.

Cacciari, E., Salardi, S., Lazzari, R., Cicognani, A., Collina, A., Pirazzoli, P., Tassoni, P., Biasco, G., Corazza, G.R., and Cassio A. (1983), Short stature and celiac disease: a relationship to consider even in patients with no gastrointestinal tract symptoms, *Journal of Pediatrics,* 103:708–711.

Carroccio, A., Iacono, G., Lerro, P., Cavataio, F., Malorgio, E., Soresi, M., Baldassarre, M., Notarbartolo, A., Ansaldi, N., and Montalto, G. (1997), Role of pancreatic impairment in growth recovery during gluten-free diet in childhood celiac disease, *Gastroenterology,* 112:1839–1844.

Carroccio, A., Iannitto, E., Cavataio, F., Montalto, G., Tumminelo, M., Campagna, P., Lipari, M.G., Notarbartolo, A., and Iacono, G. (1998), Sideropenic anemia and celiac disease: one study, two points of view, *Digestive Diseases and Sciences,* 43:673–678.

Case, S. (2001), *Gluten-Free Diet: A Comprehensive Resource Guide*, Saskatchewan, Canada: Centax Books.

Cataldo, F., Ventura, A., Lazzari, R., Balli, F., Nassimbeni, G., and Marino, V. (1995), Antiendomysium antibodies and coeliac disease: solved and unsolved questions. An Italian multicentre study, *Acta Paediatrica,* 84:1125–1131.

Cataldo, F., Marino, V., Ventura, A., Bottaro, G., and Corazza, G.R. (1998), Prevalence and clinical features of selective immunoglobulin A deficiency in coeliac disease: an Italian multicentre study. Italian Society of Paediatric Gastroenterology and Hepatology (SIGEP) and "Club de Tenue" Working Groups on Coeliac Disease, *Gut,* 42:362–365.

Catassi, C., Fabiani, E., Ratsch, I.M., Coppa, G.V., Giorgi, P.L., Pierdomenico, R., Alessandrini, S., Iwanejko, G., Domenici, R., Mei, E., Miano, A., Marani, M., Bottaro, G., Spina, M., Dotti, M., Montanelli, A., Barbato, M., Viola, F., Lazzari, R., Vallini, M., Guariso, G., Plebani, M., Cataldo, F., Traverso, G., Ughi, C., Chiaravalloti, G., Baldassarre, M., Scarcella, P., Bascietto, F., Ceglie, L., Valenti, A., Paolucci, P., Caradonna, M., Bravi, E., and Ventura, A. (1996), The coeliac iceberg in Italy: a multicentre antigliadin antibodies screening for coeliac disease in school-age subjects, *Acta Paediatrica Supplement,* 412:29–35.

Catassi, C., Fabiani, E., Corrao, G., Barbato, M., De Renzo, A., Carella, A., Gabrielli, A., Leoni, P., Carroccio, A., Baldassarre, M., Bertolani, P., Caramaschi, P., Sozzi, M., Guariso, G., Volta, U., and Corazza, G. for the Italian Working Group on Coeliac Disase and Non-Hodgkin's Lymphoma (2002), Risk of Non-Hodgkin lymphoma in celiac disease, *JAMA: The Journal of the American Medical Association,* 287:1413–1419.

Cellier, C., Delabesse, E., Helmer, C., Patey, N., Matuchansky, C., Jabri, B., Macintyre, E., Cerf-Bensussan, N., and Brousse, N. (2000), Refractory sprue, coeliac disease, and enteropathy-associated T-cell lymphoma, *Lancet,* 356:203–208.

Ciacci, C., Cirillo, M., Auriemma, G., Di Dato, G., Sabbatini, F., and Mazzacca, G. (1996), Celiac disease and pregnancy outcome, *American Journal of Gastroenterology,* 91:718–722.

Clemente, M.G., De Virgiliis, S., Kang, J.S., Macatagney, R., Congia, M., and Fasano, A. (2003), New insights on celiac disease pathogenesis: gliadin-induced zonulin release, actin polymerization, and early increased gut permeability, *Gut,* 52: 218–223.

Collin, P., Salmi, J., Hallstrom, O., Oksa, H., Oksala, H., Maki, M., and Reunala, T. (1989), High frequency of coeliac disease in adult patients with type-I diabetes, *Scandinavian Journal of Gastroenterology,* 24:81–84.

Collin, P., Korpela, M., Hallstrom, O., Viander, M., Keyrilainen, O., and Maki, M. (1992a), Rheumatic complaints as a presenting symptom in patients with coeliac disease, *Scandinavian Journal of Rheumatology,* 21:20–23.

Collin, P., Maki, M., Keyrilainen, O., Hallstrom, O., Reunala, T., and Pasternack, A. (1992b), Selective IgA deficiency and celiac disease, *Scandinavian Journal of Gastroenterology,* 27:367–371.

Collin, P., Salmi, J., Hallstrom, O., Reunala, T., and Pasternack, A. (1994), Autoimmune thyroid disorders and coeliac disease, *European Journal of Endocrinology,* 130:137–140.

Collin, P., Kaukinen, K., and Maki, M. (1999), Clinical features of celiac disease today, *Digestive Diseases and Sciences,* 17:100–106.

Corrao, G., Corazza, G.R., Bagnardi, V., Brusco, G., Ciacci, C., Cottone, M., Sategna-Guidetti, C., Usai, P., Cesari, P., Pelli, M.A., Loperfido, S., Volta, U., Calabro, A., and Certo, M. (2001), Mortality in patients with coeliac disease and their relatives: a cohort study, *Lancet,* 358:356–361.

Cronin, C.C., Jackson, L.M., Feighery, C., Shanahan, F., Abuzakouk, M., Ryder, D.Q., Whelton, M., and Callaghan, N. (1998), Coeliac disease and epilepsy, *The Quarterly Journal of Medicine,* 91:303–308.

Crowe, J.P. and Falini, N.P. (2001), Gluten in pharmaceutical products, *American Journal of Health-System Pharmacy: AJHP,* 58:396–401.

Curione, M., Barbato, M., De Biase, L., Viola, F., Lo Russo, L., and Cardi, E. (1999), Prevalence of coeliac disease in idiopathic dilated cardiomyopathy, *Lancet,* 354:222–223.

Dahms, B.B. (1993), Gastrointestinal mucosal biopsy, in R. Wyllie, and J.S. Hyams (Eds.), *Pediatric Gastrointestinal Disease*, Philadelphia: W.B. Saunders, pp. 1057–1077.

Davison, S. (2002), Coeliac disease and liver dysfunction, *Archives of Diseases in Childhood,* 87:293–296.

Dieterich, W., Ehnis, T., Bauer, M., Donner, P., Volta, U., Riecken, E.O., and Schuppan, D. (1997), Identification of tissue transglutaminase as the autoantigen of celiac disease, *Nature Medicine,* 3:797–801.

Drago, S., DiPierro, M., Giambelluca, D., Iacono, G., Catassi, C., and Fasano, A., June 2002, Gliadin Induces Occludin Down-Regulation and Tight Junctions (tj) Disassembly in Human Intestine, paper presented at the 10th International Symposium on Coeliac Disease, Paris.

Ellis, H.J., Doyle, A.P., Day, P., Wiesser, H., and Ciclitira, P.J. (1994), Demonstration of the presence of coeliac-activating gliadin-like epitopes in malted barley, *International Archives of Allergy and Immunology,* 104: 308–310.

Fabiani, E., Catassi, C., Villari, A., Gismondi, P., Pierdomenico, R., Ratsch, I.M., Coppa, G.V., and Giorgi, P.L. (1996), Dietary compliance in screening-detected coeliac disease adolescents, *Acta Paediatrica Supplement,* 412:65–67.

Farthing, M.J., Rees, L.H., Edwards, C.R., and Dawson, A.M. (1983), Male gonadal function in coeliac disease: 2. Sex hormones, *Gut,* 24:127.

Fasano, A. (2002), Toxins and the gut: role in human disease, *Gut,* 50 (Supplement 3): III9–14.

Fasano, A., Not, T., Wang, W., Uzzau, S., Berti, I., Tommasini, A., and Goldblum, S.E. (2000), Zonulin, a newly discovered modulator of intestinal permeability, and its expression in coeliac disease, *Lancet,* 355:1518–1519.

Ferguson, A., Arranz, E., and O'Mahony, S. (1993), Clinical and pathological spectrum of coeliac disease — active, silent, latent, potential, *Gut,* 34:150–151.

Forssell, F. and Wieser, H. (1995), Spelt wheat and celiac disease, *Zeitschrift fur Lebensmittel Untersuchung und — Forschung,* 201: 39–49.

Fry, L. (1995), Dermatitis herpetiformis, *Baillière's Clinical Gastroenterology,* 9:371–393.

Gandolfi, L., Bocca, A.L., and Pratesi, R. (2000), Screening of celiac disease in children attending the outpatient clinic of a university hospital, *Journal of Pediatric Gastroenterology and Nutrition,* 31:S212.

Gasbarrini, A. Torre, E.S., Trivellini, C., De Carolis, S., Caruso, A., and Gasbarrini, G. (2000), Recurrent spontaneous abortion and intrauterine fetal growth retardation as symptoms of coeliac disease, *Lancet,* 256:399–400.

Gillett, H. and Freeman, H. (2000), Comparison of IgA endomysium antibody and IgA tissue transglutaminase antibody in celiac disease, *Canadian Journal of Gastroenterology,* 14:668–671.

Gobbi, G., Bouquet, F., Greco, L., Lambertini, A., Tassinari, C.A., Ventura, A., and Zaniboni, M.G. (1992), Coeliac disease, epilepsy, and cerebral calcifications, *Lancet,* 340:439–443.

Grodzinsky, E., Jansson, G., Skogh, T., Stenhammar, L., and Falth-Magnusson. K. (1995), Anti-endomysium and anti-gliadin antibodies as serological markers for coeliac disease in childhood: a clinical study to develop a practical routine, *Acta Paediatrica,* 84:294–298.

Hadjivassiliou, M., Gibson, A., Davies-Jones, G.A., Lobo, A.J., Stephenson, T.J., and Milford-Ward, A. (1996), Does cryptic gluten sensitivity play a part in neurological illness?, *Lancet,* 347:369–371.

Hoffenberg, E.J., Haas, J., Drescher, A., Barnhurst, R., Osberg, I., Bao, F., and Eisenbarth, G. (2000), A trial of oats in children with newly diagnosed celiac disease, *The Journal of Pediatrics,* 137:361–366.

Holmes, G. and Catassi C (Eds.) (2000), *Coeliac Disease,* Oxford: Health Press.

Holmes, G., Prior, P., Lane, M., Pope, D., and Allan, R. (1989), Malignancy in coeliac disease: effect of a gluten-free diet, *Gut,* 30:333–338.

Ivarsson, S.A., Carlsson, A., Bredberg, A., Alm, J., Aronsson, S., Gustafsson, J., Hagenas, L., Hager, A., Kristrom, B., Marcus, C., Moell, C., Nilsson, K.O., Tuvemo, T., Westphal, O., and Albertsson-Wikland, K. (1999), Prevalence of coeliac disease in Turner syndrome, *Acta Paediatrica,* 88:933–936.

Jabbari, M., Wild, G., Goresky, C.A., Daly, D.S., Lough, J.O., Cleland, D.P., and Kinnear, D.G. (1995), Scalloped valvuae connivenetes: an endoscopic marker of celiac sprue, *Gastroenterology,* 95:1518–1522.

Kieslich, M., Errazuriz, G., Posselt, H.G., Moeller-Hartmann, W., Zanella, F., and Boehles, H. (2001), Brain white-matter lesions in celiac disease: a prospective study of 75 diet-treated patients, *Pediatrics,* 108:e21.

Kingham, J.C. and Parker, D.R. (1998), The association between biliary cirrhosis and coeliac disease: a study of relative prevalences, *Gut,* 42:120–122.

Knivsberg, A.M. (1997), Urine patterns, peptide levels and IgA/IgG antibodies to food proteins in children with dyslexia, *Pediatric Rehabilitation,* 1(1):25–33.

Knivsberg, A.M., Reichelt, K.L., and Nodland, M. (2001), Reports on dietary intervention in autistic disorders, *Nutritional Neuroscience,* 4(1):25–37.

Knivsberg, A.M., Reichelt, K.L., Hoien, T., and Nodland, M. (2002), A randomised, controlled study of dietary intervention in autistic syndromes, *Nutritional Neuroscience,* 5:251–261.

Leonardi, S., Bottaro, G., Patane, R., and Musumeci, S. (1990), Hypertransaminasemia as a first symptom in infant coeliac disease, *Journal of Pediatric Gastroenterology and Nutrition,* 11:404–406.

Lepore, L., Martelossi, S., Pennesi, M., Falcini, F., Ermini, M.L., Ferrari, R., Perticarari, S., Presani, G., Lucchesi, A., Lapini, M., and Ventura, A. (1996), Prevalence of celiac disease in patients with juvenile chronic arthritis, *Journal of Pediatrics,* 129:311–313.

Lindquist, B.L., Rogozinski, T., Moi, H., Danielsson, D., and Olcen, P. (1994), Endomysium and gliadin IgA antibodies in children with coeliac disease, *Scandinavian Journal of Gastroenterology,* 29:452–456.

Lock,R., Pitcher, M., and Unsworth, D. (1999), IgA anti-tissue transglutaminase as a diagnostic marker of gluten sensitive enteropathy, *Journal of Clinical Pathology,* 52:274–277.

Logan, R.F.A. (1992), Descriptive epidemiology of celiac disease, in D. Branski, P. Rozen, and M.F. Kagnoff (Eds.), *Gluten-Sensitive Enteropathy,* New York: Basel, Karger, pp. 1–14.

Lubrano, E., Ciacci, C., Ames, P.R., Mazzacca, G., Oriente, P., and Scarpa, R. (1996), The arthritis of coeliac disease: prevalence and pattern in 200 adult patients, *British Journal of Rhematology,* 35:1314–1318.

Madara, J.L. and Trier, J.S. (1980), Structural abnormalities of jejunal epithelial cell membranes in celiac sprue, *Laboratory Investigations: a Journal of Technical Methods and Pathology,* 43:254–261.

Maggiore, G., De Giacomo, C., Scotta, M.S., and Sessa, F. (1986), Coeliac disease presenting as chronic hepatitis in a girl, *Journal of Pediatric Gastroenterology and Nutrition,* 5:501–503.

Mäki, M., Hallstrom, O., Verronen, P., Reunala, T., Lahdeaho, M.L., Holm, K., and Visakorpi, J.K. (1988), Reticulin antibody arthritis and coeliac disease in children, *Lancet,* 1:479–480.

Marsh, M.N. (1988), Studies on intestinal lymphoid tissue. Xi. The immunopathology of cell-mediated reactions in gluten sensitivity and other enteropathies, *Scanning Microscopy,* 2:1663–1665.

Meyer, D., Stavropolous, S., Diamond, B., Shane, E., and Green, P.H. (2001), Osteoporosis in a North American adult population with celiac disease, *The American Journal of Gastroenterology,* 96:112–119.

Michaelsson, G., Gerdén, B., Hagforsen, E., Nilsson, B., Phil-Lundin, I., Kraaz, W., Hjelmquist, G., and Loof, L. (2000), Psoriasis patients with antibodies to gliadin can be improved by a gluten-free diet, *British Journal of Dermatology,* 142:44–51.

Molberg, O., Mcadam, S.N., Korner, R., Quarsten, H., Kristiansen, C., Madsen, L., Fugger, L., Scott, H., Noren, O., Roepstorff, P., Lundin, K.E., Sjostrom, H., and Sollid, L.M. (1998), Tissue transglutaminase selectively modifies gliadin peptides that are recognized by gut-derived T cells in coeliac disease, *Nature Medicine,* 4: 713–717.

Mora, S., Barera, G., Beccio, S., Proverbio, M.C., Weber, G., Bianchi, C., and Chiumello, G. (1999), Bone density and bone metabolism are normal after long-term gluten-free diet in young celiac patients, *The American Journal of Gastroenterology,* 94:398–403.

Murray, J.A., Herlein, J., and Goeken, J. (1997), Multicenter comparison of serologic tests for celiac disease in the USA: results of phase 1 serological comparison, *Gastroenterology,* 112:A389.

Mylotte, M., Egan-Mitchell, B., Fottrell, P.F., McNicholl, B., and McCarthy, C.F. (1974), Family studies in coeliac disease, *The Quarterly Journal of Medicine,* 71:359–369.

Not, T., Horvath, K., Hill, I.D., Partanen, J., Hammed, A., Magazzu, G., and Fasano, A. (1998), Celiac disease risk in the USA: high prevalence of antiendomysium antibodies in healthy blood donors, *Scandinavian Journal of Gastroenterology,* 33:494–498.

Patey, N., Scoazec, J.Y., Cuenod-Jabri, B., Canioni, D., Kedinger, M., Goulet, O., and Brousse, N. (1997), Distribution of cell adhesion molecules in infants with intestinal epithelial dysplasia (tufting enteropathy), *Gastroenterology,* 113:833–843.

Pellecchia, M.T., Scala, R., Filla, A., De Michele, G., Ciacci, C., and Barone, P. (1999), Idiopathic cerebellar ataxia associated with celiac disease: lack of distinctive neurological features, *Journal of Neurology, Neurosurgery and Psychiatry,* 66:32–35.

Pender, S.L., Tickle, S.P., Docherty, A.J., Howie, D., Wathen, N.C., and MacDonald, T.T. (1997), A major role for matrix metalloproteinases in T cell injury in the gut, *Journal of Immunology,* 158:1582–1590.

Pietzak, M., Wolfe, A., Rongey, C., Kaufman, F., Fisher, L., Devoe, D., Pitukcheewanont, P., Krantz, J., Berti, I., Gerarduzzi, T., Thorpe, M., Kryszak, D., Horvath, K., and Fasano, A. (2002a), Celiac disease in children with type I diabetes in Southern California, *Pediatric Research,* 51(4): 141A, abstract #820.

Pietzak, M., Wolfe, A., Rongey, C., Monarch, E., Thomas, D., Bergwerk, A., Duh, G., Naon, H., Quiros, A., Sinatra, F., Yamaga, A., Pitukcheewanont, P., Kaufman, F., Fisher, L., Devoe, D., Krantz, J., Berti, I., Gerarduzzi, T., Thorpe, M., Kryszak, D., Horvath, K., and Fasano, A. (2002b), Prevalence of celiac disease in children and adults belonging to at-risk groups in Southern California, paper presented at the 10th International Symposium on Coeliac Disease, Paris, France, June 2002b.

Prader, A., Tanner, J.M., and von Harnack, G.A. (1969), Catch-up growth in coeliac disease, *Acta Paediatrica Scandinavica,* 58:311.

Rolny, P., Sigurjonsdottir, H.A., Remotti, H., Nilsson, L.A., Ascher, H., Tlaskalova-Hogenova, H., and Tuckova, L. (1999), Role of immunosuppressive therapy in refractory sprue-like disease, *The American Journal of Gastroenterology,* 94:219–225.

Ruenala, T. and Collin, P. (1997), Diseases associated with dermatitis herpetiformis, *The British Journal of Dermatology,* 136:315–318.

Rujner, J., Wisniewski, A., Gregorek, H., Wozniewicz, B., Mlynarski, W., and Witas, H.W. (2001), Coeliac disease and HLA-DQ2 (DQA1*0501 and DQB1*0201) in patients with Turner syndrome, *Journal of Pediatric Gastroenterology and Nutrition,* 32:114–115.

Russo, P.A., Chartrand, L.J., and Seidman, E. (1999), Comparative analysis of serologic screening tests for the initial diagnosis of celiac disease, *Pediatrics,* 104:75–78.

Sblattero, D., Berti, I., and Trevisiol, C. (2000), Human tissue transglutaminase ELISA: a powerful mass screening diagnostic assay for celiac disease, *The American Journal of Gastroenterology,* 98:1253–1257.

Scala, E., Giani, M., and Pirrotta, L. (1999), Urticaria and adult celiac disease, *Allergy,* 54:1008–1009.

Schulzke, J.D., Bentzel, C.J., Schulzke, I., Riecken, E.O., and Fromm, M. (1998), Epithelial tight junction structure in the jejunum of children with acute and treated celiac sprue, *Pediatric Research,* 43:435–441.

Smith, D.M. and Miller, J. (1979), Gastro-enteritis, coeliac disease and enamel hypoplasia, *British Dental Journal,* 147:91–95.

Stevens, F.M., Lloyd, R., Egan-Mitchel, B., Mylotte, M.J., Fottrell, P.F., Wright, R., McNicholl, B., and McCarthy, C.F. (1975), Reticulin antibodies in patients with coeliac disease and their relatives, *Gut,* 16:598–602.

Sulkanen, S., Halttunen, T., Laurila, K., Kolho, K.L., Korponay-Szabo, I.R., Sarnesto, A., Savilahti, E., Collin, P., and Maki, M. (1998), Tissue transglutaminase autoantibody enzyme-linked immunosorbent assay in detecting celiac disease, *Gastroenterology,* 115:1322–1328.

Swinson, C.M., Slavin, G., Coles, E.C., and Booth, C.C. (1983), Coeliac disease and malignancy, *Lancet,* 1:111–115.

Thain, M.E., Hamilton, J.R., and Erlich, R.M. (1974), Coexistence of diabetes mellitus and celiac disease, *The Journal of Pediatrics,* 85:527–529.

Troncone, R. and Ferguson, A. (1991), Anti-gliadin antibodies, *Journal of Pediatric Gastroenterology and Nutrition,* 12:150–158.

Troncone, R., Greco, L., Mayer, M., Paparo, F., Caputo, N., Micillo, M., Mugione, P., and Auricchio, S. (1996), Latent and potential coeliac disease, *Acta Paediatrica Supplement,* 412:10–14.

Troncone, R., Maurano, F., Rossi, M., Micillo, M., Greco, L., Auricchio, R., Salerno, G., Salvatore, F., and Sacchetti, L. (1999), IgA antibodies to tissue transglutaminase: an effective diagnostic test for celiac disease, *The Journal of Pediatrics,* 134:166–171.

Vaidya, A., Bolanos, J., and Berkelhammer, C. (1999), Azathioprine in refractory sprue, *The American Journal of Gastroenterology,* 94:1967–1969.

Vajro, P., Fontanella, A., Mayer, M., De Vincenzo, A., Terracciano, L.M., D'Armiento, M., and Vecchione, R. (1993), Elevated serum aminotransferase activity as an early manifestation of gluten-sensitive enteropathy, *Journal of Pediatrics,* 122:416–419.

Valdimarsson, T., Franzen, L., Grodzinsky, E., Skogh, T., and Strom, M. (1996a), Is small bowel biopsy necessary in adults with suspected celiac disease and IgA anti-endomysium antibodies?, *Digestive Diseases and Sciences,* 41:83–87.

Valdimarsson, T., Lofmano, O., Toss, G., and Strom, M. (1996b), Reversal of osteopenia with diet in adult coeliac disease, *Gut,* 38:322–327.

Vanderhoof, J.A. (1997), Diarrheal disease in infants and children, in P.E. Hyman (Volume Ed.) and M. Feldman (Series Ed.) *Gastroenterology and Hepatology: The Comprehensive Visual Reference. Volume 4: Pediatric GI Problems*, Philadelphia: Churchill Livingstone, pp. 6.1–6.17.

Ventura, A., Magazzu, G., and Greco, L. (1999), Duration of exposure to gluten and risk for autoimmune disorders in patients with celiac disease. SIGEP Study Group for Autoimmune Disorders in Celiac Disease, *Gastroenterology,* 117:297–303.

Verkasalo, M., Kuitunen, P., Leisti, S., and Perheentupa, J. (1987), Growth failure from symptomless coeliac disease, *Helvetica Paediatrica Acta,* 47:489–495.

Vitoria, J., Arrieta, A., Arranz, C., Ayesta, A., Sojo, A., Maruri, N., and Garcia-Masdevall, M. (1999), Antibodies to gliadin, endomysium, and tissue transglutaminase for the diagnosis of celiac disease, *Journal of Pediatric Gastroenterology and Nutrition,* 29:571–574.

Index*

* Page numbers in italics refer to figures and tables.

C

D

E

F

G

H

I

N

O

P

Q

R

S

Milton Keynes UK
Ingram Content Group UK Ltd.
UKHW040445071024
449327UK00020B/998

9 780367 454197